URBAN WATER SUPPLY MANAGEMENT TOOLS

URBAN WATER SUPPLY MANAGEMENT TOOLS

Larry W. Mays, Ph.D., P.E., P.H.

Editor-in-Chief

Department of Civil and Environmental Engineering
Arizona State University
Tempe, Arizona

McGraw-Hill

New York Chicago San Francisco Lisbon London Madrid Mexico City Milan New Delhi San Juan Seoul Singapore Sydney Toronto

The McGraw-Hill Companies

Cataloging-in-Publication Data is on file with the Library of Congress.

1 2 3 4 5 6 7 8 9 0 DOC/DOC 0 9 8 7 6 5 4 3

ISBN 0-07-142836-4

The sponsoring editor for this book was Larry S. Hager, the editing supervisor was Caroline Levine, and the production supervisor was Pamela A. Pelton. The art director for the cover was Margaret Webster-Shapiro. This book was set in the HB1 design in Times Roman by Wayne A. Palmer of McGraw-Hill Professional's Hightstown, N.J. composition unit.

Printed and bound by RR Donnelley.

This book is printed on recycled, acid-free paper containing a minimum of 50% recycled, deinked paper.

CONTENTS

Contributors xi
Preface xiii

Chapter 1. Water Demand Analysis ***Benedykt Dziegielewski and Eva Opitz*** **1.1**

1.1 Definition and Measurement of Water Use / *1.1*
1.2 Public-Supply Water Use / *1.2*
1.3 Sampling of Water Users / *1.3*
1.3.1 Types of Sampling Plans / *1.3*
1.3.2 Example of Sample Size Determination for Continuous and Proportional Data / *1.5*
1.4 Development of Data Sets / *1.7*
1.4.1 Data Scales / *1.7*
1.4.2 Data Arrangements / *1.8*
1.5 Water-Use and Service Area Data / *1.10*
1.5.1 Water-Use Data / *1.10*
1.5.2 Service Area Data / *1.11*
1.6 Components of Water Demand / *1.13*
1.6.1 Sectors of Water Users / *1.13*
1.6.2 Seasonal and Nonseasonal Components / *1.15*
1.6.3 End Uses of Water / *1.19*
1.7 Water-Use Relationships / *1.20*
1.7.1 Average Rates of Water Use / *1.21*
1.7.2 Modeling of Water Use / *1.22*
1.8 Analysis of Water Savings / *1.30*
1.8.1 Statistical Estimation of Savings / *1.31*
1.8.2 Time Series Analysis of Conservation Effects / *1.33*
1.8.3 End-Use Accounting System / *1.34*
1.9 Summary / *1.36*
1.10 References / *1.37*

Chapter 2. Water Pricing and Drought Management ***Messele Z. Ejeta and Larry W. Mays*** **2.1**

2.1 Introduction / *2.1*
2.2 Background of Water Conservation / *2.3*
2.2.1 Drought Management Options / *2.3*
2.2.2 Price Elasticity of Water Demand / *2.4*
2.2.3 Demand Models / *2.6*
2.2.4 The Need for a Risk-Based Approach / *2.8*
2.2.5 Drought Severity as Risk Indices / *2.9*
2.3 Risk-Price Relationship / *2.11*
2.3.1 Developing Risk-Price Relationships / *2.11*
2.4 Operation and Management Planning Under Sustained Drought Conditions / *2.17*
2.4.1 Economic Aspects of Water Shortage / *2.18*
2.4.2 Damage Assessment / *2.20*

2.4.3 Operation and Management / 2.22
2.4.4 The $\phi(\xi)$ Function / 6.23
2.4.5 Uncertainty and Risk in Demand / 2.23
2.4.6 Operation and Management Strategy / 2.26
2.4.7 Risk Evaluation Procedure under Sustained Drought Conditions / 2.26
2.5 Summary and Conclusions / 2.32
2.6 References / 2.32

Chapter 3. Computer Programs for Integrated Management
Messele Z. Ejeta and Larry W. Mays **3.1**

3.1 Introduction / 3.1
3.2 Integrated Hydrosystems Management / 3.2
3.2.1 Definition / 3.2
3.2.2 History / 3.3
3.2.3 Importance / 3.5
3.3 Computer Programs for Integrated Management / 3.6
3.3.1 Development of Hydrosystems Simulation Computer Programs / 3.7
3.3.2 Optimization Formulations / 3.8
3.3.3 Interfacing Optimization and Simulation Computer Programs / 3.10
3.3.4 Computer-Based Information Systems: Supervisory Control Automated Data Acquisition (SCADA) / 3.14
3.3.5 Prospects of Computer Programs for Integrated Hydrosystems Management / 3.14
3.4 DSSs As Tools for Integrated Hydrosystems Management / 3.15
3.4.1 Definition of DSSs / 3.15
3.4.2 Basic Structure of DSSs / 3.15
3.4.3 Examples of DSSs for Integrated Hydrosystems Management / 3.16
3.5 State of Practice and Prospects of Hydrosystems Computer Programs / 3.21
3.5.1 State of Practice / 3.21
3.5.2 Prospects / 3.22
3.6 Summary and Conclusions / 3.23
3.7 References / 3.25

Chapter 4. Optimal Design of Water Distribution Systems
Kevin E. Lansey **4.1**

4.1 Overview / 4.1
4.2 Problem Definition / 4.1
4.3 Mathematical Formulation / 4.2
4.4 Optimization Methods / 4.3
4.4.1 Branched Systems / 4.3
4.4.2 Looped Pipe Systems via Linearization / 4.5
4.4.3 General System Design via Nonlinear Programming / 4.6
4.4.4 Stochastic Search Techniques / 4.7
4.5 Applications / 4.8
4.6 Summary / 4.11
4.7 References / 4.11

Chapter 5. Optimal Operation of Water Systems ***Fred E. Goldman, A. Burcu Altan Sakarya, and Larry W. Mays*** **5.1**

5.1 Introduction / 5.1
5.1.1 Background / 5.1
5.1.2 Previous Models for Water Distribution System Optimization / 5.2
5.2 Optimal Pump Operation—Mathematical Programming Approach / 5.3
5.2.1 Model Formulation / 5.3
5.2.2 Solution Methodology—Mathematical Programming Approach / 5.5

5.2.3 North Marin Water District / 5.7
5.2.4 Summary and Conclusions—Mathematical Programming Approach / 5.12
5.3 Optimal Pump Operation—Simulated Annealing Approach / 5.12
5.3.1 Model Formulation / 5.12
5.3.2 Solution Methodology—Simulated Annealing Approach / 5.13
5.3.3 Applications—Simulated Annealing Approach / 5.18
5.3.4 Summary and Conclusions—Simulated Annealing Approach / 5.21
5.4 Optimal Operation of Chlorine Booster Stations / 5.22
5.4.1 Model Formulation / 5.22
5.4.2 Application / 5.23
5.4.3 Summary and Conclusions—Chlorine Booster Station Operation / 5.23
5.5 Challenges for the Future / 5.25
5.6 References / 5.26

Chapter 6. Reliability and Availability Analysis of Water Distribution Systems ***Kevin Lansey, Larry W. Mays, and Y. K. Tung*** **6.1**

6.1 Failure Modes for Water Distribution Systems / 6.1
6.1.1 Distribution Repair Definitions / 6.2
6.1.2 Failure Modes / 6.3
6.1.3 Reliability Indices / 6.4
6.2 Component (Mechanical) Reliability Analysis / 6.5
6.2.1 Failure Density, Failure Rate, and Mean Time to Failure / 6.6
6.2.2 Availability and Unavailability / 6.9
6.3 Methodology for Reliability and Availability Analysis for Water Distribution Networks / 6.11
6.3.1 Reliability of a System / 6.11
6.3.2 Methodology / 6.12
6.3.3 Application of Methodology / 6.17
6.4 Other Approaches to Assessment of Reliability / 6.23
6.5 Reliability-based Design Optimization Models / 6.26
6.5.1 Framework for Reliability-based Design Framework / 6.26
6.5.2 Reliability-based Optimization Model for Operation Considering Uncertainties of Water Quality / 6.27
6.6 References / 6.28

Chapter 7. Performance Indicators as a Management Support Tool ***Helena Alegre*** **7.1**

7.1 Introduction / 7.1
7.2 Concept of Performance Indicator, Context Information, and Utility Information / 7.2
7.3 Users, Benefits, and Scope of Application of Performance Indicators / 7.3
7.4 State of the Art of Performance Assessment / 7.5
7.4.1 Overview / 7.5
7.4.2 Influence of the Growing Private Participation in Undertakings Management / 7.7
7.4.3 Objective-oriented Management and Benchmarking / 7.7
7.4.4 Lessons Arising from the U.K. Privatization Process / 7.7
7.4.5 The IWA PI System / 7.9
7.4.6 The World Bank Benchmarking Toolkit / 7.10
7.4.7 The Asian Development Bank Data Book / 7.10
7.4.8 The Water Utility Partnership for Capacity Building in Africa / 7.10
7.4.9 The Six Scandinavian Cities Group / 7.11
7.4.10 The Dutch Contact Club for Water Companies / 7.12
7.4.11 An Engineering Approach for Performance Assessment / 7.12
7.4.12 Status of Performance Assessment in the United States / 7.13
7.4.13 Other Initiatives / 7.14
7.4.14 Summary of Performance Assessment Projects for Water Supply (WS) and Wastewater (WW) Services / 7.15

7.5 The IWA PI System for Water Supply Services / *7.15*
7.5.1 Highlights of the IWA PI System / *7.15*
7.5.2 Listing of the IWA PIs and Guidance on Their Relative Importance / *7.15*
7.5.3 Data Reliability and Accuracy / *7.19*
7.5.4 Organization of the IWA PI Manual / *7.20*
7.5.5 A Partial In-depth Look into the IWA PI System / *7.26*
7.5.6 The SIGMA Lite Software / *7.30*
7.5.7 International Field Test of the IWA PI System (2000–2003) / *7.31*
7.6 Implementation of a PI System / *7.36*
7.6.1 Phases of Implementation / *7.36*
7.6.2 Definition of the Strategic Performance Assessment Policy / *7.36*
7.6.3 Establishment of the PI System / *7.37*
7.6.4 Data Collection, Validation, and Input / *7.38*
7.6.5 PI Assessment and Global Reporting / *7.38*
7.6.6 Results Interpretation / *7.38*
7.6.7 Definition and Implementation of Improvement Measures / *7.39*
7.7 Example of the Use of SIGMA Lite for Data Input and Indicators Assessment / *7.39*
7.7.1 Scope of the Example / *7.39*
7.7.2 Starting to Use SIGMA Lite / *7.40*
7.7.3 Context Information / *7.40*
7.7.4 PI Selection and Data Input / *7.40*
7.7.5 PI Assessment and Reporting / *7.41*
7.8 Example of the Use of the IWA PI System for Water Losses Control / *7.41*
7.8.1 Scope of the Example / *7.41*
7.8.2 Objective Definition and Form of Use of the Indicators / *7.42*
7.8.3 Appointment of the PI Team and Definition of the Reference PI System / *7.42*
7.8.4 Selection of the Indicators / *7.42*
7.8.5 Assessment of the Water Balance / *7.47*
7.8.6 Assessment of the Indicators / *7.50*
7.8.7 Results Interpretation / *7.51*
7.9 Final Remarks / *7.51*
7.10 Acknowledgments / *7.52*
7.11 References / *7.52*
7.12 Endnotes / *7.53*

Chapter 8. Climate Change Effects and Water Management Options
Larry W. Mays **8.1**

8.1 Introduction / *8.1*
8.1.1 The Climate System / *8.1*
8.1.2 Definition of Climate Change / *8.3*
8.1.3 Climate Change Prediction / *8.3*
8.1.4 Droughts / *8.5*
8.2 Climate Change Effects / *8.5*
8.2.1 Hydrologic Effects / *8.5*
8.2.2 River Basin/Regional Runoff Effects / *8.10*
8.3 Water Management Options / *8.13*
8.4 Uncertainties / *8.13*
8.5 References / *8.14*

Chapter 9. Water Supply Safety and Security: An Introduction
Larry W. Mays **9.1**

9.1 The Water Supply System: A Brief Description / *9.1*
9.2 Why Water Supply Systems? / *9.2*
9.2.1 The Threats / *9.2*

9.3 Prior to September 11. 2001 / *9.4*
9.4 Response to September 11. 2001 / *9.6*
9.4.1 Public Health, Security and Bioterrorism Preparedness and Response Act ("Bioterrorism Act"), (PL 107-188), June 2002 / *9.6*
9.4.2 Homeland Security Act (PL 107-296), November 25, 2002 / *9.7*
9.4.3 USEPA's Protocol / *9.8*
9.5 Vulnerability Assessment / *9.8*
9.5.1 Definition of Vulnerability Assessment / *9.8*
9.5.2 Security Tools / *9.8*
9.5.3 Security Vulnerability Assessment Using VSAT / *9.9*
9.5.4 Security Vulnerability Assessment for Small Drinking Water Systems / *9.9*
9.6 Emergency Response Planning: Response, Recovery, and Remediation Guidance / *9.11*
9.7 Information Sharing (*www.waterisac.org*) / *9.11*
9.8 Reliability Assessment / *9.11*
9.8.1 State Enumeration Method / *9.12*
9.8.2 Path Enumeration Method / *9.13*
9.8.3 Conditional Probability Approach / *9.15*
9.8.4 Fault-Tree Analysis / *9.15*
9.9 References / *9.23*
Appendix 9.A What Are Some Points to Consider in a Vulnerability Assessment / *9.26*
Appendix 9.B Security Vulnerability Self-Assessment for Small Water Systems / *9.30*
Appendix 9.C Water Utility Response, Recovery, and Remediation Guidlines / *9.40*

Index I.1

CONTRIBUTORS

Helena Alegre *National Civil Engineering Laboratory, Lisbon, Portugal* (CHAP. 7)

Benedykt Dziegielewski *Southern Illinois University, Carbondale, Illinois* (CHAP. 1)

Messele Z. Ejeta *California Department of Water Resources, Sacramento, California* (CHAPS. 2, 3)

Fred E. Goldman *GTA Engineering, Inc., Phoenix, Arizona* (CHAP. 5)

Kevin Lansey *Department of Civil Engineering and Engineering Mechanics, University of Arizona, Tucson, Arizona* (CHAPS. 4, 6)

Larry W. Mays *Department of Civil and Environmental Engineering, Arizona State University, Tempe, Arizona* (CHAPS. 2, 3, 5, 6, 8, 9)

Eva Opitz *PMCL, Carbondale, Illinois* (CHAP. 1)

A. Burcu Altan Sakarya *Department of Civil Engineering, Middle East Technical University, Ankara, Turkey* (CHAP. 5)

Y. K. Tung *Department of Civil Engineering, Hong Kong University of Science and Technology, Clear Water Bay, Kowloon, Hong Kong* (CHAP. 6)

PREFACE

For centuries humans have been concerned with the safety and security of water supply systems. Ancient Romans, the builders of magnificent water supply systems, were faced with these issues. Modern Western civilizations have tended to take for granted the safety and security of their water supply systems. The events of September 11, 2001 have forced a new focus on water infrastructure—particularly in the United States—specifically, an intensified consideration of terrorist threats to these systems.

Two very important acts that have focused on the water infrastructure and issued directives for the future were passed in 2002. The Public Health, Security, and Bioterrorism Preparedness and Response Act (PL 107-188) (June 2002), requires community water systems serving populations greater than 3300 to conduct vulnerability assessments and submit these assessments to the U.S. Environmental Protection Agency (USEPA). The Homeland Security Act (PL 107-296), November 25, 2002, directed the greatest reorganization of the federal government in decades by consolidating a host of security-related agencies into a single cabinet-level department. In October 2002 USEPA announced their Strategic Plan for Homeland Security. The goals of the plan are separated into four distinct mission areas: critical infrastructure protection; preparedness, response, and recovery; communication and information; and protection of USEPA personnel and infrastructure.

Another topic concerning water supply that is gaining more attention is the effect of global climate change on our future water supplies. These newer concerns, related to water security and climate change along with an already complex aging water infrastructure, have created an entire new set of management problems that need state-of-the-art management tools. The intent of *Urban Water Supply Management Tools* is to focus on the management tools that were presented in two previous McGraw-Hill handbooks, *Water Distribution Systems Handbook* and the *Urban Water Supply Handbook,* with additional chapters on climate change and management options and on water security and safety. This new book has been developed to focus on topics related specifically to the management tools for urban water supply systems.

First and foremost, this work is intended as a reference for those wishing to expand their knowledge of urban water supply systems, particularly the management of these systems. The book will be of value to engineers, managers, operators, and analysts involved with the various aspects of urban water supply systems management. I have attempted to cover topics that are very important for modern-day management including chapters on water demand analysis, performance indicators, reliability and availability analysis, water pricing and drought management, optimization models for operation and design, computer models for integrated management, climate change effects and management options, and water supply security and safety.

All chapter authors in this handbook are leading experts in their respective fields associated with urban water supply management, and each has published extensively in the literature. The authors were chosen because of their proven knowledge in the specific area of their contribution. Each book that I have worked on has been a part of my lifelong journey in water resources, and this handbook is no exception. I have gained more from my experiences in developing handbooks than can ever be measured in words. This book certainly will never have the impact on history that the treatises of Vitruvius and Frontinus of the Roman Empire have had, but hopefully it will provide a little insight for the present generation of urban water supply systems engineers and managers.

I dedicate this book to humanity and human welfare.

Larry W. Mays
Scottsdale, Arizona

CHAPTER 1
WATER DEMAND ANALYSIS

Benedykt Dziegielewski
Southern Illinois University
Carbondale, Illinois

Eva Opitz
PMCL
Carbondale, Illinois

1.1 DEFINITION AND MEASUREMENT OF WATER USE

From the hydrologic perspective, *water use* can be defined as all water flows that are a result of human intervention within the hydrologic cycle. A more restrictive definition of water use refers to water that is actually used for a specific purpose. Table 1.1 contains definitions of nine such uses (Solley et al., 1998). Urban water supply systems deliver water to most of these categories of use with domestic and commercial uses being almost entirely dependent on public deliveries. Several other categories such as industrial, irrigation, and public uses are also present in urban areas. Some categories are found primarily outside of urban areas or require large quantities of untreated water, and they tend to be self-supplied.

Measurements of water use are reported as water volumes per unit of time. The volumetric units include cubic meters, cubic feet, gallons, and liters, and their decimal multiples. In some cases, composite volumetric units such as acre·foot or units of water depth such as inches of rain may be used. The time periods used include second, minute, day, and year. Because the annual volumes of water use usually involve large numbers, annual water-use totals are often reported as the average daily usage rates. Two popular units for measuring total urban demands are thousand cubic meters per day (Km^3/d) and million gallons per day (mgd).

In order to make the estimates of water use easy to comprehend and to make meaningful comparisons of water use for various purposes (and various users), the annual or daily quantities are divided by some measures of size for each purpose of use. The result is an average rate of water use such as gallons per capita per day (gcd), gallons per employee per day (ged), or other unit-use coefficients.

The reported quantities of water use can be in the form of direct measurements obtained from water meters that register the volume of flow (such as displacement meters or venturi meters), or they may be estimates. Estimates of water use derived from the measurements of water levels in storages or from pumping logs are generally more accurate than those derived from related data on the volume of water-use activity. For example, the estimates of water use for industrial purposes may be obtained by multiplying the number of manufacturing employees or the value added (in dollars) by a water-use coefficient. For example, in 1995 the ratio of water use to value added by manufacture in the United States was 5783 gallons (gal) per $1000 (Dziegielewski et al., 2002a).

TABLE 1.1 Major Purposes of Water Use

Water-use purpose	Definition
Commercial use	Water for motels, hotels, restaurants, office buildings, and other commercial facilities and institutions
Domestic use	Water for household needs such as drinking, food preparation, bathing, washing clothes and dishes, flushing toilets, and watering lawns and gardens (also called residential water use)
Industrial use	Water for industrial purposes such as fabrication, processing, washing, and cooling
Irrigation use	Artificial application of water on lands to assist in the growing of crops and pastures or to maintain vegetative growth in recreational lands such as parks and golf courses
Livestock use	Water for livestock watering, feed lots, dairy operations, fish farming, and other on-farm needs
Mining use	Water for the extraction of minerals occurring naturally and associated with quarrying, well operations, milling, and other preparations customarily done at the mine site or as part of a mining activity
Public use	Water supplied from a public water supply and used for such purposes as firefighting, street washing, municipal parks, and swimming pools
Rural use	Water for suburban or farm areas for domestic and livestock needs which is generally self-supplied
Thermoelectric	Water for the process of the generation of thermoelectric power.power use

Source: Adapted from Solley et al., 1998.

1.2 PUBLIC SUPPLY WATER USE

National data on water withdrawals for public supply purposes include public and private water systems that furnish water, year-round, to at least 25 people or that have a minimum of 15 hookups (Solley et al., 1998). Nearly 55,000 community water systems serve more than 263 million people. Transient community water systems and nontransient noncommunity water systems serve another 20 million people.

Table 1.2 shows the distribution of the community systems by the size of population served. Approximately 81 percent of the population is served by 3769 systems, which deliver water to communities with more than 10,000 persons. While nearly 80 percent of public water supply systems rely on groundwater, more than one-half (58 percent) of the larger systems use surface water as their principal source of supply (calculated from table 2.1, USEPA, 1997).

In any public water supply system, water-use records can be characterized with respect to the relative needs of various customer groups (e.g., single-family residential, hotels, food processing plants), the purposes for which water is used (e.g., end uses such as sanitary needs, lawn watering, cooling), and the seasonal variation in water use. The analysis of water use can be expanded to also include the development of information on water users in the service area. Data on housing stock, household characteristics, business establishments, and other demographic and economic statistics are important because such characteristics are major determinants of water use.

The data and techniques for analyzing water demand in a public water supply system are the subject of the following sections. Such analysis must necessarily begin with a determination of quantities of water used. While total water deliveries to urban areas can be measured at one (or several) points on the water supply system, the volumes of water used for specific purposes can be obtained through an inventory or sampling of individual users.

1.3 SAMPLING OF WATER USERS

Water supply agencies and regional or state regulatory bodies usually have the ability to monitor the water use of all users or entire classes of users. In statistical terms, studies involving all users would represent the use of entire populations. However, a complete enumeration or inventory of all users may not always capture the entire population because in addition to *populations* defined in terms of users, which can be viewed as finite (or delimited), some studies may require expanded definitions of populations. For example, a study population can be defined as the monthly water-use quantities of all users over a time horizon. Because such a definition includes future water use, historical records of withdrawals constitute only a part (or a sample) of the total population. Also, in many cases, a study of an entire population must limit the number of measurements of each unit (due to cost constraints) and may not be capable of producing answers to some research questions. Because of these considerations, knowledge of water use is almost invariably based on samples or fragments of total populations. Sampling has many advantages over a complete enumeration (or inventory) of the population under study. These advantages include reduced cost, greater speed of obtaining information, and a greater scope of information that can be obtained. In addition, a greater precision of measurements can be secured by employing trained personnel to take the necessary measurements and analyze the data.

Scientific sampling designs specify methods for sample selection and estimation of sample statistics that follow the *principle of specified precision at the minimum cost,* i.e., they provide, at the lowest possible cost, estimates that are precise enough for the study objectives. *Probability sampling* refers to any sampling procedure that relies on random selection and is amenable to the application of sampling theory to validate the measurements obtained through sampling. This requires that, within the sampled population, one is able to define a set of distinct samples (where each sample consists of sampling units) with known and equal probabilities of being selected. One of these samples is then selected through a random process. In practice, the sample is most commonly constructed by specifying probabilities of inclusion for the individual units, one by one or in groups, and then selecting a sample of desired size and type. *Nonprobability sampling* refers to sampling procedures that do not include the element of random selection because samples are restricted only to a part of the population that is readily accessible, selected haphazardly without prior planning, or they consist of typical units or volunteers. The only way of examining how good the nonprobability sample may be is to know parameters for the entire population or to compare it with the probability sample statistics taken from the same population. For more information on sampling procedures, see Dziegielewski et.al. (1993), Cochran (1963), Kish (1965), or Fowler (1988).

1.3.1 Types of Sampling Plans

There are many ways of constructing a probability sample of water users. *Simple random sampling* refers to a method of selecting n sampling units out of a population of size N, such that every one of

TABLE 1.2 Community Water Systems in the United States

System description	Number of systems	Percent of systems	Population served	Percent of population
		By system size		
500 or less	31,688	59	5,148,700	2
501–3,300	14,149	26	19,931,400	8
3,301–10,000	4,458	8	25,854,100	10
10,001–100,000	3,416	6	96,709,100	37
>100,000	353	1	116,282,800	44
Total	54,064	100	263,926,100	100
		By water sources		
Groundwater	42,661	79	85,868,500	33
Surface water	11,403	21	178,057,600	67
Total	54,064	100	263,926,100	100

Note: Groundwater systems include groundwater and purchased groundwater. Surface-water systems include surface water, purchased surface water, and groundwater under the influence of surface water.
Source: USEPA, 2000.

the distinct samples (where each sample consists of n sampling units) has an equal chance of being drawn. In *stratified sampling,* the sampled population of N units is first divided into several nonoverlapping subpopulations. These subpopulations are called *strata* because they divide a heterogeneous population into homogeneous subpopulations. If a simple random sample is taken from each stratum, then the sampling procedure is described as *stratified random sampling.* In order to design a stratified random sampling plan, it is necessary to determine (1) which population characteristic (i.e., variable) should be used in stratification, (2) how to construct the strata (i.e., how many strata to use and where to set the stratification boundaries), and (3) what sample sizes should be obtained from each stratum. The statistical theory of stratified sampling offers some methods for selecting the optimal number of strata, strata boundaries, and sample sizes in advance (Cochran, 1963). However, it is usually necessary to collect and examine some data before designing a good sampling plan.

Systematic sampling is often the most expeditious way of obtaining the sample and may be used in situations where time is critically constrained. The units in the population sampled are first numbered from 1 to N in some order. To select a sample of n units, one should take the first unit at random from the first k units and every kth unit thereafter. The selection of the first unit determines the whole sample, which is often called an *every* k*th systematic sample.*

If individual sampling units are arranged according to some characteristic or variable (e.g., water use), then the systematic sample is equivalent to a stratified sample in which one sampling unit is taken from each stratum. Constructing a list of sampling units can be avoided by dividing a geographic area into subunit areas; for example, a subarea could be a county or river basin. This sampling plan, called *cluster sampling,* can result in significant cost savings. For example, a simple random sample of 600 industrial users may cover a state more evenly than 20 counties containing a sample of 30 plants each, but it will cost more because of the time devoted to travel and finding individual establishments. However, cluster sampling creates a greater risk of obtaining a nonrepresentative sample.

The *sample size* depends on the precision of measurement that is required and the variance in the parameters to be estimated. The *precision* of an estimate refers to the size of the deviations from the mean of all sample measurements obtained by repeated application of the sampling procedure. In contrast, the term *accuracy* is usually applied to indicate the deviations of the sample measurements from the true values in the population. For example, a simple random sample can be used to estimate average daily water use and variance in water use of all single-family houses in an urban area during a given year. According to sampling theory, the mean water use $\bar{y}$ obtained from the simple random sample is an unbiased estimate of the average water use $\bar{Y}$ for all houses (i.e., the population mean). Also, $\bar{Y} = N\bar{y}$ is an unbiased estimate of total water use of the population (N customers).

The *standard error* of $\bar{y}$, which describes the precision of the estimated mean value, is

$$\sigma_{\bar{y}} = \sqrt{\frac{N-n}{N}}\,\frac{S}{\sqrt{n}} \tag{1.1}$$

where S is obtained from population variance S^2 (by taking its square root). Because in practice S^2 may not be known, it must be estimated from the sample data using the following formula:

$$S^2 = \frac{\sum_{i}^{n} (y_i - \bar{y})^2}{n-1} \tag{1.2}$$

which provides an unbiased estimate of S^2 and where n is the sample size. Usually, with a population having a mean $\bar{Y}$ and a simple random sample having mean $\bar{y}$, control of the following probability condition is desired:

$$\Pr\left(\left|\frac{\bar{y}-\bar{Y}}{\bar{Y}}\right| \geq r\right) = \alpha \tag{1.3}$$

where α is a small probability (e.g., 0.05) and r is the relative error expressed as a fraction of the true population mean. By multiplying both sides of the parenthetical expression in Eq. (1.3) by $\bar{Y}$, the same condition can be restated as

$$\Pr(|\bar{y} - \bar{Y}| \geq r\bar{Y}) = \alpha \tag{1.4}$$

If, instead of the relative error r, control of the absolute error d (i.e., the absolute value of the difference between the sample mean and the population mean) in Y is desired, the formula can be written as

$$\Pr(|\bar{y} - \bar{Y}| \geq d) = \alpha \tag{1.5}$$

It is usually assumed that $\bar{y}$ is normally distributed about the population mean $\bar{Y}$, and given its standard error from Eq. (1.1), the product $r\bar{Y}$ is

$$r\bar{Y} = t\sigma_{\bar{y}} = t\sqrt{\frac{N-n}{N}} \cdot \frac{S}{\sqrt{n}} \tag{1.6}$$

where t is the value of the normal deviate corresponding to the desired confidence probability. This value is 1.64, 1.96, and 2.58 for confidence probabilities 90, 95, and 99 percent, respectively. Solving Eq. (1.6) for n gives

$$n = \frac{(tS/r\bar{Y})^2}{1 + (1/N)(tS/r\bar{Y})^2} \tag{1.7}$$

The expression in the denominator represents a finite population correction, and it should be used when n/N is appreciable. Without this correction, we can take the first approximation of the desired sample size n_o as

$$n_o = \left(\frac{tS}{r\bar{Y}}\right)^2 \tag{1.8}$$

According to this equation, n_o depends on the coefficient of variation (the ratio $S/\bar{Y}$) of the population that is often more stable and easier to guess in advance than S itself. It also depends on the error r that can be tolerated and the confidence level that is needed as captured by the value of t. For the absolute error specification as in Eq. (1.5), Eq. (1.8) is changed into

$$n_o = \left(\frac{tS}{d}\right)^2 \tag{1.9}$$

The preceding sample size relationships are illustrated in the following example.

1.3.2 Example of Sample Size Determination for Continuous and Proportional Data

A water supply agency serves 80,000 customers. The analysis of billing frequencies for the entire fiscal year indicates that average daily use per customer is 250 gal and the standard deviation is 180 gal. Using simple random sampling, how many customers must be sampled to be 95 percent confident of estimating average daily use within 2 percent of the true value?

Solution: $N = 80{,}000$, $S = 180$ gal, $\bar{Y} = 250$ gal, $\alpha = 0.05$, $t = 1.96$, and $r = 0.02$. Substituting these values into Eq. (1.8):

$$n_o = \frac{t^2 S^2}{r^2 \overline{Y}^2} = \frac{(1.96)^2\,(180)^2}{(0.02)^2\,(250)^2} = 4979$$

where n = sample size (n_o is the first approximation)
N = population size
S = population standard deviation
$\overline{Y}$ = population mean
t = confidence probability (t statistic)
r = relative error

Because n_o/N is not negligible, we need to take the finite population correction, found from Eqs. (1.7) and (1.8):

$$n = \frac{n_o}{1 + n_o/N} = \frac{4979}{1 + 4979/80{,}000} = 4687$$

The results indicate that if the average water use is unknown, to be 95 percent confident of estimating it by sampling billing records with an error of 2 percent (or 5 gal), a sample of 4687 single-family homes would be required.

In some cases, it may be necessary to obtain estimates of the percent of water users who possess a certain characteristic (e.g., use groundwater as their sources of supply) by surveying a sample of users. The sampling problem in this case is referred to as *sampling for proportions,* where the respondents are classified into two classes: groundwater users and users of other sources. In order to determine the required sample size, we must decide the margin of error d in the estimated proportion p of users who rely on groundwater and the risk α that the actual error will be larger than d. Therefore, control of the following probability condition is desired:

$$\Pr\left(|p - P| \geq d\right) = \sigma \tag{1.10}$$

where P is the true proportion of users of groundwater. Assuming simple random sampling and a normal distribution of p, the standard error of p, σ_p, is given by:

$$\sigma_p = \sqrt{\frac{N-n}{N-1}}\,\sqrt{\frac{PQ}{n}} \tag{1.11}$$

where N = population size
n = number of respondents in the sample
P = proportion of groundwater users in the population
Q = proportion of users of other sources in the population (i.e., $Q = 1 - P$)

The formula for the desired *degree of precision* is

$$d = t\sqrt{\frac{N-n}{N-1}}\,\sqrt{\frac{PQ}{n}} \tag{1.12}$$

where t is the critical value of the t distribution corresponding to the desired confidence probability (i.e., the abscissa of the normal curve that cuts off an area of α at the tails). Solving Eq. (1.12) for n gives:

$$n = \frac{t^2 PQ/d^2}{1 + (1/N)(t^2PQ/d^2 - 1)} \tag{1.13}$$

If n/N is negligible because N is large, we can take the first approximation of n_o by using an advanced estimate p for P (and q for Q) from the formula

$$n_o = \frac{t^2 pq}{d^2} \tag{1.14}$$

After obtaining n_o, we can introduce the finite population correction to the sample size from the following formula:

$$n = \frac{n_o}{1 + n_o/N} \tag{1.15}$$

The sampling plans and sample size determinations are important elements of the process of collecting water-use data. Because water users are unlikely to form a homogeneous group, stratified random sampling is the most useful procedure for obtaining representative samples of water users.

1.4 DEVELOPMENT OF DATA SETS

Water-use data are usually collected for the purpose of monitoring water use. Many states require that water users submit annual (and/or monthly) records of their water withdrawals or discharges as a part of their permitting process. These data can be used for statistical analysis of water-use trends as well as for the development of water-use models. The latter purpose would also require data on variables which influence water use such as weather, price, employment, and land use.

In economics, data on economic activities (or variables) are collected at micro- or macrolevels. Observations on individual households, families, or firms are referred to as *microdata.* National-level accounts and observations on entire industries are called *macrodata.* In water-use modeling and analysis, the corresponding types of data are sometimes referred to as *disaggregate* and *aggregate* data. Levels of aggregation may vary from the end-use level (e.g., toilet flushing, lawn watering) or the municipal level (e.g., total production or total metered use) to total withdrawals in a region or state.

In developing water-use models, it is necessary to distinguish among different types of variables and their levels of measurement. The latter are also referred to as scales. In a mathematical sense, a variable is a quantity or function that may assume any given value or set of values, as opposed to a constant that does not or cannot change or vary. Depending on the character (or type) of values that can be assumed by a variable, the following can be distinguished: (1) continuous variables, (2) discrete variables, and (3) random variables. A *continuous variable* can assume any value within the interval where it exists (e.g., monthly water use in a demand area). A *discrete variable* can assume only discrete values such as the number of customers in a demand area. Finally, a *random* (or stochastic) *variable* can take any set of values (positive or negative) with a given probability. Random variables can be discrete or continuous.

1.4.1 Data Scales

The data on a variable can be measured on different kinds of scales. Such scales are used to describe the data and variables used in the analysis. The data can be nominal (either ordinal or nonordinal) or interral (including ratio data).

Nominal, or *categorical,* data are measurements that contain sufficient information to classify and count objects. Nominal data can be classified according to ordinal or nonordinal scales. *Ordinal* scales rank data according to the *value* of the variable that is being analyzed. Objects in an ordinal scale are characterized by relative rank, so a typical relationship is expressed in terms such as *higher, greater,* or *preferred to.* Ordinal ranking of data commonly occurs in the use of surveys. For example, survey data of household income are usually ranked (or classified ordinally) by income range, such as 1 ≤ $25,000; 2 = $25,000 to $49,999; 3 = $50,000 to $74,999; etc. Notice then that ordinal rankings are hierarchical. *Nonordinal* data are ranked by variable type and therefore cannot be ranked hierarchically on a numerical scale. An example of nonordinal data comes from a survey of residential landscapes in southern California. Survey teams were asked to classify turf landscapes as 1 = Bermuda grass, 2 = other warm-season grasses, 3 = tall fescue, and 4 = cool-season grasses. Unlike the example of ordinal ranking given above, category 2 is not in any sense greater than category 1, because the categorization is based on landscape type without reference to numerical measurement of the distance between ranked objects. Finally, there are *interval* data that contain numerical values from a continuous scale. Variables with interval data are therefore most frequently called continuous variables. Because a continuous variable can assume an infinite number of values, continuous variables can theoretically be measured only over an interval, hence the name interval data. If the continuous measurement scale contains a true theoretical zero (e.g., water use), then it is called a ratio scale.

Another classification of variables is used in statistical or econometric modeling. In constructing statistical relationships, an attempt is usually made to predict or explain the effects of one variable by examining changes in one or more other variables that are known or expected to influence the former variable. In mathematics, the variable to be predicted is called the *dependent* variable, while the variables that influence it are called *independent.* However, in the analytical literature from various disciplines, a number of alternative terms are often used to describe and classify variables. Table 1.3 contains a list of such terms.

1.4.2 Data Arrangements

For modeling purposes, data (or observations) on water-use and related variables can be obtained and arranged in several ways. Depending on which type of arrangement is used, four types of data configurations can be distinguished: (1) time series data, (2) cross-sectional data, (3) pooled time series and cross-sectional data, and (4) panel (or longitudinal) data. In *time series* data, observations on all variables in the data set are taken at regular time intervals (e.g., daily, weekly, monthly, annually). In *cross-sectional* data, observations are taken at one time (either point in time or time interval) but for different units, such as individuals, households, sectors of water users, cities, or counties. *Pooled* data sets combine both time series and cross-sectional observations to form a single data matrix. Finally, *panel* data represent repeated surveys of the same cross-sectional sample at different periods of time. Databases can be built from records of water use by supplementing them with data from other sources such as random samples of water users. The possible data configurations that can be constructed from water-use histories of individual water users include

1. Time series of measurement period water-use data for individual water users
2. Time series of measurement period water-use data for all users in the sample
3. Cross-sectional water-use data for the same measurement period extending across all customers in the sample

TABLE 1.3 Alternative Terms for Dependent and Independent Variables

Dependent	Independent
Explained	Explanatory
Predicted	Predictor
Regressed	Regressor
Response	Causal
Endogenous	Exogenous
Target	Control

4. Cross-sectional water-use data aggregated for two or more measurement periods representing seasonal or annual use extending across all users in the sample
5. Pooled time series cross-sectional data for all measurement periods and all water users
6. Pooled time series cross-sectional data aggregated over two or more measurement periods for all water users

In mathematical terms, we can describe each data configuration by designating the water use of user i during measurement period t as q_{it}. If n and m represent, respectively, the number of users in the sample and the number of measurement periods, we can describe the six data configurations as

1. *Customer time series data.* Represents total water use of user i in each time period t:

$$q_{it}$$

where i is a constant and $t = 1, \ldots, m$.

2. *Aggregate time series data.* Represents total water use of all users in the sample in each time period t:

$$\sum_{i=1}^{n} q_{it}$$

where $t = 1, \ldots, m$.

3. *Cross-sectional billing period data.* Represents total water use of *each* user i during billing period t:

$$q_{it}$$

where $i = 1, \ldots, n$ and t is a constant.

4. *Seasonal (or annual) cross-sectional data.* Represents total water use of each user i during seasonal or annual period k:

$$\sum_{t=1}^{k} q_{it}$$

where $i = 1, \ldots, n$ and k is the number of billing periods in each season.

5. *Pooled time series cross-sectional data.* Represents water use of each user i in each time period t:

$$q_{it}$$

where $i = 1, \ldots, n$ and $t = 1, \ldots, m$.

6. *Pooled time series cross-sectional data with seasonal (or annual) aggregations.* Represents water use of each user i in each season comprising k billing periods:

$$\sum_{t=1}^{k} q_{it}$$

where $i = 1, \ldots, n$ and k is the number of billing periods in each season.

If observations on water use, q_{it}, are supplemented with data on variables that are believed to be predictors of water demand, then regression analysis can be applied to any of the above data configurations. Section 1.7.2 describes various water-use modeling techniques.

1.5 WATER-USE AND SERVICE AREA DATA

Water-use data from water supply agency (i.e., water utility) records can be used for examining historical trends in water use and disaggregating total use into seasons, sectors, and specific end uses within each sector.

1.5.1 Water-Use Data

Water production records are a good source of data on total water demands in the area served by a public system. Water utilities usually have one or more production meters that are generally read at least daily. These production meters are typically maintained for accuracy and therefore usually produce highly reliable measurements of water flows into the distribution system. The water treatment plants or pumping stations usually employ continuous metering of the flow of finished water to the distribution system. The data may be recorded on paper recording charts which can be used to generate a time series of total production (or production from various supply sources) at daily or hourly time intervals. These data can be used for deriving temporal characteristics of aggregate water use (e.g., peak-day, peak-hour, day-of-week). The usefulness of the production data for water-demand analysis may include, but is not limited to, (1) the analysis of unaccounted water use (comparing production with water sales data), (2) the measurement of the aggregate effect of emergency conservation campaigns on water use, or (3) the analysis of the relationship between water production and weather variability.

Customer billing data can be used for disaggregating water use into customer sectors. Typically, retail water agencies maintain individual computer records of monthly, bimonthly, quarterly, or, less often, semiannual or annual water-consumption records for all metered customers. Active computer files usually retain up to 12 or 15 past meter readings for each customer. Depending on the length of the billing cycle, the active file records can contain a 12- to 36-month history of water use. This is a valuable source of water-use data that is often not exploited to its full potential. However, billing data can suffer from the following problems: (1) unequal billing periods, (2) lack of correspondence between billing periods and calendar months, (3) estimated meter readings or incorrect meter readings due to meter misregistration, (4) unusual usage levels, (5) meter replacements and manual adjustments to meters, and (6) changes in customer occupancy. Because of these problems, sole reliance on customer billing records necessarily limits the usefulness of these data for various measurements. Although billing data are becoming more readily available on electronic media, the use of electronic media is still not routine with many utilities.

Special metering is sometimes undertaken in order to obtain water-use data for research purposes. In recent years, water utilities have begun to utilize new meter-reading technologies that greatly reduce the cost of monitoring the water use of individual customers. However, the initial investment costs of adopting these new technologies can be prohibitively high. The new technologies include automatic meter-reading (AMR) devices and electronic remote meter-reading (ERMR) devices (see Schlenger, 1991). Automatic meter-reading devices are carried by meter readers on their routes and plugged into the site meters for the automatic meter reading. The data from the automatic meter readings are stored in the AMR device and then can be downloaded to central computer systems. The ERMR devices can be used to read site meters without actually visiting the meter location. Various methods of remote meter reading include (1) telephone dial-outbound, (2) telephone dial-inbound, (3) telephone scanning, (4) cable television, (5) radio frequency, and (6) power-line carrier.

These new technologies may permit water agencies to obtain daily or weekly meter readings from individual accounts. More frequent readings can improve the precision of water-use measurements. However, the most useful measurements can be obtained by devices that monitor individual pulses of water use on the service line. These pulses can be correlated with water flows through individual fixtures and appliances on customer premises (toilet flushes, shower flows, etc.), thus permitting accurate measurements of all end uses of water (DeOreo et al., 1996). For example, the Research Foundation of the American Water Works Association and 12 cities in the United States and Canada sponsored a unique study of water use in 1200 households. The study utilized the latest technology in micrometering of residential flows to measure precisely the amounts of water used for individual domestic purposes such as showering, bathing, toilet flushing, clothes washing, dish washing, yard watering, and other end uses of urban residents. Water meter readings were recorded in 10-second intervals using electronic data loggers. More detailed technical information about this study can be found in the final technical report authored by DeOreo et al. (1998).

Currently, these technologies are used only by a few water agencies. However, as the technology becomes more frequently adopted, there is great potential for new information sources for water-demand analysis. Unobtrusive metering technologies that permit accurate measurements of end uses of water are particularly useful for measurement of efficiency-in-use and water conservation planning.

1.5.2 Service Area Data

Because water use is a function of demographic, economic, and climatic factors, an accurate description of the characteristics of the resident population, housing stock, economic activities, and weather patterns in the service area can serve as a basis for the analysis of water demands. This section discusses the types of information that can be used to characterize the service area and identify potential data sources.

Accurate *service area maps* are indispensable. Often water service areas do not follow political (i.e., city or county) boundaries. However, demographic and socioeconomic data are most readily available by political boundaries or by census-designated boundaries (i.e., census tracts). Therefore, in order to relate water service area data to demographic and socioeconomic characteristics for planning purposes, it is often necessary to determine the relationship between service area boundaries and political or census-designated boundaries. A service area map with overlays for census tract, zip code, or other political boundaries is most useful for this purpose. Service area maps with geographic or land-use overlays have usefulness in many other planning activities. For example, only parts of the service area may be targeted for specific activities (e.g., the service area might be disaggregated spatially into pressure zones or rate zones for the purpose of water-use forecasting and/or facility planning).

Usually, a set of maps can be obtained from the facility planning or engineering department of the water agency. These maps should indicate the historical growth of the service territory and potential future additions. Also, the maps should note areas within a given community that are partially served or are served by other water supply agencies. Maps denoting political boundaries and demographic characteristics can often be obtained from local and regional planning agencies. Mapping work is greatly simplified by geographic information system (GIS) technology, which is a computerized mapping technique. The GIS may have the service area divided into separate units (e.g., census tracts, pressure zones) and have several information bases about each separate unit (e.g., water-use characteristics, land use, socioeconomic characteristics).

Population and housing characteristics (i.e., household income, lot size, persons per household, home value) are determinants of residential water use. Therefore, it is important to obtain information on these characteristics as well as to understand their impact on water use. The conventional method of estimating *population served* by multiplying *total service connections* by *persons per connection* is not very accurate. Although this method may be acceptable to estimate the population served in the residential sector (assuming an accurate measurement of persons per connection), this is not an accurate method for total service area population served because of the confounding effect of commercial, institutional, governmental, and industrial accounts.

Knowledge of the number and type of housing units in the service area is very useful in water-demand analysis because water-use patterns differ among housing types. On both a per housing unit and a per capita basis, water use in the multifamily sector tends to be lower than in the single-family sector. This is the result of different household compositions and the fact that residents in multifamily housing, on a per unit basis, have less opportunity for outdoor water-use practices. The total number of residential *accounts* served by a water supply system is not a good indication of the number of housing *units* in the service area because of the varying number of units served by multifamily accounts. However, the number of single-family customer accounts is typically a good indication of the number of single-family housing units served by the water system. Unless the water agency maintains records on the number of multifamily living units per multifamily account, housing count data must be obtained from the demographic data developed by (1) the U.S. Bureau of the Census, (2) state departments of finance or economic development, (3) regional associations of governments, and (4) local county and city planning agencies.

Again, deviations of the water service boundaries from political boundaries must be considered when using these data. In addition to population and housing counts, some of the agencies listed above can also provide data on population characteristics (e.g., family size, age, income) and data on local housing (e.g., number of homes by type, new construction permits, vacancy rates). Table 1.4 gives examples of demographic and housing data that are included in the 2000 census files. Two primary questionnaires were used in the collection of census data—the short form which included questions that were asked of all persons and housing units (i.e., the 100 percent component) and the long form which targeted only about 20 percent of the population on additional subject items (i.e., the sample component). These data are presented by geographic and political subdivisions (i.e., states, counties, cities). For major urban areas, census data are further disaggregated into census tracts, city blocks, and block groups (but not for individual dwellings).

Information on *commercial, institutional, and industrial activities* in the service area is helpful in analyzing nonresidential water demands. There is a great diversity of purposes for which water is used in the commercial, institutional, and industrial (manufacturing) sectors. The uses of water may include sanitary, cooling and condensing, boiler feed, and landscape irrigation. The type of business activity conducted in a commercial and industrial establishment can provide useful information regarding the purposes for which water is used and therefore the types of conservation measures that might be applicable. Furthermore, data on square feet of floor space, land acreage, number of employees, number of rooms (for hotels and schools), and financial performance can also be useful information in predicting commercial and industrial water use. However, some of this information is not readily available for individual establishments or aggregated into political or census-designated boundaries. Some of the previously listed agencies maintain data on local economic activities including the number of establishments, employment, and financial performance of businesses (e.g., sales)

TABLE 1.4 Selected Examples of U.S. 2000 Census Data

Population	Housing
100% component	
Household relationship	Number of units in structure
Sex	Number of rooms in unit
Race	Tenure (owned or rented)
Age	Value of home or monthly rent paid
Ethnic/racial origin	Congregate housing (meals included in rent)
	Vacancy characteristics
20% sample components	
Social characteristics:	
Education	Year moved into residence
Ancestry	Number of bedrooms
Migration	Plumbing and kitchen facilities
Language spoken at home	Telephone in unit
	Heating fuel
Economic characteristics:	Source of water and method of sewage disposal
Labor force	Year structure built
Place of work and journey to work	Condominium status
Year last worked	Farm residence
Occupation, industry, and class of worker	Shelter costs, including utilities
Work experience in 1999	
Income in 1999	

disaggregated by industry type as denoted by the Department of Commerce Standard Industrial Classification (SIC) and the newly devised North American Industry Classification System (NAICS) codes.

Additional establishment or employment data can be obtained from local and regional planning commissions and local chambers of commerce or purchased from private firms. Some private vendors can provide customized computer databases containing information on large samples of businesses in designated geographic areas. Establishment and employment data can be analyzed to determine types of business establishments that represent a major portion of nonresidential water use either because of large employment or because of large water requirements for processing or other needs. Business types can be cross-checked with agency billing records and used for disaggregating water use into specific groups of nonresidential users.

Water used in parks, cemeteries, school playgrounds, and highway medians can account for a significant portion of total use and offer a potential for conservation. Data on *public and government facilities* can be obtained from city and regional planning departments, city park districts, street departments, and the department of transportation (for highway medians). Information that might be obtained includes land-use data for various purposes (in square feet or acres) as well as the number of employees in various facilities.

1.6 COMPONENTS OF WATER DEMAND

The quantities of water delivered by a public water supply system can be disaggregated by user sector and season. Table 1.5 gives an example of such decomposition of water demands in the urban area of southern California. This section describes analytical methods for disaggregating total urban water use.

1.6.1 Sectors of Water Users

Customer billing records can be used to obtain estimates of total metered use of water and to determine the distribution of total water use among several homogeneous classes of water users. A disaggregation of total metered use into major user sectors (such as residential, commercial, and industrial) can be developed from customer billing records by using one of the following four methods: (1) analysis of available premise (user type) categories, (2) distribution of meter sizes, (3) sampling of billing files, and (4) development of premise code data.

If individual customer files contain *customer premise categories* identifying the type of customers, then a simple computer program may be used to produce annual billing summaries by customer type. Each premise code can be assigned to one of the following homogeneous sectors of water users: (1) single-family residential, (2) multifamily complexes and apartment buildings, (3) commercial sector, (4) government and public sector, (5) manufacturing (industrial) sector, and (6) unaccounted uses. Water use in these sectors can be disaggregated into seasons and end uses if necessary.

In cases where the customer billing file does not contain premise code (or customer-type) information, an approximate separation of users by residential, commercial, and industrial categories can be performed based on the distribution of *meter sizes* with a manual classification of the largest users. For example, single-family homes and small businesses are usually serviced by $^5/_8$-inch (in) or $^3/_4$-in meters. The problem with this method is that some meter sizes, particularly the larger meters (e.g., 1-in) may overlap several customer types thus decreasing the precision of the disaggregation of water use into customer classes.

In cases where water-use data by customer class are not available and the distribution of water use by meter sizes produces unreliable estimates of water use by customer class, disaggregation can be accomplished by taking a *random sample of customer accounts.* The number of accounts in the sample (i.e., sample size) will depend on the desired accuracy of water use in different customer classes. Sampling efficiency (or precision) can be improved by taking a stratified random sample of users with complete enumeration of the large water-using customers. Possible stratified sampling procedures include (1) taking a random sample of customer accounts from each meter-size category

TABLE 1.5 Sectoral and Seasonal Disaggregation of Urban Water Use in Southern California (Most Likely Ranges Given in Parentheses)

		Percent of total annual use						
		Seasonal disaggregation				Components of outdoor use		
User sector/ subsector	Disaggregation of urban sectors	Nonseasonal (base use)	Seasonal (peak use)	Indoor use	Outdoor use	Irrigation	Cooling (AC)	Other
Residential sector								
Single-family (65–75)	34.4 (25–35)	68.7 (60–70)	31.3 (30–40)	65.4	34.6	30.8	0.0	3.8
Multifamily (80–90)	25.0 (10–20)	83.9 (75–85)	16.1 (15–25)	82.2	17.8	16.1	0.4	1.3
Total residential (70–80)	59.4 (20–30)	72.1 (65–75)	27.9 (25–35)	69.6	30.4	27.2	0.2	3.0
Nonresidential sector								
Commercial (70–80)	18.8 (20–30)	74.9 (70–75)	25.1 (25–30)	71.3	28.7	21.8	6.9	0.0*
Industrial (75–90)	6.0 (10–25)	79.5 (75–90)	20.5 (10–25)	79.5	20.5	12.3	8.2	0.0*
Public (30–50)	5.1 (50–70)	46.2 (30–50)	53.8 (50–70)	46.2	53.8	53.8	0.0†	0.0†
Other (30–60)	1.1 (40–70)	58.0 (30–60)	42.0 (40–70)	58.0	42.0	42.0	0.0†	0.0†
Irrigation (10–50)	0.4 (50–90)	34.0	66.0	0.0	100.0	100.0	0.0	0.0
Total nonresidential	31.4	70.0	30.0	67.4	32.6	26.9	5.7	0.0
Unaccounted use	9.2	100.0	0.0	100.0	0.0			
Total urban use	100.0	74.0	26.0	71.7	28.3	24.6	1.9	1.8

*Other uses in these sectors are included under landscape irrigation.

†Cooling and other uses are included under landscape irrigation.

Source: Dziegielewski et al. (1990).

(sample size within each stratum can be proportional to the total water use within each stratum); (2) using the same approach as in procedure 1 except excluding the upper strata (e.g., meter sizes greater than 2 in) from the sampled population and performing complete enumeration of the largest users; or (3) separating all accounts into two categories based on meter size, with the smallest meter sizes representing the residential sector and the remaining meters representing the nonresidential sector, and then taking a random sample from each category.

The samples of the customer accounts can then be assigned manually into customer classes by visual inspection of customer record (account name) and/or by telephone verification of customer accounts. Depending upon the sample sizes, this can be a time- and resource-intensive exercise.

Regardless of which sampling approach is selected, analysis of the sample should produce the following two estimates: (1) proportion (or percent) of total customers by customer class, and (2) proportion (or percent) of total metered water use by customer class. It is also desirable to calculate the precision of the estimates based on the sample variance.

If sufficient time and resources are available, the classification of all customer accounts into appropriate user types is a worthwhile undertaking. The classification would require: (1) adding a data field (or using existing unassigned fields) to the customer computer file; (2) developing a set of nonoverlapping customer classes and precisely defining each class; (3) determining the customer class for each existing customer by classifying all customers during meter reading or surveying (independently) all customers and requesting them to classify their premises on water bills; and (4) adding customer classification categories to the application forms for new connections.

1.6.2 Seasonal and Nonseasonal Components

Within each user sector, water use can be separated into its seasonal and nonseasonal components. *Seasonal use* can be defined as an aggregate of end uses of water, such as lawn watering or cooling, that varies from month to month in response to changing weather conditions (or due to other influences that are seasonal in nature). *Nonseasonal use,* on the other hand, can be defined as an aggregate of end uses of water, such as toilet flushing or dishwasher use, that remain relatively constant from month to month because these uses are not sensitive to weather conditions or other seasonal influences. Often, seasonal and nonseasonal components of water use are taken to represent the outdoor (or exterior) and indoor (or interior) water uses, respectively. Such an assumption is imprecise because some uses that occur inside the buildings can be seasonal (e.g., humidifier use or evaporative cooler), and some outdoor uses can be nonseasonal (e.g., car washing in warmer climates). The difficulties in classifying various end uses into outdoor and indoor categories must be kept in mind when water use is divided into seasonal and nonseasonal components.

Monthly water use data can be used to derive estimates of seasonal and nonseasonal water use. The terms *seasonal* and *nonseasonal* relate to the method of characterizing a monthly time series of water-use records. This method is sometimes referred to as the *minimum-month* method because it uses the month of lowest use to represent the nonseasonal component of water use. With the minimum-month method, the percent of annual use in a given year that is considered seasonal is calculated from the formula

$$S_P = 100 - (M_P \cdot 12) \tag{1.16}$$

where S_P is the percent of annual use that is seasonal and M_P is the percent of annual use during the minimum month.

The best representation of how much water is used during a given calendar month is the aggregation of daily pumpage information which records how much water enters the distribution system every day. Monthly water use information for customer groups is more difficult to obtain because of the effects of monthly and bimonthly billing cycles. When water utilities summarize the amount of water sold in a given month (both in aggregate or by customer group), this information typically represents the amount of water billed in a given month rather than the amount of water used. Bimonthly billing cycles (which indicate the amount of water used over a 2-month period) further confound calendar month water use. Differences between the amount of water billed in a given month and the actual water consumed occurs because (1) accounts are read in different months (e.g., some accounts are read in January, March, May, etc., and others are read in February, April, June, etc.), and (2) meter readings are typically recorded on any given day within a month.

In order to get a better representation of monthly water-use patterns, it is necessary to allocate water use from monthly and bimonthly billing cycle records into water use during specific calendar months. Thus, the primary purpose of the allocation (or smoothing) techniques is to adjust water consumption billing records which are read on either a monthly or bimonthly basis (e.g., even accounts read on even months and odd accounts read on odd months in the bimonthly case) into calendar month consumption for purposes of further analysis. Data smoothing procedures are performed on two levels. First, the information provided by water utilities on the amount of water billed to a customer group in a given month is smoothed to represent the amount of water actually consumed by a customer group in a given month. The smoothing procedure varies depending on whether a bimonthly or monthly billing cycle is in effect. Second, account-level water-use records are smoothed so that estimates of calendar month water use can be determined. Both types of procedures are described below.

Figure 1.1 presents a graphical representation of the procedure used to smooth aggregate sales data which utilize a *monthly billing cycle.* A monthly billing practice involves reading water meters of individual customers in approximately 1-month-long time intervals. Meters are read every working day, and all meters read during, for example, the month of February, are billed and recorded as the February water use ($Q_{\mathrm{Feb4}}{}^{b}$). In reality, only a portion of the billed water use, $Q_{\mathrm{Feb}}{}^{b}$, actually occurred during the calendar month of February. Theoretically, $Q_{\mathrm{Feb}}{}^{b}$ represents water use of individual customers during n monthly periods (where n is the number of billed customers) ending between the first and last meter-reading date in February. Therefore, for individual customers, 1-month-long periods of water use would fall between January 1 and February 28.

Assuming that all users in a given customer group (e.g., single-family, commercial) are relatively homogeneous with respect to water use and that the effects of weather on water use during the two consecutive calendar months are not substantially different (i.e., that water use of the individual customers is evenly distributed throughout the period between meter-reading dates), the calendar month water use during month N can be estimated as

$$Q_N{}^c = 0.5Q_N{}^b + 0.5Q_{N+1}{}^b \tag{1.17}$$

where Q^c is the amount of water used during the calendar month and Q^b is the amount of water billed during the calendar month. This equation indicates that water actually used during the calendar month of, for example, March ($Q_{\mathrm{Mar}}{}^{c}$) would comprise one-half of the consumption billed in March ($Q_{\mathrm{Mar}}{}^{b}$) and one-half of the consumption billed in April ($Q_{\mathrm{Apr}}{}^{b}$).

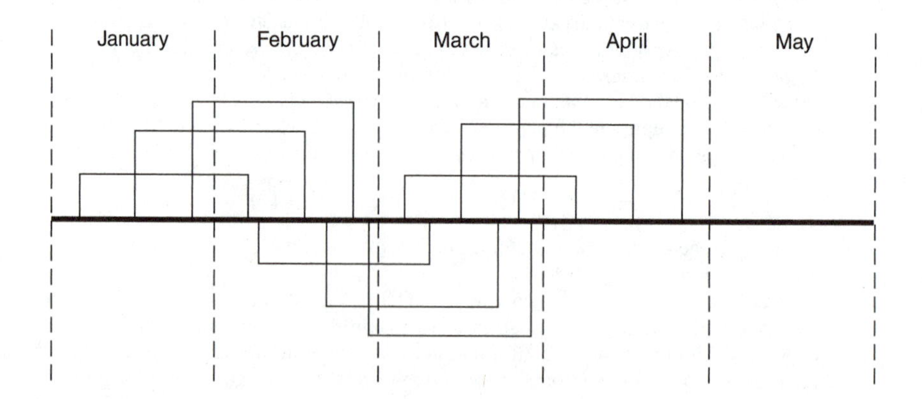

FIGURE 1.1 Allocation of monthly water-use billing records into calendar month consumption. Estimate of calendar month water-use ($Q_N{}^c$): February, $Q_{\mathrm{Feb}}{}^c = x\,Q_{\mathrm{Feb}}{}^b + y\,Q_{\mathrm{Mar}}{}^b$; March, $Q_{\mathrm{Mar}}{}^c = x\,Q_{\mathrm{Mar}}{}^b + y\,Q_{\mathrm{Apr}}{}^b$; Nth month, $Q_N{}^c = x\,Q_N{}^b + y\,Q_{N+1}{}^b$. $Q_{\mathrm{Mar}}{}^b$ = water use billed to customer during the month of March; x and y = proportions of $Q_N{}^b$ and $Q_{N=1}{}^b$ billed water use allocated to $Q_N{}^c$ and $Q_{N=1}{}^c$.

Whereas Eq. (1.17) allows the calculation of calendar month water use, a variant of this equation can be used to calculate the average water use per account in a given month when the number of billed accounts varies between the two consecutive months:

$$q_N = \frac{Q_N}{A_N}\,\frac{A_N}{A_N + A_{N+1}} + \frac{Q_{N+1}}{A_{N+1}}\,\frac{A_{N+1}}{A_N + A_{N+1}} \tag{1.18}$$

where q_N = average water use per account in any given month
Q_N = amount of water billed in any given month
A_N = number of accounts billed in any given month

Figure 1.2 presents the procedure that was used to allocate aggregate monthly sales data produced as a result of a bimonthly meter reading cycle. In this procedure, although individual meters are read every 2 months, customer billing is performed during each calendar month. As a result, all customers within a given sector are divided into two groups, referred to as group A and group B. Meters of

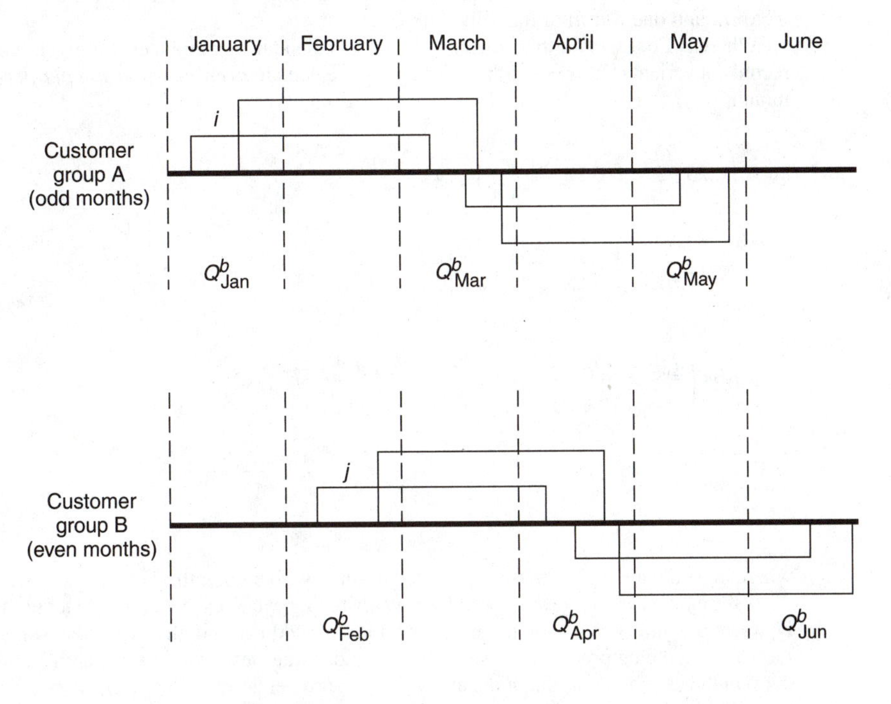

FIGURE 1.2 Allocation of bimonthly water-use billing records into calendar month consumption. Estimate of calendar month water use ($Q_N{}^c$): March, $Q_{\text{Mar}}{}^c = 0.25Q_{\text{Mar}(A)}{}^b = 0.50Q_{\text{Apr}(B)}{}^b + 0.25Q_{\text{May}(A)}{}^b$; April, $Q_{\text{Apr}}{}^c = 0.25Q_{\text{Apr}(A)}{}^b + 0.50Q_{\text{May}(B)}{}^b + 0.25Q_{\text{Jun}(a)}{}^b$; Nth month, $Q_N{}^c = 0.25Q_N{}^b + 0.50Q_{N+1}{}^b + 0.25Q_{N+2}{}^b = [(Q_N + Q_{N+2})/2 + Q_{N+1}]/2$. $Q_{\text{Mar}}{}^b$ = water use billed during the month of March (includes only customers in group A); $Q_{\text{Apr}}{}^b$ = water use billed during the month of April (includes only customers in group B).

customers in group A are read and billed during odd months (January, March, May, etc.), while in group B, meters are read and billed during even months (February, April, June, etc.).

Because some accounts are billed in any given month, aggregate monthly sales data will reflect, in any given month, only approximately one-half of the true number of accounts. For example, water use billed and recorded as the March consumption ($Q_{\text{Mar}}{}^b$) includes only customers of group A whose individual 2-month consumption period falls between January 1 and March 31. By analogy, water use billed and recorded as the April consumption ($Q_{\text{Apr}}{}^b$) includes only customers of group B whose individual 2-month consumption period falls between February 1 and April 30.

Again, assuming that water users in both groups are homogeneous with respect to water use (because they represent the same user sector) and that weather effects during the consecutive calendar months are not drastically different (i.e., that water use of the individual customer is evenly distributed throughout his or her 2-month consumption period), we can estimate total water use during any calendar month using the following formula:

$$Q_N{}^c = 0.25Q_N{}^b + 0.5Q_{N+1}{}^b + 0.25Q_{N+2}{}^b \tag{1.19}$$

This relationship is derived from Fig. 1.2. It indicates that water actually used during the calendar month of March would comprise one-fourth of consumption billed in March, one-half of that billed in April, and one-fourth of that billed in May.

Whereas Eq. (1.19) allows the calculation of calendar month water use given bimonthly billing records, a variant of this equation can be used to calculate *average water use per account* in a given month:

$$q_N = \left(0.25\frac{Q_N}{A_N} + 0.25\frac{Q_{N+2}}{A_{N+2}}\right)\frac{0.5A_N + 0.5A_{N+2}}{0.5A_N + A_{N+1} + 0.5A_{N+2}} + 0.5\frac{Q_{N+1}}{A_{N+1}}\frac{A_{N+1}}{0.5A_N + A_{N+1} + 0.5A_{N+2}} \tag{1.20}$$

$$= 0.25\left(\frac{Q_N}{A_N} + \frac{Q_{N+2}}{A_{N+2}}\right)\frac{A_N + A_{N+2}}{A_N + 2A_{N+1} + A_{N+2}} + 0.5\frac{Q_{N+1}}{A_{N+1}}\frac{2A_{N+1}}{A_N + 2A_{N+1} + A_{N+2}} \tag{1.21}$$

where q_N is the average water use per account in any given month.

In the case where both *monthly* and *bimonthly* billing cycles exist within a water utility, total monthly water use in a given month can be obtained by adding smoothed calendar water use from the monthly smoothing procedure to smoothed calendar water use from the bimonthly smoothing procedure. In the case of determining the average water use per account in any given month given the existence of both monthly and bimonthly billing cycles, the following weighting procedure can be used:

$$q_{N,avg} = q_{N,b}\,\text{WF}_b + q_{N,m}\,\text{WF}_m \tag{1.22}$$

where $q_{N,\text{avg}}$ = average water use per account in any given month
$q_{N,b}$ = average water use per account in any given month from bimonthly billing cycle
$q_{N,m}$ = average water use per account in any given month from monthly billing cycle
WF_b = weight factor for water use per account from bimonthly billing cycle [see Eq. (1.23)]
WF_m = weight factor for water use per account from monthly billing cycle [see Eq. (1.24)]

$$\mathrm{WF}_b = \frac{0.5A_{N,b} + A_{N+1,b} + 0.5A_{N+2,b}}{(0.5A_{N,b} + A_{N+1,b} + 0.5A_{N+2,b}) + (0.5A_{N,m} + 0.5_{N+1,m})}$$

$$= \frac{A_{N,b} + 2A_{N+1,b} + A_{N+2,b}}{(A_{N,b} + 2A_{N+1,b} + A_{N+2,b}) + (A_{N,m} + A_{N+1,m})} \tag{1.23}$$

$$\mathrm{WF}_m = \frac{0.5A_{N,m} + 0.5A_{N+1,m}}{(0.5A_{N,b} + A_{N+1,b} + 0.5A_{N+2,b}) + (0.5A_{N,m} + 0.5A_{N+1,m})}$$

$$= \frac{A_{N,m} + A_{N+1,m}}{(A_{N,b} + 2A_{N+1,b} + A_{N+2,b}) + (A_{N,m} + A_{N+1,m})} \tag{1.24}$$

In addition to the allocation of aggregate billing data into calendar month, water consumption records for individual customers can also be allocated to calendar months. For example, if a single account's meter is read on March 15 and then again on May 15, a smoothing procedure can be used to standardize individual account billing cycle data into calendar month use. The meter-reading cycle can be monthly, bimonthly, or even trimonthly. The composition of the equation applied to this type of smoothing is as follows:

$$Q_n^c = \sum_{i=(n-x)}^{i=(n+y)} N_i \overline{Q_i^b} \tag{1.25}$$

where n = nth calendar month
x = number of months prior to month n that fall within the billing period
y = number of months beyond month n that fall within the billing period
i = summation index (ith month)
Q_n^c = quantity of water allocated to calendar month n
N_i = number of days in ith month that are proportioned to consumption in month n
$\overline{Q_i^b}$ = average daily water consumption for billing period which represents ith month

This smoothing procedure is applied to the water billing histories for each account. In this procedure, each account will be smoothed in accordance with the current read date and the prior read date. That is, the smoothing procedure must first look at the current and prior read dates. Next, consumption for each record is allocated into calendar months by the fraction of water use that belongs to each month encountered in the billing period [see Eq. (1.25)]. Finally, consumption is summed for each account by month.

1.6.3 End Uses of Water

A meaningful assessment of the efficiency of water use cannot be made without breaking down the seasonal and nonseasonal uses of water into specific end uses. Precise measurements of the quantities of water used for showering, toilet flushing, and other purposes require installation of flow-recording devices on each water outlet found on customer premises. Because such measurements are very costly, engineering estimates are often used. However, such estimates are of limited validity because they tend to rely on many assumptions and often ignore the physical and behavioral settings in which water use takes place. Analytical methods for quantifying the significant end uses of water are described in Sec. 1.7.

1.7 WATER-USE RELATIONSHIPS

Reasonably precise estimates of water use can be obtained by disaggregating the total delivery of water to urban areas into two or more classes of water use and determining separate average rates of water use for each class. The *disaggregate estimation of water* use can be represented as the product of the number of users (or the demand driver count) and a constant average rate of usage.

$$Q_t = \sum_c N_{t,c} q_c \tag{1.26}$$

where $N_{t,c}$ represents the number of customers in a homogeneous user sector c at time t, and q_c is the unit-use coefficient (or average rate of water use per customer) in that sector.

Gains in accuracy of disaggregated estimates are possible because the historical records often show that the variance of average rate of water use within some homogeneous sectors of water users is smaller than the variance of the aggregate use. Although in some nonresidential sectors the variance of q_c is large, it is often more than offset by the smaller variance in the residential sectors with a disproportionately greater number of customers.

The *sectoral disaggregation* of total urban demands may also be extended spatially and temporally. With the added dimensions of disaggregation, Eq. (1.26) would be expanded to the form:

$$Q_t = \sum_c \sum_s \sum_g N_{t,c,s,g}\, q_{c,s,g} \tag{1.27}$$

where s denotes the disaggregation of water use according to its seasonal variations (e.g., annual, seasonal, monthly) and g represents the spatial disaggregation of water use into various geographic areas, such as pressure districts or land-use units, which are relevant for planning purposes. An example of a unit-use coefficient $q_{c,s,g}$ could be average use in a single-family residential sector during the summer season in a pressure district.

The level of disaggregation is limited by the availability of data on average rates of water use in various sectors and the ability to obtain accurate estimates of driver counts (N) for each disaggregate sector. The latter are typically obtained from planning agencies who maintain data on population, housing, and employment for local areas.

The last three decades have produced numerous studies of the *determinants of urban water demand.* The advancements in theory were followed by the development of water use models which recognized that both the level of average daily water-use and its seasonal variation can be explained adequately by selected demographic, economic, and climatic characteristics of the study area. Such advanced models retain a high level of disaggregation; however, they allow the average rates of water use and the number of users (i.e., drivers) to change in response to changes in their determining factors:

$$Q_t = \sum_c \sum_s \sum_g N_{t,c,s,g}\, q_{t,c,s,g} \tag{1.28}$$

where

$$N_{t,c,s,g} = f(Z_i) \tag{1.29}$$

and

$$q_{t,c,s,g} = f(X_j) \tag{1.30}$$

where Z_i and X_j are, respectively, determinants of the number of water users (e.g., residents, housing units, or employees) and determinants of the average rates of water use (such as average water use per person, per household, or per employee). Table 1.6 gives examples of determinants of the number of users and their average rates of water use.

The remainder of this section describes the available data on the average rates of water use and reviews a number of empirical water-use models.

1.7.1 Average Rates of Water Use

The first step in the analysis of water demands in an area served by a public water supply system is to determine average annual rates of water use. The simplest rate is the gross per capita water use which is determined by dividing the total annual amount of water delivered to the distribution system by the estimated population served. Other rates are obtained by dividing the metered water use by various urban sectors by the number of users or customers in each sector. In addition to the average annual use, average rates during high-use and low-use seasons can be estimated.

The average rates of use are sometimes compared among service areas in order to assess the relative efficiency of water use. However, the aggregate nature of these rates precludes any meaningful comparisons among various service areas. These rates vary from city to city as a result of differences in local conditions that are unrelated to the efficiency in water use. For example, per capita use may range from 50 to 500 gal per person per day [189 to 1893 liters per day (L/day).] Table 1.7 shows the distribution of per capita rates among 392 water supply systems serving approximately 95 million people in the United States. The mean per capita use in this sample was 175 gcd (662 L/day) with a standard deviation of 72 gcd (273 L/day) (AWWA, 1986).

Generally, high per capita rates are found in water supply systems servicing large industrial or commercial sectors. Therefore, more meaningful comparisons would require the disaggregation of total use into homogeneous sectors of water users. Table 1.8 shows average use rates per housing unit in the residential sector for selected cities. Nonresidential water-use rates in gallons per employee per day (ged) are shown in Table 1.9.

TABLE 1.6 Determinants of Urban Water Demand

Determinants of demand drivers, Z_i	Determinants of average rates of use, X_j
Natural birth rates	Air temperature and precipitation
Net migration	Type of urban landscapes
Family formation rates	Housing density (average parcel size)
Availability of affordable housing	Water-use efficiency
Economic growth and output	Household size and composition
Labor participation rates	Median household income
Urban growth policies	Price of water service
	Price of wastewater disposal
	Industrial productivity

TABLE 1.7 Average per Capita Rates of Water Use

Range of per capita use, gcd	Number of systems	Percent of systems
50–99	30	7.7
100–149	132	33.7
150–199	133	33.9
200–249	51	13.0
250–299	19	4.8
300+	27	6.9
Total	392	100.0

Source: Source: AWWA (1986).

TABLE 1.8 Average Rates of Water Use in Single-Family and Multiple-Unit Buildings

		Average use per dwelling unit, gpd	
Area/State	Year	Multiple-unit buildings	Single-family homes
A. Metered sales data			
National City, Calif.	1987	226	309
San Diego, Calif.	1990	261	
Santa Monica, Calif.	1988	231	376
Torrance, Calif.	1987	183	383
Anaheim, Calif.	1987	284	526
Beverly Hills, Calif.	1987	241	917
Camarillo, Calif.	1987	246	437
Fullerton, Calif.	1988	246	566
Los Angeles, Calif.	1987	264	408
Cape Coral, Fla.	1990	123	
Boston, Mass.	1990	167	
Framingham, Mass.	1992	152	
Newton, Mass.	1992	199	
Las Vegas, Nev.	1989	290	
B. Samples of buildings			
New York City, N.Y.	1990	248	
Seattle, Wash.	1988	107	
Baltimore, Md.	1965	218	
Washington, D.C.	1981	197	
Springfield, Ill.	1985	117	
Los Angeles, Calif.	1976	165	
San Francisco Bay, Calif.	1976	183	
Central Valley, Calif.	1975	144	
San Diego, Calif.	1991	137	
Pasadena, Calif.	1991	187	
Newton, Mass.	1992	155	
Bailment, Mass.	1990	121	
Framingham, Mass.	1992	164	

1.7.2 Modeling of Water Use

Water-use models are usually obtained by fitting theoretical functions to the data sets described in Sec. 1.1. The selection of appropriate data and estimation techniques depends on the desired characteristics of the final model. For example, time series models of aggregate data can be used for developing models of aggregate use for near-term forecasting. Generally, time series aggregate models can be expected to provide predictions of water use that are less reliable than those obtained from pooled time series cross-sectional observations on water use of individual customers from homogeneous groups of users.

The most commonly used technique for developing water-use models is *regression analysis.* A simple regression model that captures the relationship between two variables can be written as

$$y_i = \alpha + \beta X_i + \varepsilon_i \tag{1.31}$$

where y_i = water use of customer i

α = intercept term of the equation and the component of the effect of X upon y that is constant regardless of the value of X

β = slope coefficient of the equation and the component of the effect of X upon y that changes depending upon the value of X

ε_i = error term for the ith customer, and $i = 1, 2, \ldots, n$, which measures the difference between the estimated value of y and the true observed value of y

This model also assumes that X, the independent variable, influences y, the dependent variable, while the dependent variable does not influence the independent variable in any way. Equation (1.31) decomposes water use y_i into *explained* and *unexplained* components, where the explained component is expressed as a function of a systematic force X. The unexplained component is expressed as random noise. In other words, in Eq. (1.31), $\alpha + \beta X$ is the deterministic component of y, and ε is the stochastic or random component.

TABLE 1.9 Average Rates of Nonresidential Water Use from Establishment-Level Data

Category	SIC code	Use rate, ged	Sample size
Construction	—	31	246
General building contractors	15	118	66
Heavy construction	16	20	30
Special trade contractors	17	25	150
Manufacturing	—	164	2790
Food and kindred products	20	469	252
Textile mill products	22	784	20
Apparel and other textile products	23	26	91
Lumber and wood products	24	49	62
Furniture and fixtures	25	36	83
Paper and allied products	26	2614	93
Printing and publishing	27	37	174
Chemicals and allied products	28	267	211
Petroleum and coal products	29	1045	23
Rubber and misc. plastics products	30	119	116
Leather and leather products	31	148	10
Stone, clay, and glass products	32	202	83
Primary metal industries	33	178	80
Fabricated metal products	34	194	395
Industrial machinery and equipment	35	68	304
Electronic and other electrical equipment	36	95	409
Transportation equipment	37	84	182
Instruments and related products	38	66	147
Misc. manufacturing industries	39	36	55
Transportation and public utilities	—	50	226
Railroad transportation	40	68	3
Local and interurban passenger transit	41	26	32
Trucking and warehousing	42	85	100
U.S. Postal Service	43	5	1
Water transportation	44	353	10
Transportation by air	45	171	17
Transportation services	47	40	13
Communications	48	55	31
Electric, gas, and sanitary services	49	51	19
Wholesale trade	—	53	751
Wholesale trade—durable goods	50	46	518
Wholesale trade—nondurable goods	51	87	233
Retail trade	—	93	1044
Building materials and garden supplies	52	35	56
General merchandise stores	53	45	50
Food stores	54	100	90
Automotive dealers and service stations	55	49	198
Apparel and accessory stores	56	68	48

TABLE 1.9 (*Continued*)

Category	SIC code	Use rate, ged	Sample size
Retail trade			
Furniture and home furnishings stores	57	42	100
Eating and drinking places	58	156	341
Miscellaneous retail	59	132	161
Finance, insurance, and real estate	—	192	238
Depository institutions	60	62	77
Nondepository institutions	61	361	36
Security and commodity brokers	62	1240	2
Insurance carriers	63	136	9
Insurance agents, brokers, and service	64	89	24
Real estate	65	609	84
Holding and other investment offices	67	290	5
Services	—	137	1878
Hotels and other lodging places	70	230	197
Personal services	72	462	300
Business services	73	73	243
Auto repair, services, and parking	75	217	108
Miscellaneous repair services	76	69	42
Motion pictures	78	110	40
Amusement and recreation services	79	429	105
Health services	80	91	353
Legal services	81	821	15
Educational services	82	117	300
Social services	83	106	55
Museums, botanical, zoological gardens	84	208	9
Membership organizations	86	212	45
Engineering and management services	87	58	5
Services, NEC	89	73	60
Public administration	—	106	25
Executive, legislative, and general	91	155	2
Justice, public order, and safety	92	18	4
Administration of human resources	94	87	6
Environmental quality and housing	95	101	6
Administration of economic programs	96	274	5
National security and international affairs	97	112	2

Source: Planning and Management Consultants, Ltd., 1994, unpublished data.

In *ordinary least-squares* (OLS) regression analysis, the parameters α and β are estimated by fitting a regression line to water-use data so that the sum of squared residuals of y ($\Sigma\varepsilon_i^2$) away from the line is minimized. The method of least squares dictates that one choose the regression line where the sum of the squared deviations of the points from the line is a *minimum,* resulting in a line that "fits" the data as well as possible.

In order for OLS to yield valid results, the residuals (ε) must meet the five assumptions of simple linear regression:

1. *Zero mean.* $E(\varepsilon_i) = 0$ for all i, or the expected value of mean error is 0. In other words, the errors are expected to fluctuate randomly about 0 and, in a sense, cancel each other out.
2. *Common (or constant) variance.* $\text{Var}(\varepsilon_i) = \sigma^2$ for all i, which states that each error term has the same variance for each customer.

3. *Independence.* ε_i and ε_j are independent for all $i \neq j$.
4. *Independence of* X_j. ε_i and X_j are independent for all i and j, which says that the distribution of ε does not depend on the value of X.
5. *Normality.* ε_i are normally distributed for all i. This also implies that ε_i are independently and normally distributed with mean 0 and a common variance σ^2. The concept of normality is needed for inferences on parameters, but is not required to find least-square estimates.

When the five basic assumptions of the regression model are satisfied, OLS provides unbiased estimates of the regression coefficients α and β, which have minimum variance among all unbiased estimates. In other words, the least-squares estimators indeed yield the estimated straight line that has a smaller residual sum of squares than any other straight line. For this reason, OLS estimates are referred to as *best linear unbiased estimates* (BLUE). Any violation of these assumptions can reduce the validity of the OLS method. The greater the departure of the model from this set of assumptions, the less reliable is OLS. In such situations, one must use alternative estimation procedures depending on the type of violation of the above assumptions. One alternative estimation technique called generalized least squares (GLS) is described later in this section.

Multiple-regression techniques should be used to model a dependent variable instead of simple regression because (1) the dependent variable can be predicted more accurately if more than one independent variable is used and (2) if the dependent variable depends on more than one independent variable, a simple regression on a single independent variable may result in a biased estimate of the effect of this independent variable on the dependent variable. The theoretical model of multiple regression is basically the same as in simple regression. The only difference is that the dependent variable is assumed to be a linear function of more than one independent variable. For example, if there are three independent variables, the model is

$$y = \alpha + \beta_1 X_1 + \beta_2 X_2 + \beta_3 X_3 + \varepsilon \tag{1.32}$$

where X_1, X_2, X_3 = independent variables assumed to affect the dependent variable y
ε = random error term
$\alpha, \beta_1, \beta_2, \beta_3$ = estimated coefficients

Just as in the case of simple regression, the coefficients α and β_i are estimated by finding the value of each that minimizes the sum of the squared deviations for the observed values of the dependent variable from the values of the dependent variable predicted by the regression equation. Furthermore, in order to obtain least-square estimates, multiple regression must follow each of the five assumptions required by simple regression, with two added conditions. The first condition is that none of the independent variables can be an exact linear combination of any of the other independent variables. In other words, no one variable can be an exact multiple (or linear combination) of any other independent variable. For example, X_i cannot be written as aX_2. This situation is called multicollinearity. The second condition is related to degrees of freedom. Specifically, the number of observations N must exceed the number of coefficients being estimated. In practice, the sample size should be considerably larger than the number of coefficients to be estimated in order to obtain meaningful information about the underlying relationship.

A *time series analysis of monthly data* on volumes of water sold in consecutive billing periods can be used to estimate water-use models. However, the reliability of such models will depend on (1) the ability to disaggregate sales data into classes of similar users (e.g., single-family residential, multiunit residential, small commercial, large industrial); (2) the ability to separate (or account for) the seasonal effects and weather effects in the time series data; and (3) the ability of the estimation technique to deal with nonconstant error variance and correlation of model errors through time. A theoretical time series model can be written as

$$y_t = a + \sum_{i=1}^{N} b_i S_{i,t} + \sum_{j=1}^{M} c_j W_{j,t} + \sum_{k=1}^{P} d_k X_{k,t} + \sum_{r=1}^{R} e_r C_{r,t} + \mathring{a}_t \tag{1.33}$$

where y_t = aggregate volume of water sold to a homogeneous class of customers during a monthly or bimonthly billing period t where $t = 1, \ldots, T$
a = model intercept
S_i = set of N seasonal variables that capture the seasonal variability of water use ($i = 1, \ldots, N$)
W_j = set of M weather variables that capture the effect of actual weather conditions on water use ($j = 1, \ldots, M$)
X_k = set of P "trend forming" variables that capture changes in water use unrelated to seasonal and weather effects ($k = 1, \ldots, P$)
C_r = set of R conservation variables ($r = 1, \ldots, R$)
ε_t = error term
b_i, c_j, d_k, e_r = coefficients to be estimated

The selection and definition of variables to represent the four types of systematic forces that affect aggregate water use over time are very important. The seasonal component in water-use data can be captured in many ways. Three possible specifications used in modeling time series water-use data include (1) a seasonal index, (2) a discrete step function, and (3) a Fourier series of sine and cosine terms.

A *seasonal index* is usually expressed as the average fraction of total annual water use to be expected during a given calendar month. This fraction can be estimated using the time series data on water use. For example, the value of the index in July can be obtained by dividing water use during the month of July by total annual use for each calendar year and then calculating the average value of the index for all years in the data set. The process is repeated for each calendar month until all 12 values of the seasonal index are obtained. The seasonal index is then used as a simple variable to capture the seasonal component of water use in Eq. (1.33). A *discrete step function* can be represented by 12 indicator variables corresponding to individual calendar months (e.g., $M_1, \ldots, M_{12}$, where $M_1 = 1$, if the month in the data is January, and $M_1 = 0$ elsewhere). When bimonthly data are modeled, six indicator variables would be created, one for each bimonthly period. In order to avoid multicollinearity, only $M - 1$ indicators should be specified, where M denotes the number of monthly or bimonthly periods. A *Fourier series* of sine and cosine terms is a harmonic function that can be applied to the data to generate a smooth sinusoidal cycle of seasonal effects. In the case of monthly data, the Fourier series may include six sine and cosine harmonics that can be written as

$$\sum_{h=1}^{6} \left(a_h \sin \frac{2\pi hm}{12} + b_h \cos \frac{2\pi hm}{12} \right) \tag{1.34}$$

where a_h and b_h are coefficients to be estimated and m is the calendar month ($m = 1$ for January, $m = 2$ for February, etc.).

The cycle corresponding to $h = 1$ has a 12-month period. The cycles corresponding to $h = 2$ are harmonics of the 6-month period. All six harmonics represent the seasonal cycle of water use, which is periodic but not directly sinusoidal. Because the lower harmonics tend to explain most of the seasonal fluctuations, in most situations, it may be possible to omit higher-frequency harmonics in Eq. (1.34), thus representing the seasonal component as

$$\sum_{i=1}^{4} b_i S_{i,t} = b_1 \,\mathrm{SIN}(1) + b_2 \,\mathrm{COS}(1) + b_3 \,\mathrm{SIN}(2) + b_4 \,\mathrm{COS}(2) \tag{1.35}$$

where b_1, b_2, b_3, b_4 = coefficients to be estimated
SIN (1) = $\sin(2\pi m/12)$
COS (1) = $\cos(2\pi m/12)$
SIN (2) = $\sin(4\pi m/12)$
COS (2) = $\cos(4\pi m/12)$

The significance of each cycle is usually tested first. The cycles with insignificant amplitudes (i.e., b_i) can then be deleted from the equation.

Air temperature and rainfall are usually used to capture the effects of weather on water use. In most cases, these two variables will be correlated with the seasonal variables. Therefore, the weather variables should be measured as deviations from their normal values for each month (or billing period). Also, lagged weather variables can be used to take into account (1) the fact that the recorded consumption in any given month represents water use which took place during the current and the previous month and (2) the short-term memory in water use (e.g., water use in month t is affected by rainfall in month $t - 1$). The following weather variables can be included in the specification of the weather effects in Eq. (1.33): (1) deviation of monthly rainfall from monthly norms, (2) deviation of monthly average of maximum daily temperatures from monthly norms, (3) deviation of the number of days with precipitation greater than 0.01 in from monthly norms, and (4) deviation of cooling-degree days from monthly norms.

These deviations can be specified both as contemporaneous and lagged measurements. The *normal* values can be calculated for the period of the time series data or for weather data extending up to 30 years back (i.e., long-term averages).

In addition to properly measuring the seasonal and weather effects, it is necessary to include the effects of variables such as the number of customers, the price of water, the cost of wastewater disposal, and other factors such as income in residential sectors and productivity in nonresidential sectors. The changes in the *number of customers* can be incorporated by expressing the dependent variable in terms of water use per customer (by dividing total volume of water by the number of customers billed). The *price of water and wastewater disposal* should be included in the model. In modeling aggregate water-use data, it is difficult to determine what measure of price should be used. The relevant measure is the marginal price faced by an individual customer. This price is the same for all customers when a uniform rate structure is used. If increasing block rates are used, then the average price determined for an average consumption level can be used, so that only actual increases in the price of water and wastewater are captured by the price variable. All nominal values of price should be converted into constant dollars using the Consumer Price Index (CPI) for all items. *Median household income* should also be included among the variables of residential models if the data are available and expressed in constant dollars. Usually, household income statistics can be obtained for each quarter from the Internal Revenue Service. Monthly values of income can be obtained by interpolating the quarterly data. Several other variables that are known to influence water use can be omitted if the changes over time are minimal (e.g., average number of persons per customer connection, average lot size). If changes in such variables are significant, then these variables should be included in the model.

Finally, the effects of the conservation programs can be accounted for in the time series model by including an indicator variable which separates the data into pre- and postprogram periods. This variable takes on the value of 0 for all months before the program, and the value of 1 for the months after program implementation. The effects of other passive and active conservation measures such as a conservation-oriented plumbing code should be included in the model. This can be accomplished by introducing a variable that measures the cumulative number of new service connections sold after the code went into effect. The *error term* can be specified as additive or multiplicative. A logarithmic transformation of the water-use variable will result in a multiplicative error term. Such a transformation will often produce a better fit of the model than untransformed water use.

The potential problems with estimating the parameters of the time series regression model are endogeneity of the price variable (i.e., the price is related to the quantity of water used), nonconstant error variance (also known as heteroskedasticity), and autocorrelation. Most regression software packages have routines that attempt to correct for the problem of autocorrelation. Nonconstant error variance and endogeneity are problems best suited for alternative regression methods such as generalized least squares.

If customer-level monthly (or billing period) data on water use for a period of 2 to 4 years can be obtained and supplemented with information on customer characteristics and external factors such as price of water and weather, then a *pooled time series cross-sectional* (TSCS) data set can be constructed and used to estimate the parameters of a multiple-regression model. The theoretical form of the model may be written as

$$y_{it} = \beta_1 + \sum_{k=2}^{K} \beta_k X_{k,it} + \varepsilon_{it} \tag{1.36}$$

where y_{it} = monthly water use of customer i in month t (instead of months, billing periods may be used)
β_1 = intercept term
β_k = regression coefficients
X_k = set of independent variables that represent all possible systematic forces which affect that use
ε_{it} = error term

The types of independent variables of residential water-use models that may be used in Eq. (1.36) are illustrated in Table 1.10. Variables that explain use in other sectors can be found in Dziegielewski et al. (2002b). Measurements of these variables (or as many variables as possible) should be obtained for *each* customer and *each* time period using such sources of information as (1) telephone or mail surveys of customers in the sample, (2) real estate and tax assessor records, (3) aerial photographs, (4) "driveby" surveys of customer premises, (5) water and wastewater prices and rate structures, and (6) meteorological stations. Many variables will have values that are constant over time or will have only one observation in the time that is available. In the latter case, their values can be assumed constant over the period for which water-use data are obtained.

Once the pooled time series cross-sectional database is complete, an appropriate form of the functional relationship between water use and its determinants must be selected. Also, in the context of the structure of systematic forces, the analyst should consider the appropriate structure of the model error. *Residential water demand models* are often estimated using one of the following functional forms:

1. Linear model

$$y_{it} = \beta_1 + \sum_{k=2}^{k} \beta_k X_{k,it} + \varepsilon_{it} \tag{1.37}$$

in which the error is additive

2. Log-linear (or double-log) model

$$\log y_{it} = \log(\beta_1) + \sum_{k=2}^{k} \beta_k (\log X_{k,it}) + \log \varepsilon_{it} \tag{1.38}$$

with multiplicative error

3. Exponential model

$$\log y_{it} = \beta_1 + \sum_{k=2}^{k} \beta_k X_{k,it} + \varepsilon_{it} \tag{1.39}$$

with multiplicative (and/or exponential) error

Good examples of a pooled time series cross-sectional multivariate regression model can be found in Dziegielewski and Opitz (1988) and Dziegielewski et al. (1993).

Most frequently, pooled time series cross-sectional analyses use the linear functional form of the model and estimate parameters using the OLS estimation technique. However, the nature of pooled time series cross-sectional data often results in the violation of one or more of the basic regression assumptions.

TABLE 1.10 Important Explanatory Variables for Residential Water-Use Models

Variables
Family characteristic variables
1. Family size (number of persons)
2. Number of children under 18
3. Household income
4. Ownership of residence
Household fixture and appliance variables
1. Number of showers
2. Number of toilets
3. Washing machine (presence of)
4. Dishwasher (presence of)
5. Garbage disposal
Frequency of appliance use variables
1. Laundry loads per week
2. Dishwasher loads per week
Outdoor feature variables
1. Lot size
2. Lawn size
3. Total irrigated area
4. Automatic sprinkling system
5. Swimming pool
Frequency of outdoor use variables
1. Lawn and landscape watering per week
2. Car washing per week
3. Hosing of concrete (blacktop) surfaces
Price variables
1. Marginal price of water
2. Rate structure
3. Wastewater charge
Weather variables
1. Monthly average of maximum daily temperatures
2. Total monthly precipitation
3. Number of days with precipitation greater than 0.01 in
4. Cooling-degree days
Other variables
1. Age of the house
2. Type of sewerage system

Two common problems that have been encountered in practical research are heteroskedasticity and autocorrelation. *Heteroskedasticity* usually arises from cross-sectional components of the data. The variance of the error term ε is not constant across all observations. When this occurs, the scale of the dependent variable and the explanatory power of the OLS model tend to vary across observations. *Autocorrelation* is usually found in time series data. The disturbances ε_i are not independent of each other; they are correlated. Time series data often display a "memory" such that variation is not independent from one period to the next. For example, an earthquake or flood may affect water use in a particular community for many periods following the actual event. Note, however, that it does not always take such a large disturbance to produce autocorrelated errors. If either one of these problems exists, the coefficient estimates of the OLS model no longer have minimum variance among all linear unbiased estimators. In other words, the OLS estimators are no longer the best linear unbiased estimates.

Several diagnostic tests for heteroskedasticity and autocorrelation are commonly available in statistical software packages. Perhaps the simplest diagnostic check is to plot the residuals. If heteroskedasticity and/or autocorrelation are detected, it is advisable to specify an alternative to the OLS model. One such model, the *generalized least squares* (GLS) regression model, can be shown to provide the best linear unbiased estimators under these conditions. Instead of minimizing the sum of squared residuals as in OLS estimation, the GLS procedure produces a more efficient estimator by minimizing a *weighted* sum of the squared residuals. Observations whose residuals are expected to be large because the variances of their associated disturbances are known to be large are given a *smaller* weight. Observations whose residuals are expected to be large because other residuals are large are also given smaller weights (Kennedy, 1985). In order to produce coefficient estimates using GLS, the variance-covariance matrix of the disturbance terms must be *known* (at least to a factor of proportionality). In actual estimating situations, however, this matrix is usually not known. A procedure called *estimated generalized least squares* (EGLS) can then be employed to estimate the variance-covariance matrix of the disturbances. EGLS estimators are no longer linear or unbiased, but because they account for the effects of heteroskedastic and autocorrelated errors, they are thought to produce better coefficient estimates.

The EGLS estimation technique is used to estimate what are called *error components* models of pooled time series and cross-sectional data (see Kennedy, 1985). The general specification for the random-effects model can be written as

$$y_{it} = \beta_0 + \sum_{i=1}^{K} \beta_k X_{k,it} + \varepsilon^* \tag{1.40}$$

where

$$\varepsilon = u_i + v_t + \varepsilon_{it}$$

Notice that this model has an overall intercept and an error term that consists of three components. The u_i represents the extent to which the ith cross-sectional unit's intercept differs from the overall intercept. The v_t represents the extent to which the tth time period's intercept differs from the overall intercept. The u_i and v_t are each assumed to be independently and identically distributed with a mean of 0 and variance of σ_u^2 and σ_v^2, respectively. The third component, ε_{it}, represents the traditional error term that is unique to each observation. All three error components are assumed to be mutually independent. The extent to which the intercept coefficients differ across cross-sectional units and across time is assumed to be randomly distributed. Because of this, the error components model is sometimes referred to as the *random effects model*. A good example of an error components model can be found in Chesnutt and McSpadden (1990).

1.8 ANALYSIS OF WATER SAVINGS

A precise measurement of water savings that can be attributed to various demand management programs is difficult because the observed water use often shows great variability among different users and it also significantly varies over time for the same user. For example, the amounts of water used inside and outside a residential home can vary substantially from month to month and from household to household. This variability is caused by many factors, including conservation practices. Because of the great variability in water use, the observed changes in water use over time, or differences in use between individual customers or groups of customers, may be caused by influences unrelated to the customer participation in a conservation program. Therefore, the most important consideration in measuring water conservation savings is the design of a measurement procedure that is capable of correctly measuring not only the changes in water use but also separating these changes into those caused by the program and those caused by changes in weather, prices, economic factors, and other confounding factors. The precision of the measurements of water savings depends on whether the study design was capable of isolating and controlling for (1) the characteristics of the conservation program that could significantly influence the results of the estimation of water savings, (2) the characteristics of the customer groups targeted by the program that could also influence the results, and (3) the characteristics that are external to both the conservation program design and the targeted customer groups. Research in evaluation designs identified a number of factors referred to as *outside effects* or *externalities* (Dziegielewski et al., 1993).

Once a conservation practice is adopted, the baseline demand that represents water use without the practice cannot be directly measured, and the unaltered demand has to be reconstructed somehow. In practice, all study designs employ comparisons of water-use behavior (and other customer characteristics in some cases) over time and/or between groups of customers. Possible types of comparisons are illustrated by Fig. 1.3. Implementation of a conservation program divides the time continuum into two periods, namely, pretreatment conditions and posttreatment conditions. It also divides the water users into two groups—the control group of nonparticipants and the treatment group of program participants. The conditions of a valid study design are achieved by a careful selection of a sample of water users and the use of proper methods of data analysis.

There are two basic approaches for estimating water conservation savings: statistical techniques and leveraged approaches. These two approaches will be discussed.

1.8.1 Statistical Estimation of Savings

Statistical comparison methods produce estimates of conservation savings by comparing water use between a participant group and a control group (or changes in water use before and after the program). The *comparison-of-means method* is derived from the statistical theory of randomized controlled experiments which utilizes a treatment and control design. Conservation savings are estimated as the difference in the mean level of water use between the treatment group and the control group, i.e.,

$$d = \bar{q}_t - \bar{q}_c \tag{1.41}$$

where

$$\bar{q}_t = \frac{1}{n_1} \sum_{t=1}^{n_1} q_t \tag{1.42}$$

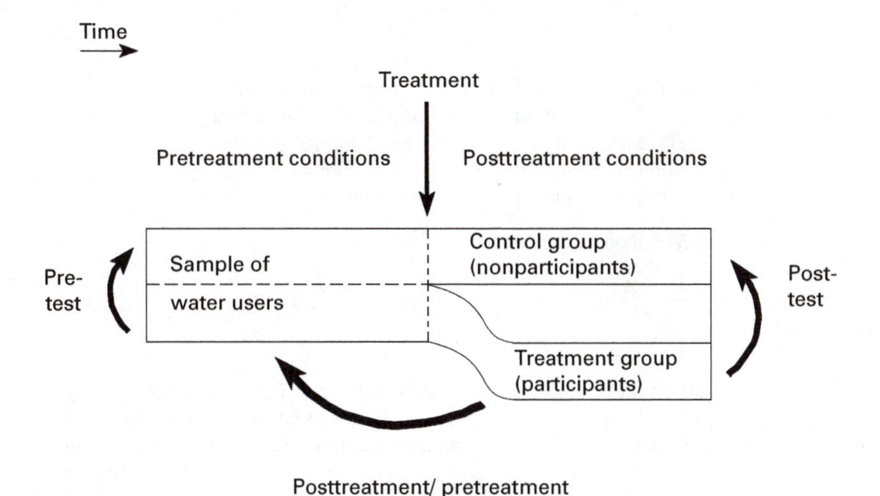

FIGURE 1.3 Evaluation designs for measuring water conservation savings.

and

$$\bar{q}_c = \frac{1}{n_2} \sum_{c=1}^{n_2} q_c \tag{1.43}$$

where d = conservation effect (difference between means)
$\bar{q}_t$ = mean water use in treatment sample
$\bar{q}_c$ = mean water use in control sample
q_t = water use of customer t in treatment sample
q_c = water use of customer c in control sample
n_1 = number of customers in treatment sample
n_2 = number of customers in control sample

The data can be used to test whether the observed difference d can be attributed to chance, or whether it is indicative that the two samples come from populations of unequal means. Given that the parent population distribution of the differences in means is unknown or not normal, and that the population standard deviation of water use in each group (σ_t and σ_c) are unknown but assumed equal, the sampling distribution of the differences in mean water use should follow a t distribution for large samples. The *central limit theorem* (and the concept of repeated sampling) implies that for large sample sizes, the sampling distribution of the difference in means between groups will approach a normal distribution. This allows statistical inference about population parameters when the population distribution is unknown or not normal. However, in order for the sampling distribution of the difference in two means to be approximated by the normal distribution, the population standard deviations (σ_t and σ_c) must be known. If σ_t and σ_c are not known but are assumed equal, one may use the sample estimates of σ_t and α_c (s_t and s_c) and the t distribution for statistical inference. For large sample sizes, the t distribution will approximate the normal distribution.

A calculated t statistic can be used to test the hypothesis that the differences between mean water use of the treatment and control samples is 0. In order to calculate a t statistic, the *standard error of the difference between the two means* must be determined using the following formula:

$$S_d = \sqrt{\frac{(n_t - 1)\, s_t^2 + (n_c - 1)\, s_c^2}{n_t + n_c - 2} \; \frac{n_c + n_t}{n_c n_t}} \tag{1.44}$$

where S_d = estimated standard error of conservation effect d
s_t = standard deviation of water use in treatment sample
s_c = standard deviation of water use in control sample
n_t = number of customers in treatment sample
n_c = number of customers in control sample

The t *statistic* is then calculated as

$$t = \frac{\bar{q}_t - \bar{q}_c}{S_d} = \frac{d}{S_d} \tag{1.45}$$

Using the properties of the t distribution, one can test the null hypothesis stating that the true population mean of the treatment group is equal to the population mean of the control group, or the alternative hypothesis stating that the population means of the treatment and control groups are different. The value of the t statistic calculated by Eq. (1.45) uses the sample estimates to infer whether the difference d is large enough to reject the null hypothesis that the true difference in means is equal to 0. To test the null hypothesis, the resultant value of t must be compared to statistical tables of the t distribution. These tables may be found in any standard statistics textbook. Depending on the type of assertion to be made about the difference between means, one may use either a one-tail or two-tail (also called one-sided or two-sided) test of significance.

The two-tail test should be used to test the hypothesis that the difference between means, d, is 0, against the alternative hypothesis that the difference in means is positive or negative. The one-sided test of significance should be used to test the null hypothesis of no difference in means against the alternative hypothesis that, after the treatment, the mean water use in the treatment group is *lower* than the mean water use in the control group.

In order to produce reliable estimates, the comparison-of-means method must satisfy two requirements: (1) the two random variables representing water use in the treatment and control groups must be drawn from the same population distribution, and (2) the distribution is normal. The first assumption is often violated. In the classic experimental design on which the comparison-of-means method is based, the experimenter has careful control over all factors that might affect the variable under consideration. Therefore, by carefully designing the samples to be used in the experiment, any difference between the treatment and control groups can be attributed solely to the "treatment." However, the water planner is not likely to have complete control over the confounding factors that affect household water use. Although statistical theory suggests that randomly assigning households into treatment and control groups will result in groups that are less likely to be systematically different in terms of water use, this does not ensure that the two groups are identical with respect to household income, average number of persons per household, yard size, and many other factors. Therefore, unless a great deal of matching or sampling work is done, water use cannot be considered a random variable. It is related, if not caused, by the uncontrolled-for factors that differ between the treatment and control groups. In other words, one runs the risk of incorrectly attributing observed changes in water use to the treatment (e.g., a retrofit), when in fact they are caused by the different average values of external factors in each group. With respect to the second assumption, it must be stressed that empirical distributions of water use have most often been found not to follow a normal distribution. Typically, distributions of water use show a long right-hand tail, and thus do not conform to the symmetric bell-shaped appearance associated with the normal distribution. The violation of this assumption is not a fatal flaw, however, if a normalizing transformation of the data is used. For example, taking the log of water use should at least pull in the right-hand tail and minimize the leverage of "contaminated" or outlying data points or they can be screened out by rejecting values greater than $x + 3s$.

When adhering to a strict experimental design, the comparison-of-means method is more likely to produce meaningful and reliable results in situations where (1) the expected conservation effect is large when compared to mean water use, (2) the variance in water use is small compared to the conservation effect, (3) the mean and variance in water use are very similar (in terms of size) for both groups prior to treatment, and (4) the sample sizes in the treatment and control groups are large. The comparison-of-means method can produce reliable and informative results if used in conjunction with experimental designs and large sample sizes.

Multivariate regression models represent the most sophisticated method of comparing water-use data over time or between groups of customers while controlling for the effects of a large number of external factors. One can choose from a variety of regression methods depending on the types of available data and the acceptable level of estimation complexity. The statistical models described in Section 1.7.2 can be used for this purpose.

1.8.2 Time Series Analysis of Conservation Effects

A time series of the volumes of water sold in consecutive billing periods can be used to measure conservation effects of full-scale programs while controlling for external influences other than the conservation program. The reliability of the estimates will depend on (1) the ability to disaggregate sales data into classes of similar users (e.g., single-family residential, small commercial), (2) the ability to separate (or account for) the seasonal effects and weather effects in the time series data, and (3) the ability of the estimation technique to deal with nonconstant error variance and correlation of model errors through time. The effects of a conservation program under investigation can be measured by including an indicator variable which separates the time series data into pre- and postprogram periods. However, it is also important to capture the effects of other passive and active conservation measures which are adopted by water customers independently of the program under evaluation.

If customer-level monthly (or billing period) data on water use for a period of 2 to 4 years can be supplemented with information on customer characteristics, and such external factors as price of water and weather conditions, then a pooled time series cross-sectional data set can be constructed and used to estimate the parameters of a multiple-regression model. The important explanatory variables will depend on the customer class. In the residential sectors they usually include information on family characteristics, household features and appliances, frequency of appliance use, outdoor features, and frequency of outdoor uses. This measurement technique can produce very accurate estimates of actual water savings for some programs, especially those targeting the residential sector.

The major drawback of the multivariate regression models is that they are relatively expensive and time-consuming. They require large amounts of data and a lot of expertise from the analyst. They also need large sample sizes and are less appropriate for nonresidential water users.

1.8.3 End-Use Accounting System

The accumulating experience in the evaluation of conservation programs indicates that it is very difficult to obtain measurements of water savings with a high level of precision using a single best method. The most precise estimates can be achieved by taking advantage of the strong features of the statistical methods with engineering methods. The known strengths, weaknesses, and biases of each approach can be used to narrow down the confidence bands surrounding the actual water savings.

Engineering (or mechanical) estimates are obtained using laboratory measurements or published data on water savings per device or conservation practice. These data can be combined with assumptions regarding the magnitude of factors expected to impact on the results of the conservation program in order to generate estimates of program savings. However, the resultant estimates can be very sensitive to the underlying assumptions and relationships. For example, the savings resulting from the installation of an ultralow-flush toilet replacing a standard toilet will depend on assumptions regarding flushing volumes and frequency of flushing. The resultant savings can range from 19.5 gal per person per day (3.9 gal $\times$ 5 flushes per person per day) to 7.6 gal per person per day (1.9 gal $\times$ 4 flushes per person per day). The high estimate is almost 3 times greater than the low estimate. The validity of the assumptions used in the above example can easily come under attack, since they rely on subjective conclusions and a great deal on the professional judgment of the engineer or analyst.

Despite their obvious shortcomings, engineering estimates may be considered appropriate for providing preliminary estimates of potential savings when field measurements are not available. They can also be used to verify statistical estimates by setting limits on a possible range of savings. However, they become most useful in leveraged techniques where they can be used to augment and strengthen statistical models.

The most promising method of leveraging information involves the use of information from one approach within the procedures of another. For example, engineering estimates or special metering measurements can be used as independent variables in statistical models. Statistical models of urban water demands do not accommodate the needs of planning and evaluation of demand management programs because of an inability to disaggregate water demands down to the end-use level. Because many demand-side programs target specific end uses, the absence of end-use water demand models severely impairs the development of effective demand management policies. Without adequate end-use models, the effects of various demand management programs cannot be measured precisely. In order to enhance the ability of water planners to formulate, implement, and evaluate various demand management alternatives, it is necessary to disaggregate the usually observed sectoral demands during a defined season of use into the applicable end uses. Only such a high level of disaggregation will permit water planners to make all necessary determinations in estimating water savings of various programs.

The first step is to disaggregate the observed water demands into their specific components or end uses. Figure 1.4 illustrates how water demand of a homogeneous sector of water users can be disaggregated into its seasonal and end-use components. A rational representation of each end use is made using a *structural end-use equation* of the following form:

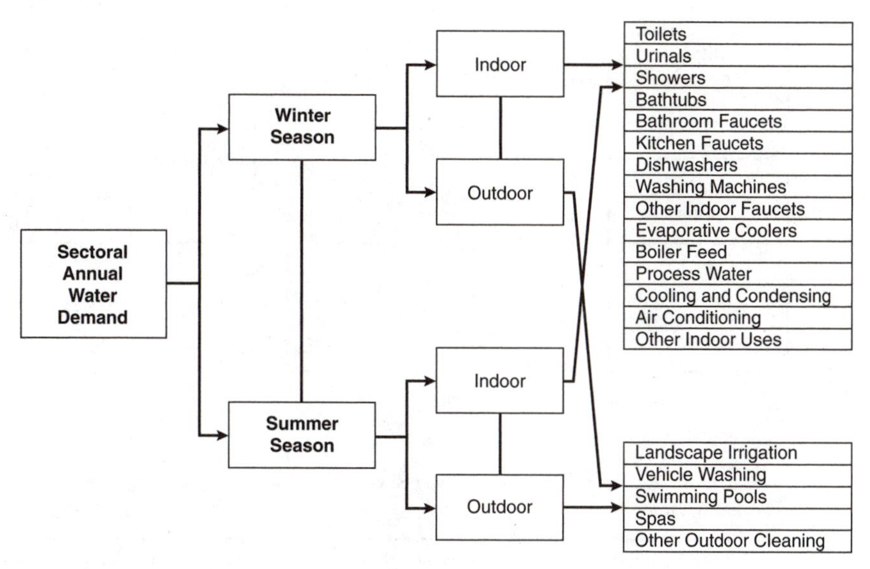

FIGURE 1.4 Disaggregation of annual water demands.

$$q = [(M_1S_1 + M_2S_2 + M_3S_3)U + KF]A \qquad (1.46)$$

where q = average quantity of water in a given end use
M_1, M_2, M_3 = efficiency classes of end use (design parameters)
S_1, S_2, S_3 = fractions of end uses within efficiency class
U = usage rate (or intensity of use)
K = average flow rate of leaks
F = fraction of end uses with leaks (incidence of leaks)
A = presence of end use in a given sector of users

A graphical representation of this structural end-use relationship is given in Fig. 1.5. An application of Eq. (1.46) to the toilet end use in the residential sector would require the knowledge of all end-use parameters. An example of the uses of this equation for analyzing the toilet end use is presented in Table 1.11. Other end uses and effects of improvements in their efficiency can be estimated using similar parameters and data.

The structural end-use relationship [Eq. (1.46)] is dictated by the need to distinguish between changes in water demand caused by active and passive demand management programs from the changes caused by other factors. The structure of end uses which exists in the service area at any point in time will not remain constant over the planning horizon. It will change in response to changes in the determinants of water use such as income, household size, and housing density. The effects of various interventions of demand management programs must be counted relative to the baseline forecasts of water demands, which capture the effects of the relevant external factors. The parameters of each end use are affected by the external factors. For example, changes in price will cause a decrease in the incidence of leaks in the short run and will affect the distribution of end uses among the classes of efficiency in the long run. The other two parameters in the end-use equation (i.e., intensity and presence) also will be affected by changes in price.

For example, changes in price will cause a decrease in the incidence of leaks in the short run and will affect the distribution of end uses among the classes of efficiency in the long run. The other two parameters of the end-use equation (i.e., intensity and presence) also will be affected by changes in price. The ideal forecasting model would be capable of predicting the parameters of the structural end-use

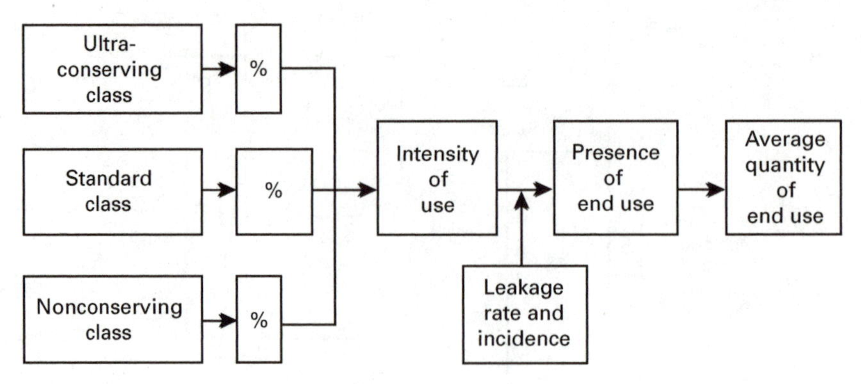

FIGURE 1.5 Structural end-use relationships.

TABLE 1.11 Example of Estimating Water Savings with End Use Accounting System (Toilet End Use)

End-use parameter	Before change	After change	Net effect (savings)
Inefficient class rate	5.50	5.50	
Inefficient class fraction	0.35	0.25	−0.10
Standard class rate	3.50	3.50	
Standard class fraction	0.55	0.50	−0.05
Efficient class rate	1.60	1.60	
Efficient class fraction	0.10	0.25	+0.15
Intensity of use, fpd	14.00	14.00	
Presence of end use	1.00	1.00	
Leakage rate, gpd	20.00	20.00	
Incidence of leaks	0.15	0.15	
Average quantity, gpd	59.10	52.35	−6.75

equations as a function of such influencing variables as price, income, household size, housing density, and weather. For example, in the case of lawn irrigation end use, the parameters of Eq. (1.46) should be modeled as functions of the explanatory variables.

The structure of the end-use equation allows the planner to estimate the net effects of long-term conservation programs by tracking the values of end-use parameters over time. It also accommodates the handling of such issues as interaction and overlapping of multiple programs, customer-initiated conservation effects, and the relationships between long-term and short-term (e.g., drought emergency) programs. Each of these effects creates problems in measuring the effectiveness of efficiency improvement programs.

1.9 SUMMARY

Urban water supply systems are designed to deliver water "on demand" as the system operators have no direct control over the quantity of water taken out by the customers. Accordingly, water demands are usually taken as given quantities that have to be matched with supplies. However, statistical studies of historical rates of water use have shown that water use fluctuates in response to various influencing factors. As a result, the per capita use rates in urban areas can fluctuate from year to year in response to changing weather conditions and other conditions. The usage rate also exhibits strong long-term trends caused by changes in the mix of housing types, commercial activities, new growth, prices of water, and other factors. Because these influencing forces are likely to operate in the future, future demand cannot be assumed to be a simple product of the projected population and the historical rate of per capita water use. The adoption of water conservation measures, price of water, housing density, urban growth policies, and types of landscaping are important determinants of water use and can be viewed as instruments for managing water demands.

Analytical techniques described in this chapter can help water planners to isolate and quantify the effects of many explanatory factors, and the impacts on water demand of anticipated changes in these factors. An even greater challenge will be the development of models that are able to estimate the long-term effects of water-demand management programs and policies with a high degree of confidence.

1.10 REFERENCES

American Water Works Association. 1986. *1984 Water Utility Operating Data.* Denver, Colo.

Chesnutt, T. W., and C. N. McSpadden. 1990. *A Model-Based Evaluation of the Westchester Water Conservation Program.* Metropolitan Water District of Southern California, Los Angeles.

Cochran, W. G. 1963. *Sampling Techniques.* John Wiley, New York.

DeOreo, W. B., J. P. Heaney, and P. Mayer. 1996. Flow Trace Analysis to Assess Water Use. *Journal AWWA,* 88(1):79–90.

DeOreo, W. B., P. W. Mayer, E. M. Opitz, B. Dziegielewski, J. C. Kiefer, W. Y. Davis, and J. O. Nelson. 1998. *Residential End Uses of Water: Final Report.* AWWA Research Foundation, Denver, Colo.

Dziegielewski, Benedykt, and Eva Opitz. 1988. *Phoenix Emergency Retrofit Program: Impacts on Water Use and Consumer Behavior.* Phoenix Water and Wastewater Department, Phoenix.

Dziegielewski, Benedykt, Eva Opitz, and Dan Rodrigo. 1990. *Seasonal Components of Urban Water Use in Southern California.* Metropolitan Water District of Southern California, Los Angeles.

Dziegielewski, Benedykt, Eva M. Opitz, Jack C. Kiefer, and Duane D. Baumann. 1993. *Evaluation of Urban Water Conservation Programs: A Procedures Manual.* American Water Works Association, Denver, Colo.

Dziegielewski, B., S. Sharma, T. Bik, X. Yang, and H. Margono. 2002a. *Analysis of Water Use Trends in the United States: 1950–1995. Project Completion Report.* Southern Illinois University, Carbondale, Ill.

Dziegielewski, B., S. Sharma, T. Bik, X. Yang, H. Margono, and R. Sa. 2002b. *Predictive Models of Water Use: Analytical Bibliography.* Southern Illinois University, Carbondale, Ill.

Fowler, Floyd J. Jr. 1988. *Survey Research Methods.* Sage Publications, Newbury Park, Calif.

Kennedy, P. 1985. *A Guide to Econometrics.* MIT Press, Cambridge, Mass.

Kish, Leslie. 1965. *Survey Sampling.* John Wiley, New York.

Schlenger, Donald. 1991. Current Technologies in Automatic Meter Reading. *Waterworld News* 7(3):14–17.

Solley, Wayne B., Robert R. Pierce, and Howard A. Perlman. 1998. *Estimated Use of Water in the United States in 1995.* U.S. Geological Survey Circular 1200, Washington, D.C.

USEPA. 1997. *Community Water System Survey,* Vol. II.

USEPA. 2000. *Factoids: Drinking Water and Ground Water Statistics for 2000.* EPA 816-k01-004. *www.epa.gov/safewater/data/00factoids.pdf.*

CHAPTER 2
WATER PRICING AND DROUGHT MANAGEMENT

Messele Z. Ejeta
California Department of Water Resources
Sacramento, California

Larry W. Mays
Department of Civil and Environmental Engineering
Arizona State University
Tempe, Arizona

2.1 INTRODUCTION

Droughts continue to rate as one of the most severe weather-induced problems around the world. Global attention to natural hazard reduction includes drought as one of the major hazards. Changnon (1993) gave seven lessons or truths that have emanated out of studying the major droughts from 1932 to 1992 in the United States. These lessons are summarized below:

1. A major drought is a pervasive condition affecting most portions of the physical environment as well as the socioeconomic structure.
2. Droughts are a major but unpredictable part of the climate of all parts of the United States. Moreover, they occur infrequently, and this results in a decay in the attention to drought preparedness and mitigation.
3. Responses and adjustments to drought problems can be sorted into two classes: (*a*) short-term fixes and (*b*) long-term improvements.
4. Although many long-term adjustments have been made as a result of the major droughts of the last 60 years, many factors make today's society generally more vulnerable to drought than ever before.
5. Agriculture, in general, cannot escape from experiencing major drought losses in the future, even with healthier crop strains and increased irrigation.
6. Opportunities for improvement in water management exist and could make the nation's water resources more impervious to drought. However, many water-related problems are localized and at the substate scale and often do not get needed attention.
7. Drought is ubiquitous: everything and everybody is affected, and yet no one (everyone) is in charge.

The shortage of water supply during drought periods is such a significant factor for the general welfare that its effect cannot be easily undermined. Domestic water supply shortages during these periods in particular have been crucial in some cases, and as a result, various measures targeting different means of reducing water demand during such periods were initiated by different water supply agents. These measures, which may be considered semiempirical to empirical, included water metering, leak detection and repair, rate structures, regulations on use, educational programs, drought contingency planning,

TABLE 2.1 A Topology of Drought Management Options

- I. Demand Reduction Measures
  1. Public education campaign coupled with appeals for voluntary conservation
  2. Free distribution and/or installation of particular water-saving devices:
     - 2.1 Low-flow showerheads
     - 2.2 Shower flow restrictors
     - 2.3 Toilet dams
     - 2.4 Displacement devices
     - 2.5 Pressure-reducing valves
  3. Restrictions on nonessential uses:
     - 3.1 Filling of swimming pools
     - 3.2 Car washing
     - 3.3 Lawn sprinkling
     - 3.4 Pavement hosing
     - 3.5 Water-cooled air conditioning without recirculation
     - 3.6 Street flushing
     - 3.7 Public fountains
     - 3.8 Park irrigation
     - 3.9 Irrigation of golf courses
  4. Prohibition of selected commercial and institutional uses:
     - 4.1 Car washes
     - 4.2 School showers
  5. Drought emergency pricing:
     - 5.1 Drought surcharge on total water bills
     - 5.2 Summer use charge
     - 5.3 Excess use charge
     - 5.4 Drought rate (special design)
  6. Rationing programs:
     - 6.1 Per capita allocation of residential use
     - 6.2 Per household allocation of residential use
     - 6.3 Prior use allocation of residential use
     - 6.4 Percent reduction of commercial and institutional use
     - 6.5 Percent reduction of industrial use
     - 6.6 Complete closedown of industries and commercial establishments with heavy uses of water
- II. System improvements
  1. Raw water sources
  2. Water treatment plant
  3. Distribution system:
     - 3.1 Reduction of system pressure to minimum possible levels
     - 3.2 Implementation of a leak detection and repair program
     - 3.3 Discontinuing hydrant and main flushing
- III. Emergency water supplies
  1. Interdistrict transfers:
     - 1.1 Emergency interconnections
     - 1.2 Importation of water by trucks
     - 1.3 Importation of water by railroad cars
  2. Cross-purpose diversions:
     - 2.1 Reduction of reservoir releases for hydropower production
     - 2.2 Reduction of reservoir releases for flood control
     - 2.3 Diversion of water from recreation water bodies
     - 2.4 Relaxation of minimum streamflow requirements

TABLE 2.1 *(Continued)*

3. Auxiliary emergency sources:
3.1 Utilization of untapped creeks, ponds, and quarries
3.2 Utilization of dead reservoir storage
3.3 Construction of a temporary pipeline to an abundant source of water (major river)
3.4 Reactivation of abandoned wells
3.5 Drilling of new wells
3.6 Cloud seeding

Source: Dziegielewski, 1986.

water recycling and reuse, and pressure reduction. Such efforts are collectively termed water conservation, although there has not been a uniform definition among authors.

On the other hand, different researchers and scientists have tried to develop more scientific methods for water conservation during drought periods. These methods have been aimed at water conservation through price increases of the water supply to the customers. The results elucidated the fact that water is more of a commodity than it is a public resource. However, the several models developed so far which relate reduction in demand for water due to the increase in its price, through price elasticity, used different variables that range from the income of the customers to hydrologic conditions. The relations developed used regression analysis, and as a result the differences and the variations of the variables considered are significant enough that the estimated demand is subject to uncertainty. Thus the demand may be better expressed by an estimated value and a probability distribution.

The basics of the *price-elasticity approach* presumes that the demand can be adjusted to the available supply. This may happen on an average basis; however, the demand has a random distribution about the available supply. By similar reasoning, the available supply corresponding to a given return period of weather conditions may have a random distribution about the expected value. All the aforementioned uncertainties call for risk evaluation to determine the probability that the demand exceeds the available supply, for water supply project planning. Conversely, the price of water supply for a given tolerable risk level can be determined.

This chapter first discusses various efforts reported in the literature for water conservation and then culminates with the new idea of a *risk-based approach*. The different water-conservation practices are briefly discussed, giving coverage of the price-elasticity formulation. The basic reasons that make it necessary for a risk-based approach are described. Some risk-level indices, which have been used for the evaluation and prediction of a drought period, are given and their limitations are explained. A method for evaluation of the damage associated with certain levels of drought severity is developed. This new approach relates the risk, the price, and the return period. It is found through this relationship that risk is sensitive to the return period and to the price changes.

2.2 BACKGROUND OF WATER CONSERVATION

2.2.1 Drought Management Options

Experiences from past droughts have shown that the action of water managers can greatly influence the magnitude of the monetary and nonmonetary losses from drought. There have been a variety of *drought management options* that have been undertaken in response to anticipated shortages of water, which can be categorized as (1) demand reduction measures, (2) improvements in efficiency in water supply and distribution system, and (3) emergency water supplies (Dziegielewski, 1986). A topology of drought management options is given in Table 2.1.

Not only is water conservation necessary during drought periods, but the economic merits are also important to consider. In the United States, federal mandates urge that opportunities for water conservation be included as a part of the economic evaluation of proposed water supply projects

(Griffin and Stoll, 1983). Water conservation during drought periods, however, requires important attention because our demand of water may exceed the available resources in the demand environment. Conservation may be achieved through different activities. According to the U.S. Water Resources Council (1979a), these activities include, but are not limited to,

1. Reducing the level and/or altering the time pattern of demand by metering, leak detection and repair, rate structure changes, regulations on use (e.g., plumbing codes), education programs, drought contingency planning
2. Modifying management of existing water development and supplies by recycling, reuse, and pressure reduction
3. Increasing upstream watershed management and conjunctive use of ground and surface water (Griffin and Stoll, 1983)

The effort to conserve water started out with metering rather than providing a flat rate. Both domestic and sprinkling demands reduced significantly as a result of the introduction of water meters (Hanke, 1970). Grunewald et al. (1976) stated: "Traditionally, water utility managers have adjusted water quantity [rather] than prices as changes in demand occurred."

In general, some of the major measures followed for water-conservation efforts with references are as follows:

1. Use restrictions (no car washing, or hosing down of sidewalks, alternate-day lawn and garden watering and the like) (Moncur, 1989).
2. Increasing rate structures, also called inverted-block rates, inclining-block rates, increasing blocks, inverted pyramid rates (Jordan, 1994).
3. A lump-sum charge and a commodity charge per unit volume imposed in addition to the normal rates (Carver and Boland, 1980).
4. Allowing the market process to operate, that is, adopting marginal cost pricing, even for normal periods, rather than averaging price (Moncur, 1989).
5. Attempting to decrease the amount used by industries by trying to utilize existing technology to design and install production processes using less water per unit of output (Grebenstein and Field, 1979).
6. Reducing withdrawals for production processes by recycling (Grebenstein and Field, 1979).
7. Passing water-conservation acts, requiring builders to install ultralow-flow fixtures in all new projects (Jordan, 1994).
8. Forcing the public to be bound to treated wastewater for new recreational use. Officials in Phoenix, for instance, considered not allowing new recreational lakes unless treated wastewater was used (Maddock and Hines, 1995).

Increasing the price of domestic water supply has been a focus of several studies. These studies were conducted to analyze the effect of urban water pricing and how it contributes to water conservation during a drought period (Agthe and Billings, 1980; Moncur, 1989). However variations have been observed in the approaches followed. According to Jordan (1994), water pricing is an effective way of conserving water, compared to the other measures mentioned above. An increase in the price of water contributes to water conservation because of the fact that customers have limited money. For every percent increase in the price, there is some decrease in the demand, which is explained through price elasticity. A significant number of studies have been undertaken in different regions to determine price elasticity associated with pricing. Section 2.2.2 explains price elasticity.

2.2.2 Price Elasticity of Water Demand

The elasticity of demand is the responsiveness of consumers' purchases to varying price. The most frequently used elasticity concept is *price elasticity,* which is defined as the percentage change in quantity taken if price is changed 1 percent. Young (1996) states that "the price elasticity of demand for water measures the willingness of consumers to give up water use in the face of rising prices, or conversely, the tendency to use more as price falls." Two different ways have been followed to formulate the price elasticity of demand for water: one based upon average price and the other based upon marginal price. Agthe and Billings (1980) state that the elasticity determined based upon average price overestimates the result. Therefore they recommend (as do several others) that the marginal price be used.

Howe and Linaweaver (1967) defined the *price elasticity of water* as

$$\eta_P = \frac{\Delta d}{\bar{d}} \div \frac{\Delta p}{\bar{p}} \tag{2.1}$$

where η_P = price elasticity
$\bar{d}$ = the average quantity of water demanded
$\bar{P}$ = average price
Δd = change in demand
ΔP = change in price

For a continuous demand function, the following more general formula is applicable.

$$\eta_P = \frac{dd}{d} \div \frac{dP}{P} \tag{2.2}$$

Table 2.2 is a summary of some of the values of price elasticity of water demand reported in the literature.

The use of the price elasticity of water has been applied to some cities with some important achievements having been obtained. The following schematic may depict the general trend of this principle, as derived from the conclusion reached by Jordan (1994):

$$\uparrow \text{(Price)} \rightarrow \downarrow \text{(water demand)} \ \& \uparrow \text{(revenue)} \tag{2.3}$$

An increase by less than 40 percent of the price resulted in a 10 percent decrease in the demand in Honolulu—the announced goal of the restrictions imposed in the drought episodes of 1976 to 1978 and in 1984 (Moncur, 1987). This was achieved using a price elasticity of only −0.265. In Tucson, Arizona, an inverted rate structure was claimed to have been credited with reducing public demand from about 200 gal per capita per day (gcd) to 140–160 gcd (Maddock and Hines, 1995).

The way in which water utilities are structured is probably the most important factor, which complicates the study of price elasticity. For instance, some customers who own homes or who pay water bills, more or less, react to the price change, whereas those who rent apartments or who do not pay for water bills are almost indifferent to it. Furthermore, water necessity for residential, commercial, and industrial purposes are not equally important. Because of this reason, different researchers had to study demand elasticity by categorizing water distribution systems for industrial, commercial, and residential uses. The demand patterns under these categories are not uniform. One of the most comprehensive studies on price elasticity of water demand done by Schneider and Whitlatch (1991) for six user categories (residential, commercial, industrial, government, school, and total metered) showed different results for these categories. Residential water use is further complicated by different factors: many residents who rent housing do not pay for water and as such are indifferent to demand regulations; the patterns for indoor and outdoor water demand differ quite significantly and hence necessitate different approaches of demand analysis. The climatic conditions of a given area and the time of the year are also worth mentioning. These are probably the reasons why apparently different elasticity values are reported for the eastern and the western United States and for winter and summer uses.

From the studies enumerated so far, a general conclusion is reached: demand is elastic to price increase. Almost all research has reinforced this hypothesis. However, differences exist between the elasticity values calculated for different geographic locations. For instance, Howe (1982) obtained values of −0.57 and −0.43 for the eastern and the western United States, respectively. On the other hand, no clear consistency exists in the way that elasticity is calculated: some use average price, some use marginal price, and still some include the intramarginal rate structure. Although some of the studies targeted alleviating water shortage problems during drought periods, they did not approach the problem from the perspective of risk analysis.

TABLE 2.2 Summary of Some of the Price Elasticity Values

No.	Researchers	Research area	Year	Estimated price elasticity	Estimated income elasticity	Remarks
1	Howe & Linaweaver	Eastern U.S.	1967	−0.860		
2	Howe & Linaweaver	Western U.S.	1967	−0.52		
3	Wong	Chicago	1972	−0.02	0.20	
4	Wong	Chicago suburb	1972	−0.28	0.26	
5	Young	Tucson	1973	−0.60 to −0.65		Exponential and linear models used
6	Gibbs	Metropolitan Miami	1978	−0.51	0.51	Elasticity measured with the mean marginal price
7	Gibbs	Metropolitan Miami	1978	−0.62	0.82	Elasticity measured with the average price
8	Agthe & Billings	Tucson	1980	−0.27 to −0.71		Long-run model
9	Agthe & Billings	Tucson	1980	−0.18 to −0.36		Short-run model
10	Howe		1982	−0.06		
11	Howe	Eastern U.S.	1982	−0.57		
12	Howe	Western U.S.	1982	−0.43		
13	Hanke & de Maré	Malmö, Sweden	1982	−0.15		
14	Jones & Morris	Metropolitan Denver	1984	−0.14 to −0.44	0.40 to 0.55	Linear and log-log models used
15	Moncur	Honolulu	1989	−0.27		Short-run model
16	Moncur	Honolulu	1989	−0.35		Long-run model
17	Jordan	Spalding County, Georgia	1994	−0.33		A price elasticity of −0.07 was also reported for no rate structure, but increased price level

2.2.3 Demand Models

It is important to have demand related to the drought severity. Several studies have expressed demand as a function of different variables. Mays and Tung (1992) gave a general form of demand models as

$$d = f(x_1, x_2, \ldots, x_k) + \varepsilon \tag{2.4}$$

where f is the function of explanatory variables $x_1, x_2, \ldots, x_k$ and ε is a random error (random variable) describing the joint effect on q of all the factors not explicitly considered by the explanatory variables.

Several explicit linear, semilogarithmic, and logarithmic models have been developed through different researches. Billings and Agthe (1980), for example, gave the following water demand function for Tucson, Arizona (notations modified to fit the notations adopted for this study).

$$\ln d = -7.36 - 0.267 \ln P + 1.61 \ln I - 0.123 \ln \text{DIF} + 0.0897 \ln W \tag{2.5}$$

where d = monthly water consumption of average household, 100 ft^3
P = marginal price facing average household, cents per 100 ft^3
DIF = difference between actual water and sewer use bill minus what would have been paid if all water was sold at marginal rate, \$
I = personal income per household, \$/month
W = evapotranspiration minus rainfall, in

Equation (2.5) implicitly relates demand to the hydrologic index, W. The positive coefficient of W shows that demand increases exponentially with W, which indirectly indicates increases of demand with the dryness of weather conditions. The general trend of the average demand with the return period, therefore, may be shown as given by the demand curve in Fig. 2.1. Demand increases with the return period of the drought severity because the more severe the drought, the more the customers are prompted to use more water. Different demand curves are illustrated in Fig. 2.2 for different price levels. As shown in this figure, the higher the price, the lower the demand for given hydrologic conditions.

Equation (2.5) may be rearranged as

$$d = 0.00006362 P^{-0.267} I^{1.61} (\text{DIF})^{-0.123} W^{0.0897} \tag{2.6}$$

or in more general terms,

$$d = a' P^{b'} I^{c'} (\text{DIF})^{d'} W^{e'} \tag{2.7}$$

where a', b', c', d', and e' are constants. The price elasticity of demand for Eq. (2.6) is -0.267. Therefore, changing the price while keeping the other variables constant results in different average

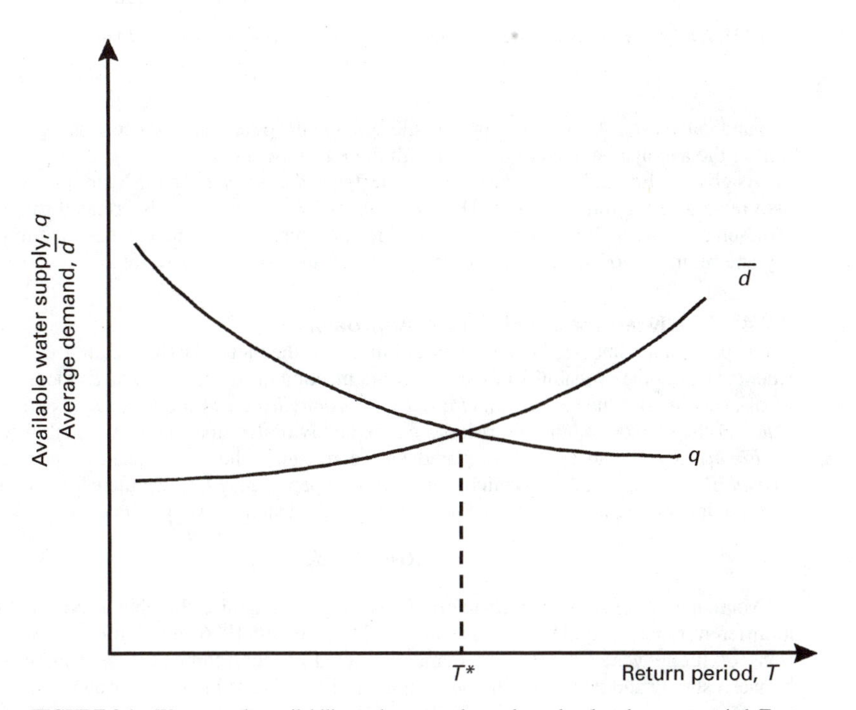

FIGURE 2.1 Water supply availability and average demand as related to the return period, T.

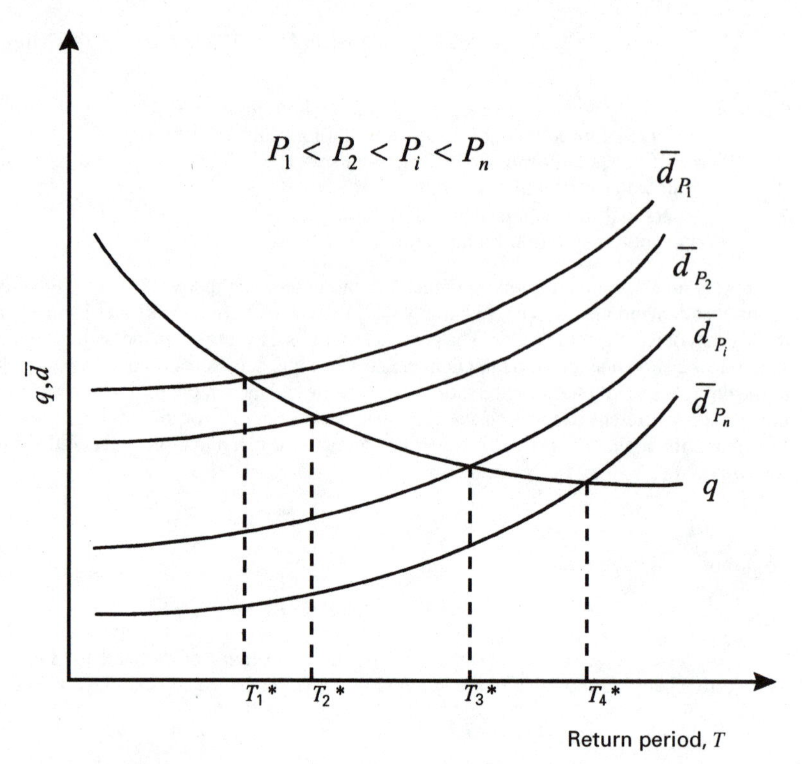

FIGURE 2.2 Water supply and average demands for different price values as related to T.

demand values, $\bar{d}_{P_i}$. Again, varying W while keeping the other variables constant gives a general relation of the average demand associated with the return period T.

As given in Eq. (2.7), it can be seen that the demand d is related to the hydrologic index W, which is also related to the return period. The available supply (flow) q is also related to the return period (Hudson and Hazen, 1964). Thus the general relationships between demand and return period and supply and return period which are shown in Fig. 2.1 are based on these trends.

2.2.4 The Need for a Risk-Based Approach

A few past studies analyzed risk only by defining it as the monetary (financial) loss. They did not consider the risk as the probability of the supply not meeting the demand. To make these two connotations of risks distinctive, the terms *financial risk* and *probabilistic risk* are introduced and used differently. Many of the previous studies on risk did not explicitly define financial and probabilistic risks.

Financial risk can be simply stated as the monetary loss associated with a certain damage. *Probabilistic risk,* which is explicitly used in this paper, may be formulated as the probability $p(\ )$ that the demand d_{Pi} at price P_i for level i exceeds the available supply q_T, expressed as

$$\text{Risk} = p(d_{Pi} > q_T) \tag{2.8}$$

Municipal water supply shortage problems have been manifesting themselves in different regions at different times for a long time. A study by Dixon et al. (1996) for California showed that projections of future water supply and demand (including environmental uses) indicate that the gap between supply and demand will widen to 4.1 million acre·ft in average water years and 7.4 million acre·ft in drought-water years by 2020. In 1977 in Fairfax County, Virginia, the drought was so severe that drastic measures such as the closing of schools and businesses were actively being considered (Sheer, 1980).

Two major groups of actions are undertaken by water agencies in order to avert some serious consequences of impending water shortages caused by droughts: (1) measures that reduce demands and (2) measures that enhance existing supplies. Developing practical methods for determining the necessary prices and devising structures of water rates that would achieve the desired reductions in water use are the most critical needs for establishing effective drought pricing policies (Dziegielewski et al., 1991).

There are different uncertainties involved with either one of these measures. In trying to reduce demands by increasing the price, uncertainty is involved in that the demand volume may not be equal to the limited available supply. This is simply because the demand depends on so many factors that cannot be totally controlled, irrespective of the price increase. On the other hand, enhancing the existing supply may cost more than the risk of not undertaking this task at all. By way of risk analysis, it is possible to optimize between the economic loss and the cost of enhancing the existing supply—such as emergency supply construction. Many scientists in different professions agree that the level of risk as a decision support system is a good indicator for sound decisions. Decisions in which the effects are portrayed relatively in the long run may finally result in adverse effects. Such effects are incurred at the expense of nonconservative risk-level designs.

Suter (1993) gave the following reasons for using the risk assessment approach for decision making:

1. Cost of estimating all environmental effects of human activities is impossibly high.
2. Regulatory decisions must be made on the basis of incomplete scientific information.

He concluded that a risk-based approach balances the degree of risk to be permitted against the cost of risk reduction and against competing risks. Lansey et al. (1989) also suggested that reliability analysis (a complement of risk analysis) be viewed as an alternative to making a decision without an analytical structure.

It has not been a common practice by responsible bodies to systematically incorporate in the decision process the risk of water supply shortages during sustained drought periods. Bruins (1993) indicated that "…governments often respond to drought through crisis management rather than preplanned programs (i.e., risk management)." Wilhite (1993) also criticized that until recently, nations had devoted little effort toward drought planning, preferring instead the crisis management approach.

A consensus among water managers and researchers regarding water supply during drought is that the key to adequate management in urban areas lies in predrought preparation, especially as it relates to conservation and planning for future water needs (Dziegielewski et al., 1991). All the above accounts prompt us to focus on the necessity of risk-based design, especially when we deal with phenomena such as drought that are very difficult to predict accurately with regard to timing and magnitude.

2.2.5 Drought Severity as Risk Indices

Every natural phenomenon with associated detrimental effects to human beings and their environment needs our keen attention to how and when it occurs. Unfortunately, the degree of severity of some such phenomena including drought is difficult to determine, as accurately as desirable, before they occur. A study by the National Research Council (1986) indicated that there is not a firm rationale or explanation of the drought mechanism. It added that though empirical relations have been documented so far, why and when these relations trigger the occurrence of significant drought is not understood.

In the absence of such rationale, it is worth studying the degree (level) of risk, such as in the case of droughts, based on the available indices. The level of risk is apparently reflected by the severity of the drought. Severe drought implies a relative shortage of the required water supply, which in turn can be expressed by a certain level of the risk that the demand is not met. Thus calibrating drought severity may be used to indirectly determine the risk level.

No single definite method has been in use as a drought severity indicator. Nonetheless there are some methods being used in different fields. According to Wilhite (1993), the simplest drought index in widespread use is the percent of normal precipitation. This, indeed, is a good approach to infer the status of the available supply. However, it does not render an obvious forecast to enable a risk management body to be prepared for a forthcoming drought period.

Sheer (1980) tried to calculate the risk that the reservoir of a water supply system becomes empty by blending together the severest hydrologic and hydraulic conditions of different time periods. Specifically, he considered a condition in which the demands were the highest, the reservoir storage the lowest, and the date when these conditions occurred the beginning of one of the worst drought years. This simulation resulted in 4 out of 26 years in which the reservoir was empty, and it is concluded that the risk is $4/26$.

Although the approach is reasonable enough to indicate what would have happened had the conditions been met, the authors hardly believe that it fully reflects the realistic situation. One simple reason is that if the actual conditions were as bad as the ones selected, the demand could be higher and might result in more years of an empty reservoir, since the demands under such conditions would be much higher over the considered time span. Another reason may be that the risk in that study is not fully analogous with the usual convention. This is to say that the risk is based on the demand exceeding the supply, which is reached long before the reservoir becomes empty.

As the best alternative, risk analysis may be viewed in relation to the uncertainty associated with the different variables. Tung (1996) points out that the most complete and ideal description of uncertainty is the probability density function of the quantity subject to uncertainty. It is, therefore, very feasible to consider the probability density function of the demands about a fixed available supply during drought and thus derive the risk as the cumulative probability function of the supply being exceeded.

To be able to calculate the risk, the level of the drought severity must be determined (forecast). There are several drought severity indices, which have been used so far. Some of them are used to assess the severity of a drought event that has already happened, while a few others are used for forecasting. The Palmer Drought Severity Index (PDSI) and the Sheer Steila Drought Index (DI) are examples of the former category, while the Surface Water Supply Index (SWSI) and the Southern Oscillation Index (SOI) are examples of indices used for drought forecasting.

Palmer (1965) expressed the severity of a drought event by developing the following equation (Steila, 1972; Puckett, 1981):

$$\mathrm{PDSI}_i = 0.897\ \mathrm{PDSI}_{i-1} + \tfrac{1}{3} Z_i \tag{2.9}$$

where Z is an adjustment to soil moisture for carryover from one month to the next, expressed as

$$Z_i = k_j[PPT_i - (\alpha_j PE_i + \beta_j G_i + \gamma_j R_i - \delta_j L_i)] \tag{2.10}$$

where PPT_i = precipitation
PE_i = potential evapotranspiration
G_i = soil moisture recharge
R_i = surface runoff (excess precipitation)
L_i = soil moisture loss for month i.

Subscript j represents one of the calendar months and i is a particular month in a series of months. The coefficients α_j, β_j, γ_j, and δ_j are the ratios for long-term averages of actual to potential magnitudes for E, G, R, and L based on a standard 30-year climatic period.

The SWSI gives a forecast of a drought event. It is a weighted index that generally expresses the potential availability of the forthcoming season's water supply (U.S. Soil Conservation Service, 1988). It is formulated as a rescaled, weighted equation of nonexcedence probabilities of four hydrologic components: snowpack, precipitation, streamflow, and reservoir storage (Garen, 1993).

$$\frac{\mathrm{SWSI} = \alpha p_{\mathrm{snow}} + \beta p_{\mathrm{prec}} + \gamma p_{\mathrm{strm}} + \omega p_{\mathrm{resv}} - 50}{12} \tag{2.11}$$

where α, β, γ, and ω are weights for each hydrologic component and add up to unity; p_i is the probability of nonexcedence (in percent) for component i; and the subscripts snow, prec, strm, and resv stand for the snowpack, precipitation, streamflow, and reservoir storage hydrologic components, respectively. This index has a numerical value for a given basin, which varies between -4.17 to $+4.17$. The following are the ranges for the index for practical purposes: $+2$ and above, -2 to $+2$, -3 to -2, -4 to -3, and -4 and below. These ranges are associated with the qualitative expressions of abundant water supply, near normal, moderate drought, severe drought, and extreme drought conditions, respectively.

The SWSI has been in use to forecast monthly surface water supply forecasts of different basins (see, for example, *U.S. Soil Conservation Service Monthly Report,* 1988). In fact, it gives a forecast of both wet and dry (drought) months. On the other hand, Wilhite (1993) reports that several scientists agree that it has been possible to forecast drought for up to 6 months in Australia by using the SOI, which is based on forecast meteorological conditions.

2.3 RISK-PRICE RELATIONSHIP

Risk can be defined as the probability that the loading exceeds the resistance (Chow et al., 1988; Mays and Tung, 1992). Analogously, the risk in water distribution systems is defined as the probability that the demand exceeds the available supply where the demand is considered as the loading and the supply as the resistance. For future planning purposes, it is not certain when a drought event of a certain severity level will occur.

In planning for urban water supply projects, therefore, it is important to determine the probability distribution parameters of the demand and the supply. Both demand and supply are related to hydrologic indices. Also, operation and management of an existing water distribution system can be handled better through a risk analysis approach when the forthcoming period's (say month's) conditions of weather or water supply availability can be predicted ahead of time.

One of the common ways to represent uncertain events such as demand and supply under drought conditions is using an appropriate probability distribution of these variables. On the other hand, both variables are related to the return period T of the drought. The available supply data of many years can be arranged in descending order of magnitude for drought indication. These arranged flow data can be plotted versus the return period T, which is a measure of hydrologic conditions, as shown in Fig. 2.1. Two basic ways can be considered for selecting the representative flow data in relation to the return period. The first one is selecting one extreme value for each unit of time, e.g., the lowest monthly flows in a period of years. The second is selecting the lowest monthly flows in a period of years (Hudson and Hazen, 1964). Both of these procedures give a general relationship between available supply (flow) and its corresponding return period, as given in Fig. 2.1.

2.3.1 Developing Risk-Price Relationships

Demand depends on many uncertain factors and consequently is uncertain for a given return period drought event. The uncertainty can be represented through a probability distribution function as illustrated in Fig. 2.3, which indicates the risk at two different return periods.

For decision purposes, the design may be fixed at the condition where the demand equals the available supply for a given price level. Beyond this point the demand exceeds the supply and there will be some associated risk. As shown in Fig. 2.2, the intersection points and the region beyond represent different values of water price and the associated risk. The illustrations in Figs. 2.1 and 2.2 show that for return periods larger than the critical return period T^* at the intersection point of supply and demand, the demand at the given price is greater than the supply. As the price decreases, the shortage volume increases thereby increasing the risk. Thus a graph of risk versus price may be plotted as shown in Fig. 2.4.

The regions beyond each of the intersection points in Fig. 2.2 have some corresponding risk levels, that is, the probability $p(\ )$ that the available supply falls below the demand corresponding to

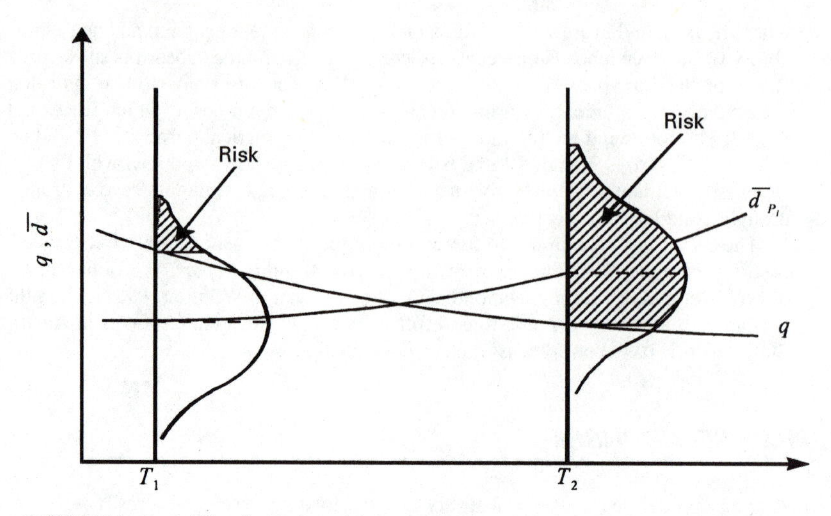

FIGURE 2.3 Probability distribution of demand at different return periods.

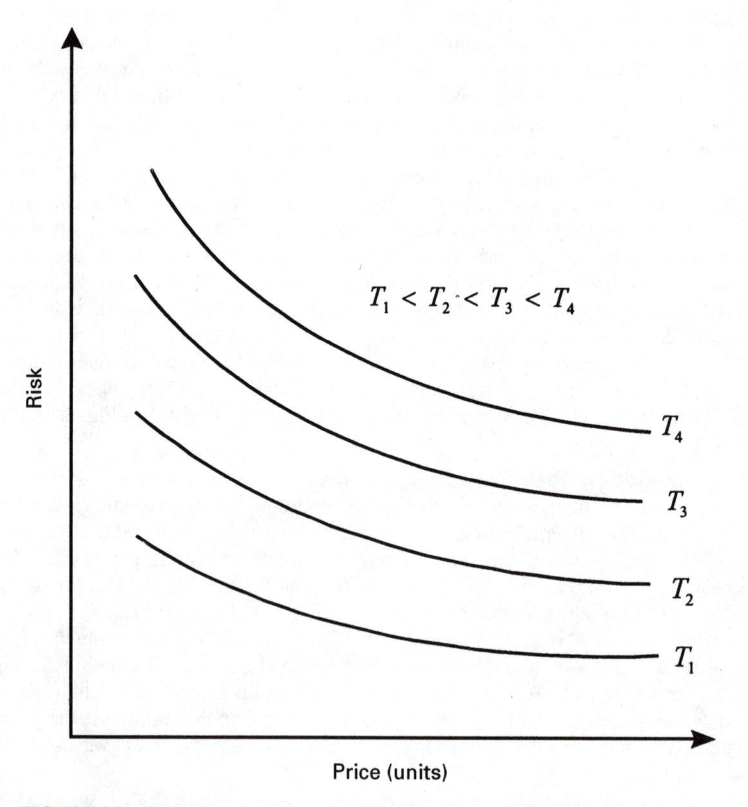

FIGURE 2.4 Risk-price relationships for different return periods.

the price adopted at level i, $i = 1, 2, 3, \ldots, n$. Such a relation can help water supply planners to determine a municipal water supply price based on a predetermined tolerable risk or to assess the risk associated with a certain price level. Although not yet demonstrated by data analysis, the price-risk relationship indicates that price is infinite at no risk and risk is close to 1.00 at zero price (Fig. 2.4).

Risk Evaluation Procedures. The general procedures for evaluating the risk of a system's loading exceeding a system's capacity are considered under two different scenarios. For water supply systems, the demand may be considered as the loading and the supply as the capacity. The two scenarios are (1) when the loading is uncertain and the capacity is certain, and (2) when both the loading and the capacity are uncertain. Risk evaluation in the first case involves consideration of the probability density function of one variable (the demand) which is computationally simpler. Risk evaluation in the second case involves composite risk evaluation.

Suppose that the probability density function of loading L is $f(L)$. The probability p that the loading will exceed a fixed and known capacity C^* is given as (Chow et al., 1988; Mays and Tung, 1992)

$$p(L > C^*) = \int_{C^*}^{\infty} f(L)\, dL \tag{2.12}$$

This relationship holds true when the capacity C is a deterministic quantity, which corresponds to the first scenario. Analogously, if the probability density function of demand d_{P_i} at price level P_i is $f(d_{P_i})$, the risk of demand d_{P_i} at price level P_i exceeding the supply q_T for return period T is expressed as

$$\text{Risk}_{(P_i,T)} = \int_{q_T}^{\infty} f(d_{P_i})\, dd \tag{2.13}$$

Using this definition for risk, the risk-price relationship may be developed for each T. The higher the price the lower the demand is, and consequently the lower the risk.

When the capacity is also uncertain but may be represented by a probability density function $g(C)$, i.e., the second scenario, the composite risk is used. The general formula for risk in this case is (Fig. 2.5).

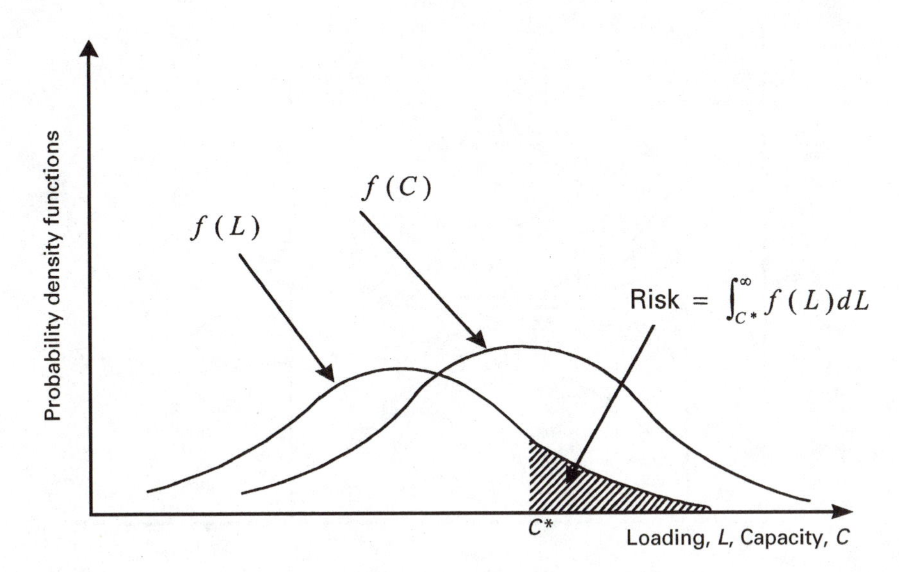

FIGURE 2.5 Probability distribution functions of loading and capacity.

$$\text{Risk} = \int_{-\infty}^{\infty} \left[\int_{C^*}^{\infty} f(L)\, dL \right] g(C)\, dC \tag{2.14}$$

Again in a similar analogy, the corresponding composite risk where both demand and supply are considered to be uncertain (for a given price and return period) is expressed as (Fig. 2.6)

$$\text{Risk}_{(P_i,T)} = {}_p(\bar{d}_{P_i} > \bar{q}_T)$$

$$= \int_{-\infty}^{\infty} \left[\int_{\bar{q}_T}^{\infty} f(q_T)\, dq \right] f(d_{P_i})\, dd \tag{2.15}$$

A similar relationship as the one shown in Fig. 2.4 can also be developed for the composite risk from these relationships. In both cases [Eqs. (2.13) and (2.15)], the risk at a given price and return period is computed as the probability of the demand exceeding the supply. The difference is in the certainty of the supply in the former equation and its uncertainty in the latter one.

Methodology of Risk Evaluation. Numerical evaluations of risk using the above equations call for the approach to determine the quantitative values of different statistical parameters of the loading and/or the capacity. The risk equations consist of complex probability distribution functions, which become difficult to integrate. Because of this, alternative ways of evaluating the value of risk are often utilized. The safety margin and safety factor approaches (see Chow et al., 1988; Mays and Tung, 1992) are generally used for the computation of the risk from the probability distributions of the loading and/or the capacity. The safety margin approach is illustrated below with numerical data and the safety factor approach will be introduced in Sec. 2.4.5.

The safety margin SM is generally given as the difference between the loading and the capacity or SM = $C - L$. Thus the risk in terms of the safety margin is given as

$$\text{Risk} = p(C - L < 0) = p(\text{SM} < 0) \tag{2.16}$$

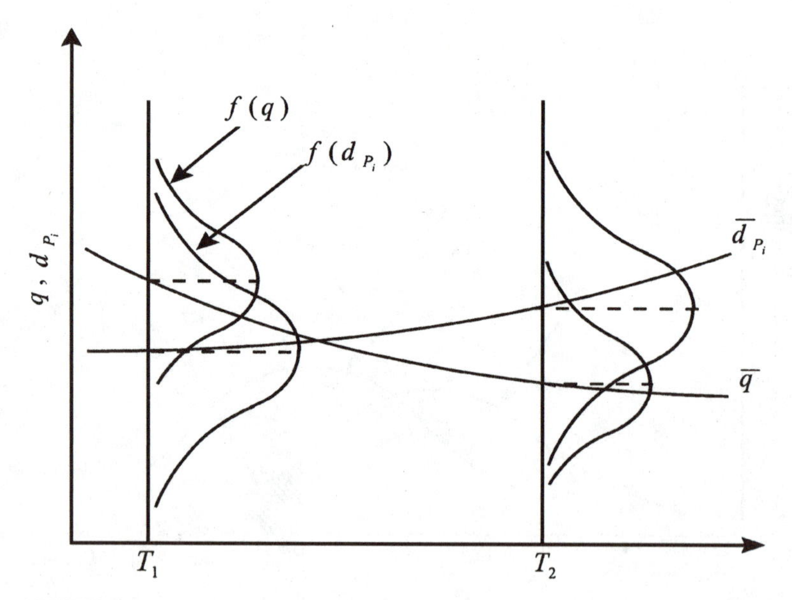

FIGURE 2.6 Probability distribution of both demand and supply at different return periods.

If C and L are independent random variables, the mean value and the standard deviation of SM are given, respectively, as

$$\mu_{SM} = \mu_C - \mu_L \tag{2.17}$$

$$\sigma_{SM}^2 = \sigma_C^2 + \sigma_L^2 \tag{2.18}$$

By taking water demand and the available supply as the loading and the capacity, respectively, the risks at different price levels for different return periods can be easily computed. Using the safety margin approach,

$$\text{Risk} = p[(d_{Pi} - q_T) < 0] = p(\text{SM} < 0) \tag{2.19}$$

$$\mu_{SM} = \mu_q - \mu_{dP} \tag{2.20}$$

$$\sigma_{SM}^2 = \sigma_q^2 + \sigma_{dP}^2 \tag{2.21}$$

where μ_q is the mean supply and μ_{dP} is the mean demand at price level P, respectively. Assuming that the safety margin is normally distributed, the risk is expressed as

$$\text{Risk} = p\left(\frac{\text{SM} - \mu_{SM}}{\sigma_{SM}} < \frac{0 - \mu_{SM}}{\sigma_{SM}}\right) = p\left(z < \frac{-\mu_{SM}}{\sigma_{SM}}\right) = p\left(z < \frac{-(\mu_q - \mu_d)}{(\sigma_q^2 + \sigma_d^2)^{1/2}}\right)$$

$$= \Phi_z\left(-\frac{\mu_{SM}}{\sigma_{SM}}\right) = \Phi_z\left(\frac{-\mu_q - \mu_d}{\sigma_q^2 + \sigma_d^2}\right)^{1/2} \tag{2.22}$$

where z is the standard normal variable with mean 0 and standard deviation 1.

However, before using these equations it is further required that the mean and the standard deviation estimates of demand and/or supply must be estimated. The expected value of demand at different price levels can be estimated using the price elasticity formula. Its standard deviation, on the other hand, can be estimated from the first-order analysis of uncertainty of the demand model (equation). If a dependent variable Y is a function of independent variables $\mathbf{X}$ ($\mathbf{X} = X_1, X_2, \ldots, X_k$) such that $Y = g(\mathbf{X})$, the first-order approximation of Y is given as

$$Y \approx g(\bar{x}) + \sum_{i=1}^{k} \left[\frac{\delta g}{\delta X_i}\right]_{\bar{x}} (X_i - \bar{x}_i) \tag{2.23}$$

in which $\bar{\mathbf{x}} = (\bar{\mathbf{x}}_1, \bar{\mathbf{x}}_2, \ldots, \bar{x}_k)$, a vector containing the means of k random variables (Mays and Tung, 1992). The variance of Y, Var[Y] or σ_Y^2, is estimated by the following equation, which can be derived from the first-order analysis of uncertainty of Eq. (2.23).

$$\sigma_Y^2 = \text{Var}[Y] \approx \sum_{i=1}^{k} \alpha_i^2 \sigma_i^2 + 2 \sum_{1<}^{k} \sum_{j}^{k} a_i a_j \,\text{Cov}[X_i, X_j] \tag{2.24}$$

where $a_i = [\delta g / \delta X_i]_{\bar{x}}$ and σ_i^2 is the variance corresponding to random variable X_i. When the X_i's are independent random variables, Cov $[X_i, X_j] = 0$.

The foregoing discussion in general indicated that for a given return period for design, water supply planners can decide the price of the water supply for an affordable risk level or can determine the risk at a given affordable water price. The flowchart given in Fig. 2.7 summarizes the basic steps used to develop the risk-price-return period relationships.

Risk Evaluation Example. Based on the safety margin analysis given by Eqs. (2.19) to (2.22) and the price elasticity of demand definition [Eq. (2.1)], it is possible to determine the risk values for a given return period and different price levels. For a given return period of drought, the expected demand when the price is increased by a certain amount can be determined by Eq. (2.25). Table 2.3

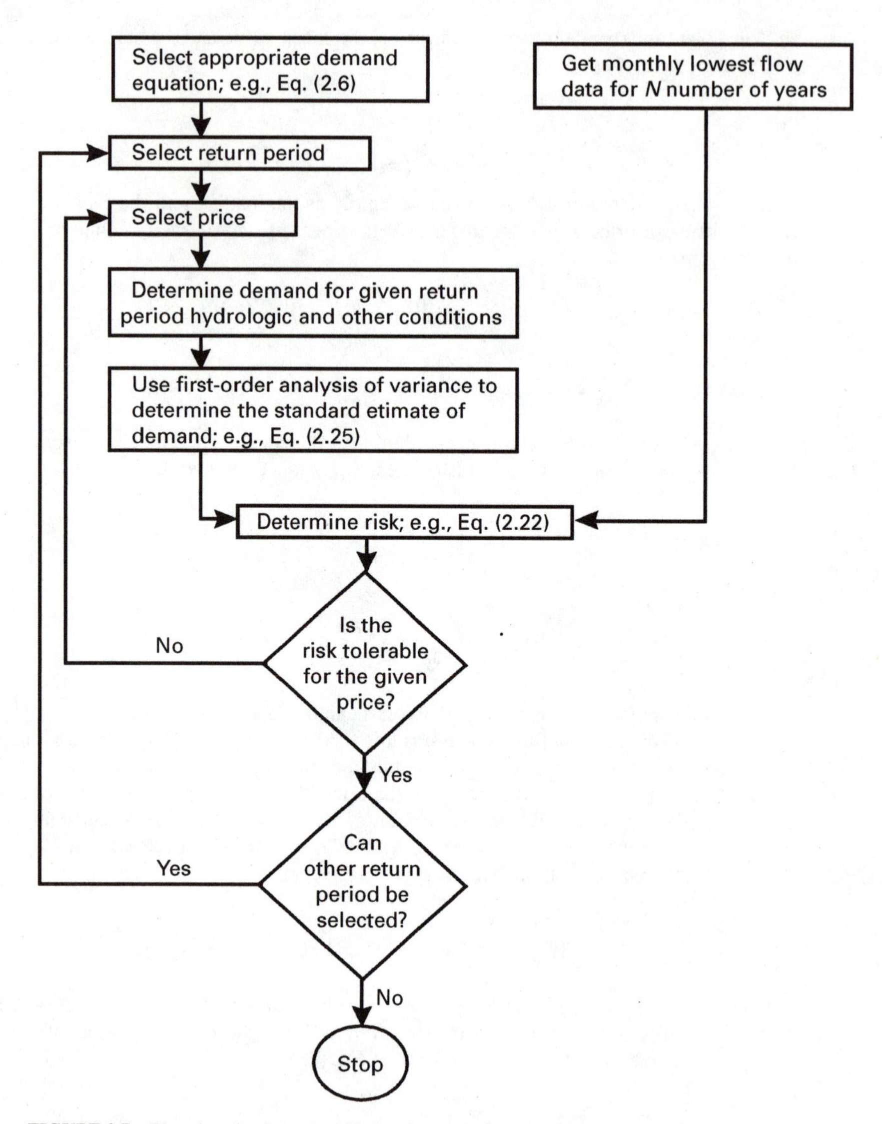

FIGURE 2.7 Flowchart for the proposed planning procedure.

TABLE 2.3 (Hypothetical) Data of Demand and Supply for Different Return Periods

Return period T, years	Demand at price level P_1, units	Available flow q, units
1	8.0	12.0
5	8.5	11.0
10	9.5	9.5
25	11.0	8.0
50	13.0	7.0

lists demand for an initial price level P_i and also the available supply for different return periods. Equation (2.1) for the price elasticity of demand is rearranged to solve for d_i as

$$d_i = \frac{d_{i-1}\,[1 + \eta_P(P_i - P_{i-1}) / (P_i + P_{i-1})]}{1 - \eta_P(P_i - P_{i-1}) / (P_i + P_{i-1})} \tag{2.25}$$

Equation (2.25) is used to determine the demand at a given price level and a given return period. Price increases of up to 200 percent and a price elasticity of -0.5 are used to compute the demand reduction due to the increases in the price for each of the return periods. The risks associated with different price levels and different return periods are determined based on approximate estimates of the standard error for supply as 2.0 units and for demand as 4.0 units, for which σ_{SM} equals 4.47. The results thus obtained are given in Table 2.4 and also plotted as shown in Fig. 2.8. The plots show that the risk is not significantly sensitive to the price change for small price increases. The plot in Fig. 2.9 shows how sensitive risk is to the return period T. It is inferred from these two plots that planning and/or overcoming the shortage of water supply during drought periods requires a strong commitment to increase the price sufficiently.

2.4 OPERATION AND MANAGEMENT PLANNING UNDER SUSTAINED DROUGHT CONDITIONS

The price-elasticity formulation indicates that when the available supply is less than the demand, the latter can be adjusted to the former by increasing the price. In other words, for a drought event of severity index greater than the one at which the demand equals the available supply (Fig. 2.10), it is possible to force the demand curve down to the supply curve by increasing the price. However, the fact that it has not been easy to forecast drought conditions well ahead of time and the uncertainty in its magnitude and length requires operation and management of water supply systems that will attempt to smooth out the effect of the drought. Such operation and management efforts will be based on data of a short time interval. The efforts in effect are a supplement to the planning procedure already mentioned above. The planning basically turns out to be a one-time decision, while operation and management especially under sustained drought conditions involve routine decisions. The severity of the drought could be so high that emergency water supply construction projects may be

TABLE 2.4 Risk Values for Different Return Periods and Price Increases of up to 200%

	Return period, T, years				
Price (unit)	1	5	10	25	50
1.0	0.183	0.288	0.500	0.749	0.910
1.2	0.147	0.236	0.425	0.674	0.864
1.4	0.123	0.198	0.367	0.614	0.813
1.6	0.102	0.169	0.326	0.564	0.770
1.8	0.090	0.147	0.295	0.516	0.726
2.0	0.079	0.131	0.264	0.480	0.674
2.2	0.069	0.119	0.245	0.448	0.655
2.4	0.064	0.109	0.224	0.421	0.622
2.6	0.058	0.102	0.212	0.394	0.599
2.8	0.054	0.093	0.198	0.378	0.568
3.0	0.050	0.087	0.187	0.359	0.544

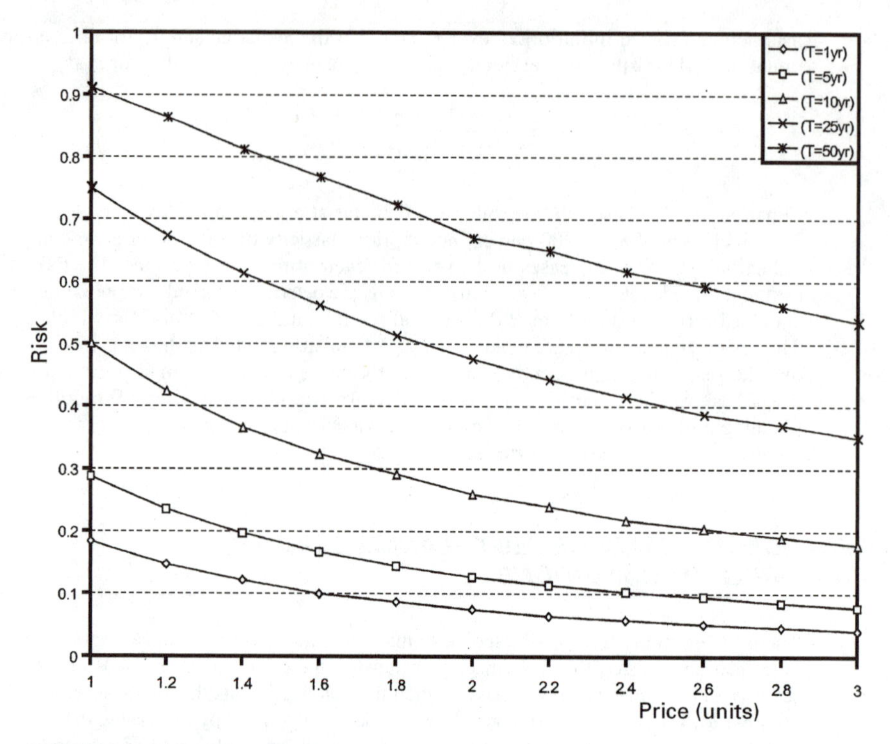

FIGURE 2.8 Risk-price-return period relationships.

considered. An estimate of the damage that would result from a sustained drought period is required and, most of all, may be used to determine if an emergency water supply should be implemented.

2.4.1 Economic Aspects of Water Shortage

The shortage of water supply during drought periods results in different types of losses in the economy including, but not limited to, agricultural, commercial, and industrial. In agriculture, lack of water supply results in crop failures; in commerce, it may result in a recession of the business; and in industry, it may result in underproduction of commodities. The loss in each production or service sector depends on the purpose of the sector. For instance, the economic impact of drought on agriculture depends on the crop type (Easterling, 1993). There is no single common way of assessing the economic impact of drought on any one of the sectors. Evaluating and comparing what actually happens during a drought period with what would have happened had there been no drought may be one way of assessing the effects of drought (Dixon et al., 1996). Dixon et al. (1996) adopted the concept of willingness-to-pay to value changes in well-being. They define *willingness-to-pay* as the maximum individuals would have been willing to pay to avoid the drought management strategies imposed by water agencies.

On the other hand, since water is supplied during a drought period at a greater price, it can be viewed as a revenue generator. Therefore, when the demand exceeds the available supply, the revenue collected by the water supply agency will be less than what could have been collected had there been more supply than that actually available. In other words, if the demand exceeds the supply, the problem is not only limited to lack of water but there will also be economic loss since the customers would pay for more supply if there were enough. Depending on the risk level, it is possible to reach a decision of whether supply augmentation is necessary or the pressure for more demand could be tolerated with the available supply.

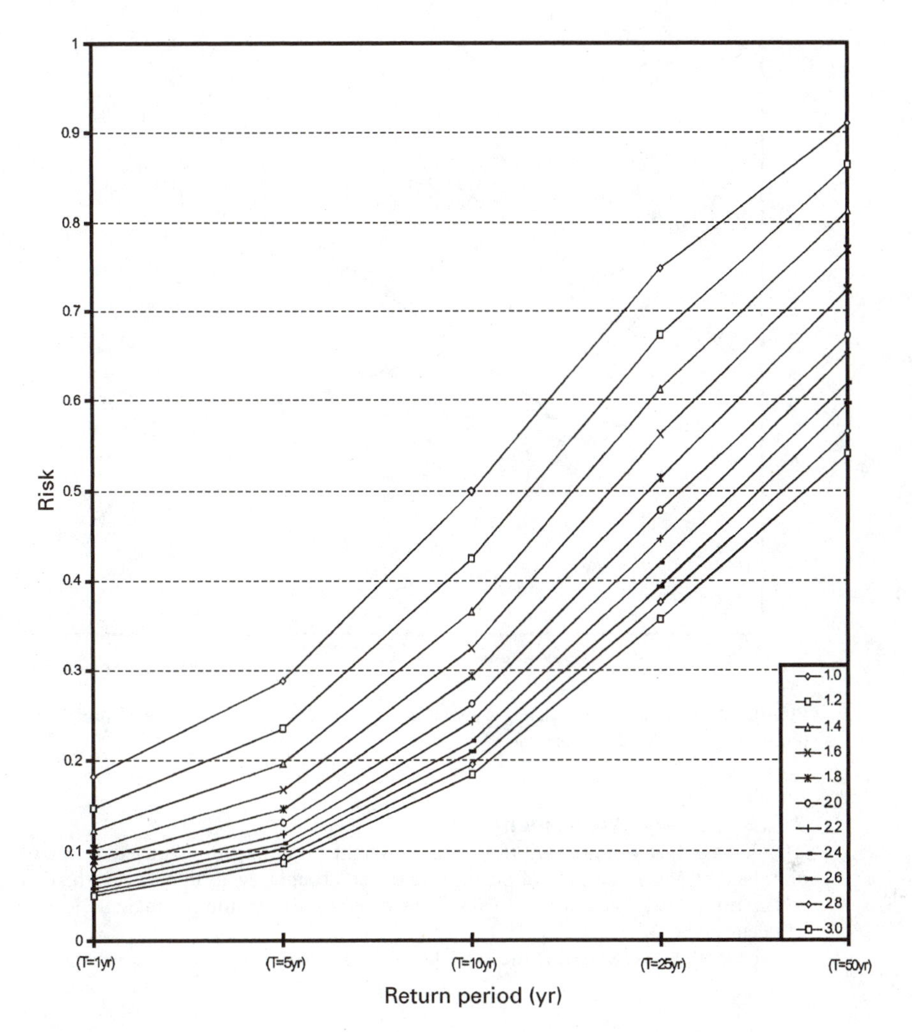

FIGURE 2.9 Risk-price-return period relationships. The legend indicates the price in units.

Some water shortage relief efforts can be undertaken so that emergency water supplies may be made available to the users. This can be implemented by well drilling, trucking in potable supplies, or transporting water through small-diameter emergency water lines. In such cases, it may be required that the emergency supply construction costs be paid by the users (Dziegielewski et al., 1991). The estimation of the expected financial loss can be used to determine and inform the users of its extent and advise them of the necessity, if any, of paying for the emergency supply construction costs.

If the option for emergency supply construction is justified, then the design needs to take into consideration the possibilities of optimization. The construction can be designed such that the financial risk and the cost of construction are at optimum. Figure 2.11 illustrates this optimization process.

The economic loss (damage) can be calculated with the help of Eqs. (2.29) and (2.30) (given in Sec. 2.4.2), and the cost of emergency construction must be determined from the physical conditions at the disposal of the water supply agency.

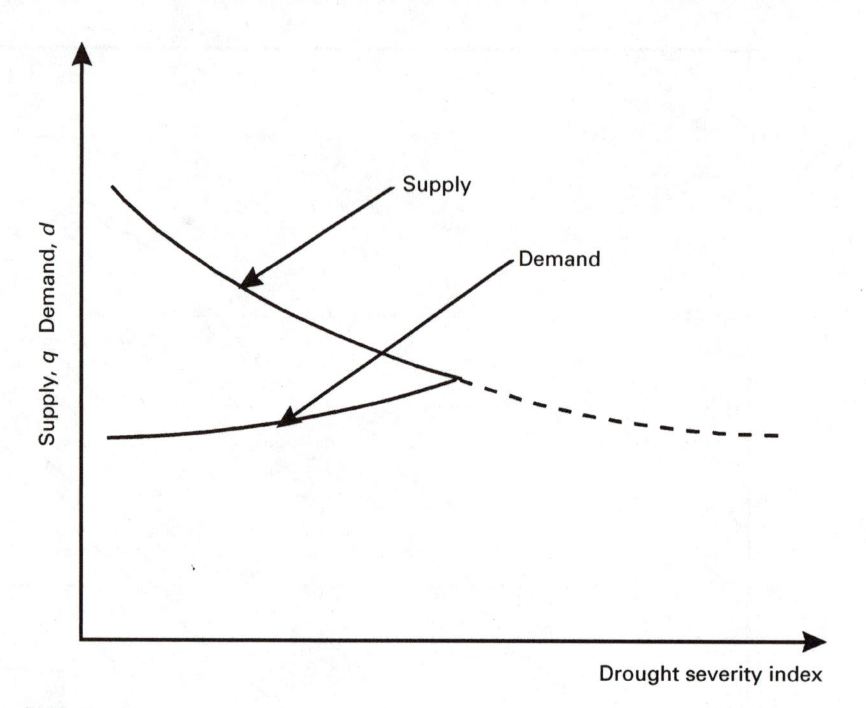

FIGURE 2.10 Water demand during a sustained drought period as adjusted to the available supply (the broken line shows the adjustment).

2.4.2 Damage Assessment

The damage that would result if a certain drought event occurred is used as one of the main decision factors. Since the time of occurrence of the drought event that causes the damage is difficult to determine, only the expected value is assessed by associating its magnitude with its probability of occurrence.

Chow et al. (1988) define the expected annual damage cost D_T for the event $x > x_T$ as

$$D_T = \int_{x_T}^{\infty} D(x)\, f(x)\, dx \tag{2.26}$$

where $f(x)\, dx$ is the probability that an event of magnitude x will occur in any given year and $D(x)$ is the damage cost that would result from that event. The event x in this case can be assumed as the demand and x_T can be the available supply during a drought event of return period T.

Breaking down the expected damage cost into intervals,

$$\Delta D_i = \int_{x_{i-1}}^{x_i} D(x)\, f(x)\, dx \tag{2.27}$$

from which the finite difference approximation is obtained as

$$\Delta D_i = \frac{D(x_{i-1}) + D(x_i)}{2} \int_{x_{i-1}}^{x_i} f(x)\, dx$$

$$= \frac{D(x_{i-1}) + D(x_i)}{2} [p(x \geq x_{i-1}) - p(x \geq x_i)] \tag{2.28}$$

Thus the annual damage cost for a structure designed for a return period T is given as

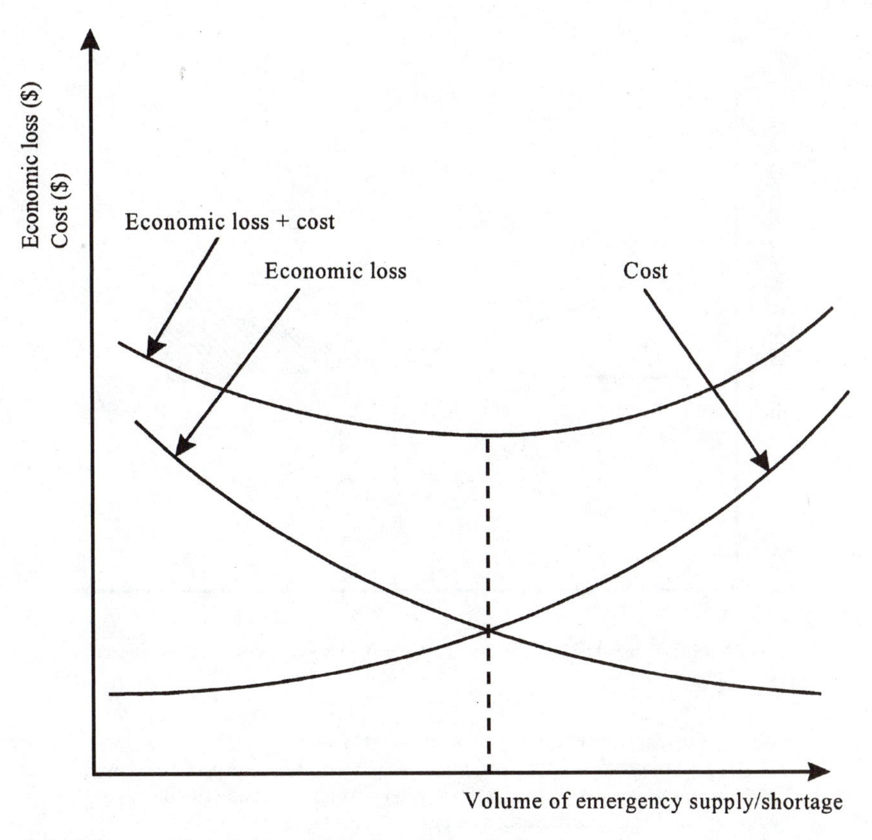

FIGURE 2.11 Optimization for emergency water supply construction.

$$D_T = \sum_{i=1}^{\infty} \frac{D(x_{i-1}) + D(x_i)}{2} [p(x \geq x_{i-1}) - p(x \geq x_i)] \tag{2.29}$$

To determine the annual expected damage in Eq. (2.29), the damage that results from drought events of different severity levels must be quantified.

The magnitude of the drought (in monetary units) may be obtained by estimating the volume of water shortage that would result from that drought. In other words, not having the water results in some financial loss to the water supply customer. The resulting financial loss to the customer from a certain drought event is thus considered as the damage from that drought event.

As was shown in Fig. 2.1, after the critical return period T^*, the divergence between the demand and the supply increases with the return period. Expressing the demand and the supply as a function of return period T of drought events enables one to estimate the annual expected water supply shortage volume as

$$S_V = \int_{T^*}^{T} [d(T) - q(T)]\, dt \tag{2.30}$$

The shortage volume S_V is illustrated by the shaded area in Fig. 2.12. The shortage volume for a drought event of a higher return period than the critical one results in a higher shortage volume and consequently a higher associated damage. The relationship between the shortage volume and the

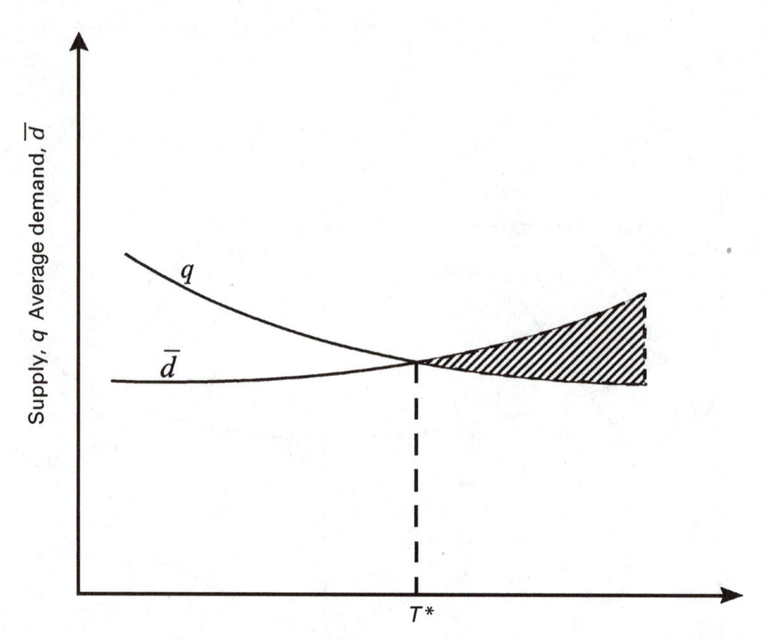

FIGURE 2.12 Demand and supply showing water shortage volume when demand exceeds supply.

associated damage generally depends on several factors including the water-use category—residential, industrial, commercial, agricultural, and so on. To use the procedure presented herein for assessing the damage that results from a certain water shortage volume, the damage given by Eq. (2.29) must be developed for a specific user category.

2.4.3 Operation and Management

For operation and management of an existing municipal water supply system during a sustained drought period, administrative decisions may be based on short-time forecasting of the hydrologic conditions. A forecast of, say 1 month ahead of the available supply, helps the supply managers to preadjust the expected demand to the forecasted available supply by increasing the price. In other words, the expected demand can be, in principle, suppressed to the forecasted available supply by increasing the price. Howe (1993) points out that since price presumably affects the quantities users demand, price can be used to adjust demand to the available supply. The basic factor in the decision will be the damage that would occur if the adjustment were not undertaken. This is the reason why we need to focus on the assessment of such damages.

It may be easily conceived from the above reasoning that it is possible to express price increase as some function of damage. If dP is an elementary increase in price due to a certain level of drought, the following general relationship may be formulated:

$$dP = \phi(\xi) \tag{2.31}$$

where ξ is an implicit variable for the drought severity level.

The amount of decrease in the demand attained as a result of the increase in the price may be determined from the concept of price elasticity of demand for water, which is rewritten in finite difference form as

$$\Delta d = \frac{\eta_P(d_{x_{i-1}} + d_{x_i})/2}{(P_{x_{i-1}} + P_{x_i})/2} \Delta P \tag{2.32}$$

The equation for the increased price P_{xi} is obtained from Eq. (2.32) as

$$P_{x_i} = \frac{P_{x_{i-1}} + P_{x_{i-1}}(d_{x_i} - d_{x_{i-1}})/\eta_P(d_{x_{i-1}} + d_{x_i})}{1 - (d_{x_i} - d_{x_{i-1}})/\eta_P(d_{x_{i-1}} + d_{x_i})} \tag{2.33}$$

Also Eq. (2.31) can be written in finite difference form as

$$\Delta P = P_{x_i} - P_{x_{i-1}} = \phi(\xi) \tag{2.34}$$

$$P_{x_i} = P_{x_{i-1}} + \phi(\xi) \tag{2.35}$$

A close-up look at Eq. (2.33) indicates that the price P_{x_i} at drought event level x_i is greater than the price $P_{x_{i-1}}$ at drought event level x_{i-1}, as expected. To achieve this, the price must increase from $P_{x_{i-1}}$ to P_{x_i} by the amount $\phi(\xi)$, as shown by Eq. (2.35). Thus by increasing the price, the supply deficiency of water during sustained drought periods can be overcome or minimized. In fact, the price can be forced to rise to the level that limits the demand of water to that amount which is available. Doing so will theoretically enable us to adjust the portion of the demand curve beyond the critical drought severity index (Fig. 2.11) down to the supply curve. However, this may not be readily accepted by the customers and thus arises the uncertainty. In essence, there will result a positively skewed distribution tendency of the customers for the demand, and hence the analysis of the associated uncertainty comes into picture.

2.4.4 The $\phi(\xi)$ Function

To fully make use of Eq. (2.33) or (2.35) for water demand abatement through price increase, an explicit form of the drought function, $\phi(\xi)$, in which ξ is the drought severity index, must be determined. Different approaches have been followed to develop indices for a drought event. Presuming that the SWSI is one of the alternatives available to forecast a drought severity level, ϕ(SWSI) will be used herein. The subscripts of d and P may be substituted by the numerical values of SWSI. For instance, if a drought month of SWSI $= -2.00$ is forecast to follow a normal month of SWSI $= 0.00$, the price $P_{-2.00}$ can be determined based on Eq. (2.33), with the price during the normal month $P_{0.00}$ known. Since the SWSI depends on p_{snow}, p_{prec}, p_{strm}, p_{resv}, the following general relation between SWSI and the variables may be conceived.

$$\text{SWSI} = \psi(p_{\text{snow}}, p_{\text{prec}}, p_{\text{strm}}, p_{\text{resv}}) \tag{2.36}$$

Apparently, then,

$$\phi(\xi) \equiv \phi(\text{SWSI}) = \phi'(p_{\text{snow}}, p_{\text{prec}}, p_{\text{strm}}, p_{\text{resv}}) \tag{2.37}$$

Once a fully explicit model is developed for Eq. (2.37), it becomes possible to recompute the expected demand using Eq. (2.25). As an alternative for this, the following equation may also be derived from Eq. (2.32):

$$P_{x_i} = P_{x_{i-1}} + \frac{(d_{x_i} - d_{x_{i-1}})\,(P_{x_i} + P_{x_{i-1}})/2}{\eta_P(d_{x_i} + d_{x_{i-1}})/2} \tag{2.38}$$

The second term on the right-hand side in Eq. (2.38) is equivalent to the $\phi(\xi)$ function mentioned earlier. It is to be noted that Eq. (2.33) gives an explicit equation to determine the price at drought event level x_i, while Eq. (2.38) gives a term equivalent to the $\phi(\xi)$. The SWSI can be used to indicate if a drought may occur and to determine its severity level if it occurs. It is to be recalled that, as indicated by Eq. (2.33), the demand at drought event level x_i can be adjusted to the estimated available supply q_{xi} by increasing the price from its value at drought event level x_{i-1} to a new value at drought event level x_i.

2.4.5 Uncertainty and Risk in Demand

Although it is presumed that demands can be adjusted to the available supply, there is uncertainty. Demand is a variable and may not meet the available supply irrespective of the increase in the price.

Hobbs (1989) points out that future demands are random because they depend upon weather, consumer tastes and preferences, household income, water rates, and level of economic development. These reasons naturally cause the demand to have some positively skewed probabilistic distribution.

Some organizations and researchers have used different probability distributions for demand. Charles Howard and Associates (in 1984) and Norrie (in 1983) used a gamma distribution for demand for Seattle (Hobbs, 1989). Also, it may be possible that the statistics of the distribution of the demand about the available supply is not uniform at different drought severity levels.

A general trend of the supply with the drought severity index and the distribution of the demand about the supply may be represented as shown in Fig. 2.13. A general gamma probability density function given by the following equation (Montgomery and Runger, 1994) is assumed.

$$f_x(x; \lambda, r) = \frac{\lambda^r x^{r-1} e^{-\lambda x}}{\Gamma(r)} \qquad x > 0 \qquad \lambda > 0, \qquad \text{and} \qquad r > 0 \tag{2.39}$$

where

$$\Gamma(r) = \int_0^{\infty} x^{r-1} e^{-x}\, dx \qquad r > 0 \tag{2.40}$$

Taking the SWSI as the drought severity level indicator and the demand as the variable x in the gamma function given in Eq. (2.40), a general relationship between the supply q, the SWSI, and the density function of the demand $f(d)$ can be given as illustrated in Fig. 2.13. Negatives of the SWSI values normally adopted are used in this figure to indicate the increase of severity with the index in absolute terms.

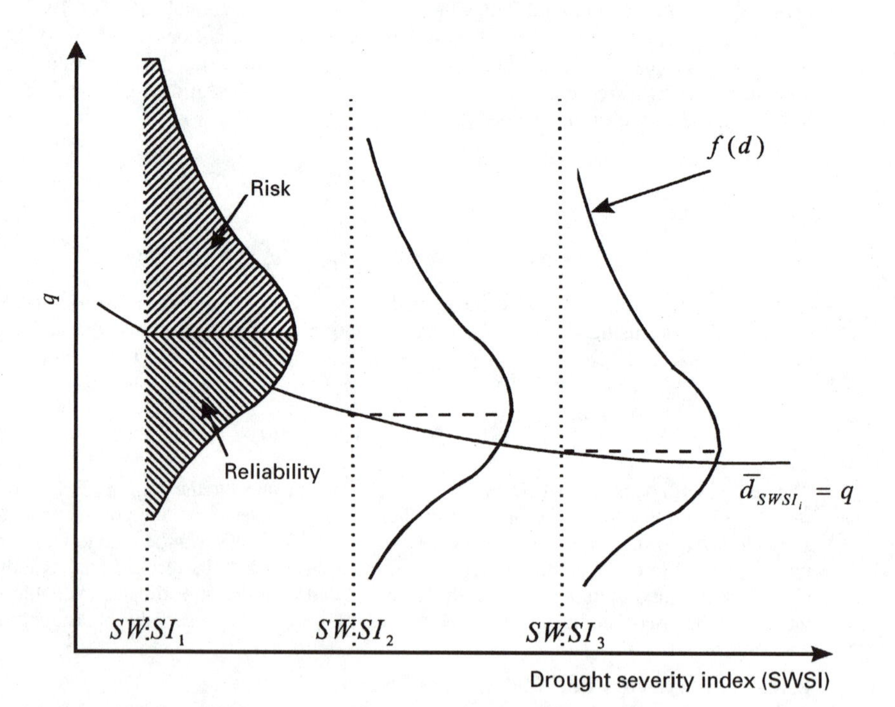

FIGURE 2.13 Expected water demand as adjusted to the available supply under sustained drought conditions and its probability distribution.

Let d_{x_i} be the random demand at drought severity level x_i, and by implication let q_{x_i} be the corresponding known available supply at drought severity level x_i. Tung (1996) defines *reliability* as the probability that the resistance is greater than the loading. In a similar analogy, the reliability of a water supply system may be defined as the probability that the available supply is greater than the expected demand. Thus, the reliability of supply is the probability that q_{x_i} is greater than d_{x_i}, expressed as

$$R = p(q_{x_i} > d_{x_i}) \tag{2.41}$$

and the risk is

$$\text{Risk} = 1 - p(q_{x_i} > d_{x_i}) = p(d_{x_i} > q_{x_i}) \tag{2.42}$$

where R is the reliability that the available supply is greater than the estimated demand and p is the probability. As illustrated in Fig. 2.13, a higher demand above the available supply implies a higher risk and a lower reliability. The risk defined by Eq. (2.42) can also be expressed in terms of the *safety factor,* SF, which may be defined as

$$\text{SF} = \frac{q_{x_i}}{d_{x_i}} \tag{2.43}$$

where the corresponding risk formula is

$$\text{Risk} = p(\text{SF} < 1) \tag{2.44}$$

For different drought severity indices, different risk–safety factor relationships can be developed. This is illustrated in Figs. 2.14 and 2.15. Once such relationships are developed, it is easier for water supply managers to decide the tolerable risk for a given drought severity index. It is to be noted here that the reliability analysis is just complementary to the risk analysis, whereas the safety factor approach is simply an alternative to the safety margin approach discussed in Methodology of Risk Evaluation in Sec. 2.3.1.

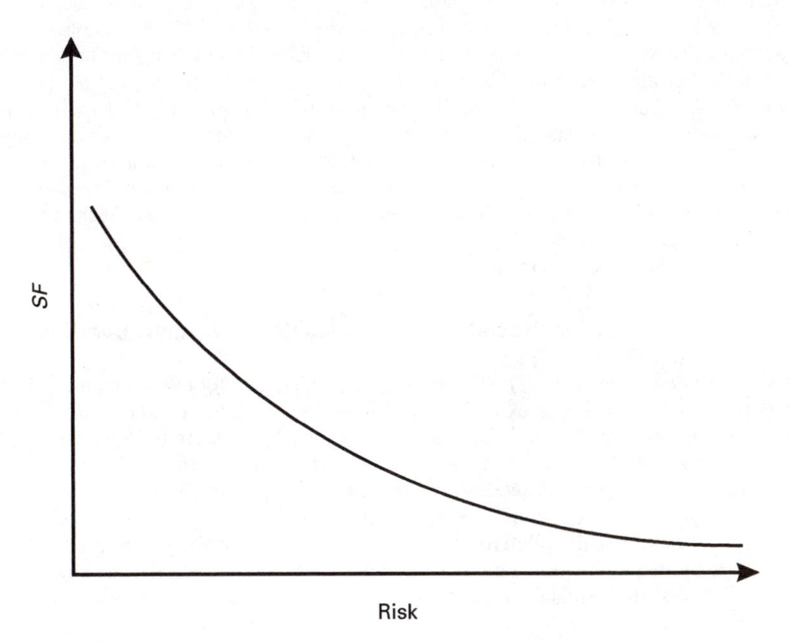

FIGURE 2.14 General illustration of risk–safety factor relationship.

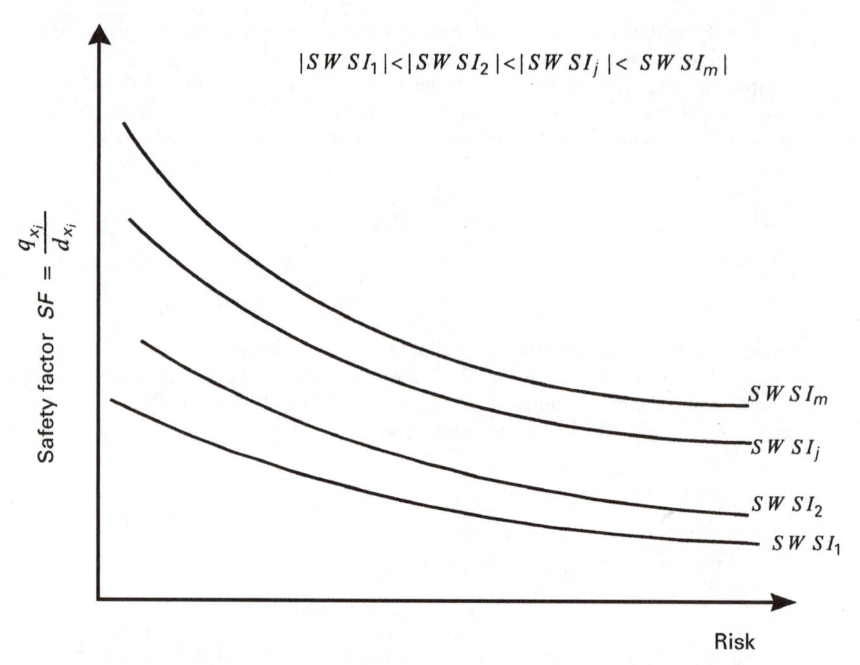

FIGURE 2.15 Risk–safety factor relationship for different SWSIs.

2.4.6 Operation and Management Strategy

It is indicated in the foregoing sections that urban water supply operation and management during sustained drought periods requires preparation at least by the water supply agents. Sound preparation procedures entail good strategy to be used. Most of all, collection of enough data affecting the water supply during a forthcoming period of time enables the supply agents to be prepared better for smoothing out the effect from a forecast drought event. Such efforts must be undertaken continuously during a sustained drought period. The flowchart shown in Fig. 2.16 will help water supply operation and management during drought periods. As indicated in the flowchart, the important data to forecast are the forthcoming period's (say month's) weather conditions or available flow and the expected demand. A comparison between the expected demand (after some necessary price increase) and the available flow clearly indicates if there will be a drought event. If the expected demand is found to be greater than the available flow, it implies that a drought event exists and hence necessary measure(s) should be sought.

2.4.7 Risk Evaluation Procedure under Sustained Drought Conditions

Developing the Procedure. The flowchart given in Fig. 2.16 for evaluating the risk under sustained drought conditions entails an explicit method of risk assessment. A certain gamma distribution must be determined among a possibly infinite number of such distributions. Specifically, the parameters of the distribution, λ and r, must be selected. The chi-square distribution, perhaps the most widely used gamma distribution, is selected for this example. The value of λ for chi-square distribution equals ½, whereas an r value of 1 is selected from possible values of ½, 1, 1½, …

Figure 2.17 shows the relative values of the available supply q_{x_i} and the expected demand d_{x_i} which are plotted on a chi-square distribution with $\lambda = ½$ and $r = 1$. The shaded area shows the probability that the demand exceeds the supply, or equivalently, the probability that the safety

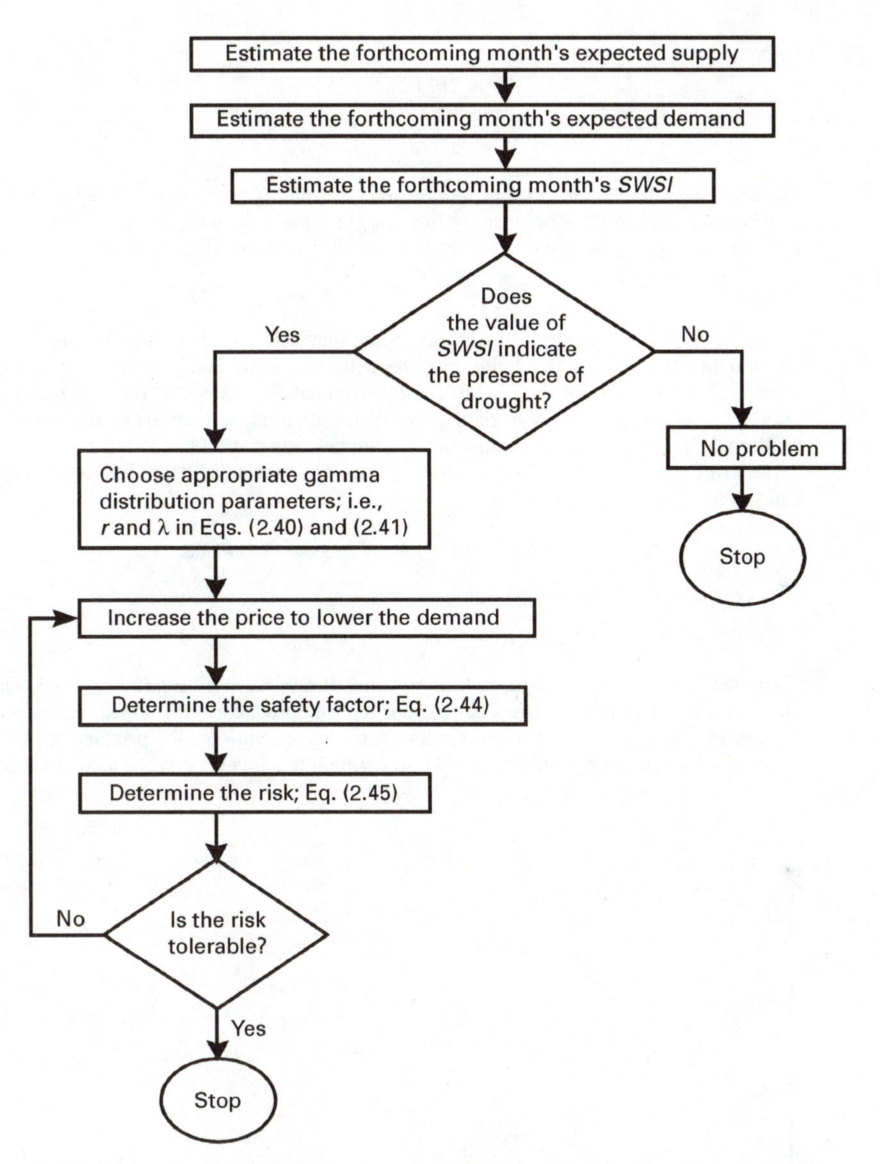

FIGURE 2.16 Flowchart for adjusting demand to available supply under sustained drought conditions.

factor SF is less than 1. The area under the chi-square distribution that lies to the right of the d_{x_i} limit (or the $\chi_{p/2,\,n-1}{}^2$ limit) is simply $p/2$, which in essence represents a probability value. Also the area that lies to the left of the $\chi_{1-p/2,\,n-1}{}^2$ limit is $p/2$. The letter n represents the sample size used in determining the expected average demand, and the value $(n - 1)$ represents the degrees of freedom for the error in estimating the expected demand. Therefore, the risk which is represented in the figure by the shared area is one-half of $1 - p$, i.e., Risk $= \frac{1}{2}(1 - p)$. Equivalently, the risk can be derived from Eq. (2.43) as

$$\text{Risk} = p\left(\frac{q_{x_i}}{d_{x_i}} < 1\right) = p\left(\frac{d_{x_i}}{q_{x_i}} > 1\right)$$
$$= p(d_{x_i} - q_{x_i} > 0) \tag{2.45}$$

Since it is assumed that the expected demand is distributed as chi-square and the available supply is considered deterministic, the quantity ($d_{xi} - q_{xi}$) is also distributed as chi-square. As shown in Fig. 2.17, the $\chi_{p/2,\,n-1}{}^2$ value where $d_{xi} = q_{xi}$ is $\chi_{0.5,\,n-1}{}^2$. Therefore, the risk can be computed as

$$\text{Risk} = p(d_{x_i} - q_{x_i} > \chi_{0.5,\,n-1}{}^2) \tag{2.46}$$

Figure 2.18 gives a simplification procedure that will help determine the risk value for a given demand level. Figure 2.18*a* represents the actual demand level on the chi-square distribution. However, to use some of the readily available probability values of the chi-square distribution at varying levels of the given variable, the distribution must start from the origin (Fig. 2.18*c*). Subtracting q_x from d_x displaces the actual demand distribution to the left by q_x, as shown in Fig. 2.18*b*. Adding $\chi_{0.5,\,n-1}{}^2$ to $d_x - q_x$ results in the desired final distribution. Referring to this figure and simplifying Eq. (2.46) results in

$$\text{Risk} = p(d_{x_i} - q_{x_i} + \chi_{0.5,\,n-1}{}^2) - p(\chi_{0.5,\,n-1}{}^2) \tag{2.47}$$

where $p(\chi_{0.5,\,n-1}{}^2) = 0.5$. Therefore,

$$\text{Risk} = p(d_{x_i} - q_{x_i} + \chi_{0.5,\,n-1}{}^2) - 0.5 \tag{2.48}$$

It may be noted here that the maximum value of risk given by Eq. (2.48) is 0.5. This is in agreement with the definition of the safety factor and the procedure developed in this work. Referring back to Fig. 2.10 shows that under sustained drought conditions, the primary objective is to lower the demand down to the available supply, lowering it to a level below the available supply not being

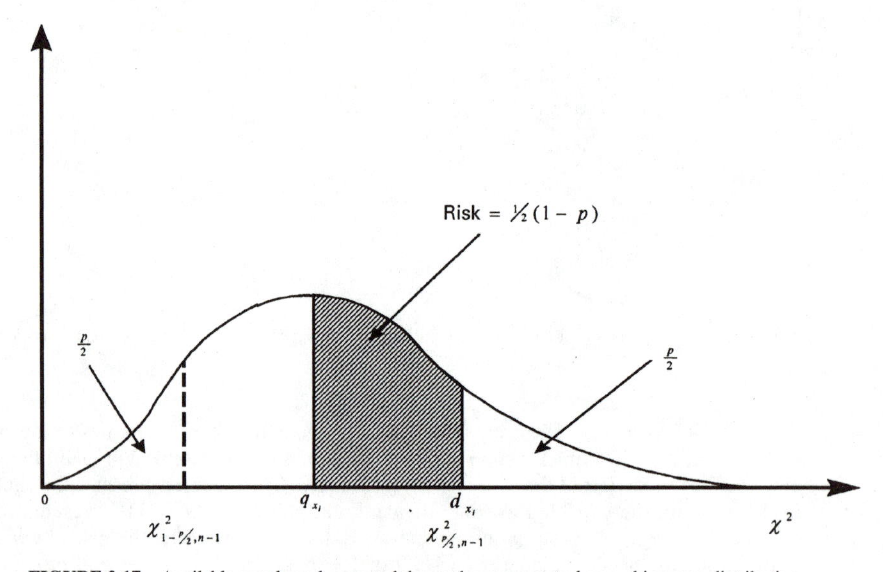

FIGURE 2.17 Available supply and expected demand as represented on a chi-square distribution.

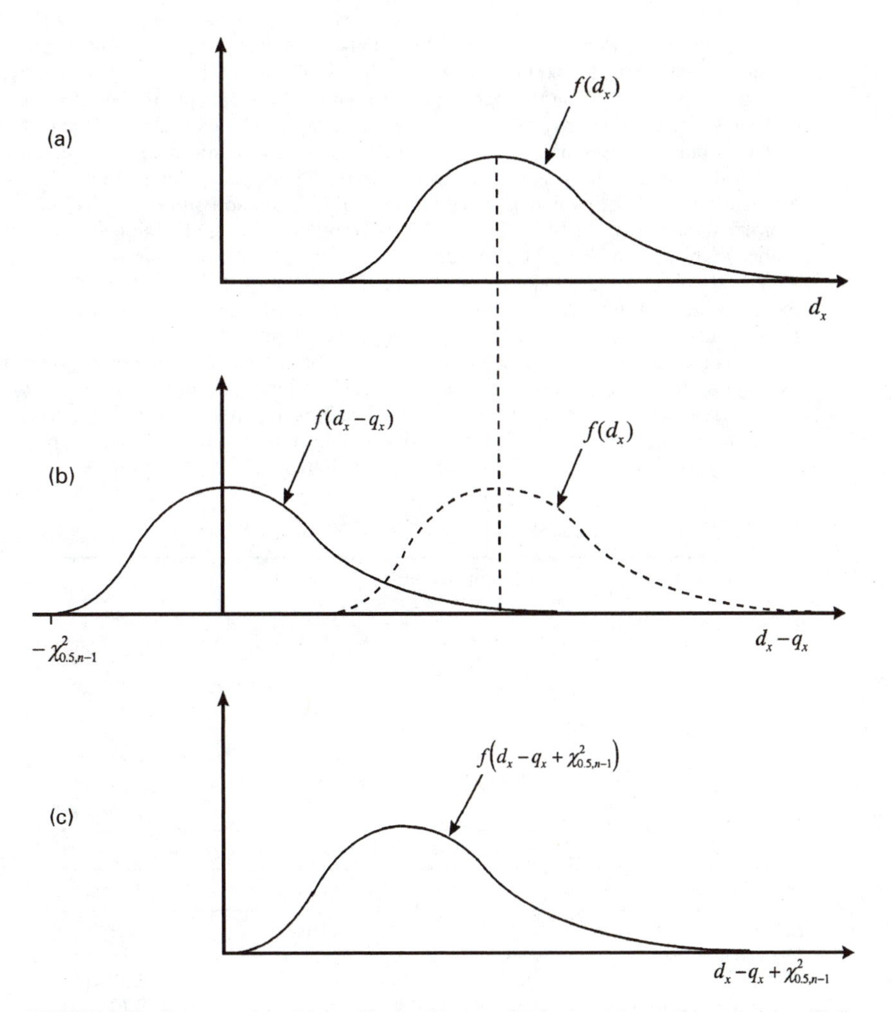

FIGURE 2.18 Representation of demand by the chi-square distribution: (*a*) general representation, (*b*) displaced distribution so that the mean demand is put at the origin, and (*c*) demand distribution in (*b*) displaced to the right by $\chi_{0.5,n-1}^{2}$.

the primary focus. However, should the latter be considered, a little modification can be applied to Eq. (2.48) simply by dropping the constant 0.5 for $\chi_{0.5,\,n-1}^{2}$.

Risk Evaluation Example Using Data from the City of Phoenix. To show numerical evaluation of risk, data was obtained from the city of Phoenix (Table 2.5). As indicated in the table, there is no uniform water price for the city. The price arrangement is divided into two categories: inside city and outside city uses. The costs for both categories are further divided into three seasons: low months (December, January, February, March); medium months (April, May, October, November); and high months (June, July, August, September) (Kiefer, 1994). In view of the procedure developed in this work that is sought for drought conditions, the high months' current price of $1.53/100 ft^3 and an applicable environmental charge of $0.08/100 ft^3 are used. In addition, the service charge of $5.16 for inside city use for a meter size of 5/8-in is selected.[1] It is to be noted that if a customer having a 5/8-in meter size doesn't use more than the 600-ft^3 limit during the months of October through May or the 1000-ft^3 limit during the months of June through September, she or he pays only the monthly service charge. In general, such a customer pays \$5.16 plus \$1.61 for every 100 ft^3 above the specified limit during a given month. Table 2.5 shows the general data from which the above information was extracted.

Based on the data for water price given in Table 2.5, the data given in Table 2.6, an assumed price elasticity of −0.267 [a value assessed for Tucson (Agthe and Billings, 1980)] and an assumed certain dry month (say SWSI = 2), the risk-price relationship is determined as given in Table 2.7. Although the data in Table 2.6 show that, in June 1997, the demand was below the available supply, a demand as high as 736.9 gallons per day (gpd) was recorded in July 1989 (Kiefer, 1994). To show the application of the foregoing procedure, an expected demand of 700 gpd is assumed as the typical demand during a given month of the selected drought period. The price elasticity concept [Eq. (2.25)] is used to lower the demand first, and then the risk at each price level is computed. With the assumption of a sample size of 20, the risk value is determined for each safety factor computed.

The risk–safety factor result obtained (Table 2.7) is drawn as shown in Fig. 2.19. It may be noted that to bring the high demand down to the available supply, it requires several price increase steps. Thus only part of the risk–safety factor data are plotted.

TABLE 2.5 City of Phoenix Water Rates (Effective March 1, 1997)

Part I. Monthly service charge		
Meter size, in	Inside city, \$	Outside city, \$
⅝	5.16	7.74
1	5.61	8.42
1½	8.88	13.32
2	9.78	14.67
3	39.06	58.59
4	47.24	70.86
6	51.33	77.00
Part II. Volume charges (above the limits)		
Months	Inside city, \$	Outside city, \$
Low months	1.01	1.52
Medium months	1.19	1.79
High months	1.53	2.30
Environmental charge	0.08	0.12

TABLE 2.6 Inside City Data for the City of Phoenix for Low-Density Residential Water Use for the Month of June 1997

Code	Conservation, million gal (A)	% of total* (B)	No. of accts. (C)	Avg. conservation, gpd (ft^3/d) (A/30C) (D)	Total production, million gal (E)	Supply share [B(E)/100] (F)	Avg. supply available, gpd (ft^3/d) (F/30C) (G)
0†	2435	24	132,584	612 (82)	11,247	2672	672 (90)

*Total water use during the same month equals 10,251,693,804 gal.
†This code is for single-family residential water use.
Source: City of Phoenix Water-Wastewater and Water Services Departments.

[1] The monthly service charge is for 600 ft^3 for the months of October through May inclusive and 1000 ft^3 for the months of June through September inclusive.

TABLE 2.7 Price-Risk Relations for a Typical Drought Month

Available monthly supply, ft³	Monthly demand, ft³	Demand subject to surcharge, ft³	Basic service charge, $	Price per additional 100 ft³, $	Total price, $	Difference $d_x - q_x$	Safety factor (SF)	Risk
2695	2800.00	1800.00	5.16	1.61	34.14	105.00	0.963	0.500
2695	2791.13	1791.00	5.16	1.65	34.78	96.13	0.966	0.500
2695	2784.28	1784.28	5.16	1.68	35.28	89.28	0.968	0.500
2695	2777.00	1777.00	5.16	1.72	35.82	82.00	0.970	0.500
2695	2769.83	1769.83	5.16	1.76	36.37	74.83	0.973	0.500
2695	2762.65	1762.65	5.16	1.80	36.93	67.65	0.976	0.500
2695	2755.49	1755.49	5.16	1.83	37.50	60.49	0.978	0.500
2695	2748.34	1748.34	5.16	1.87	38.07	53.34	0.981	0.500
2695	2741.21	1741.21	5.16	1.92	38.66	46.21	0.983	0.500
2695	2734.10	1734.10	5.16	1.96	39.26	39.10	0.986	0.500
2695	2727.00	1727.00	5.16	2.00	39.87	32.00	0.988	
2695	2719.92	1719.92	5.16	2.05	40.48	24.92	0.991	
2695	2712.85	1712.85	5.16	2.09	41.11	17.85	0.993	
2695	2705.80	1705.80	5.16	2.14	41.75	10.80	0.996	
2695	2698.77	1698.77	5.16	2.18	42.40	3.77	0.999	
2695	2691.75	1691.75	5.16	2.23	43.07	−3.25	1.001	
2695	2684.75	1684.75	5.16	2.28	43.74	−10.25	1.004	

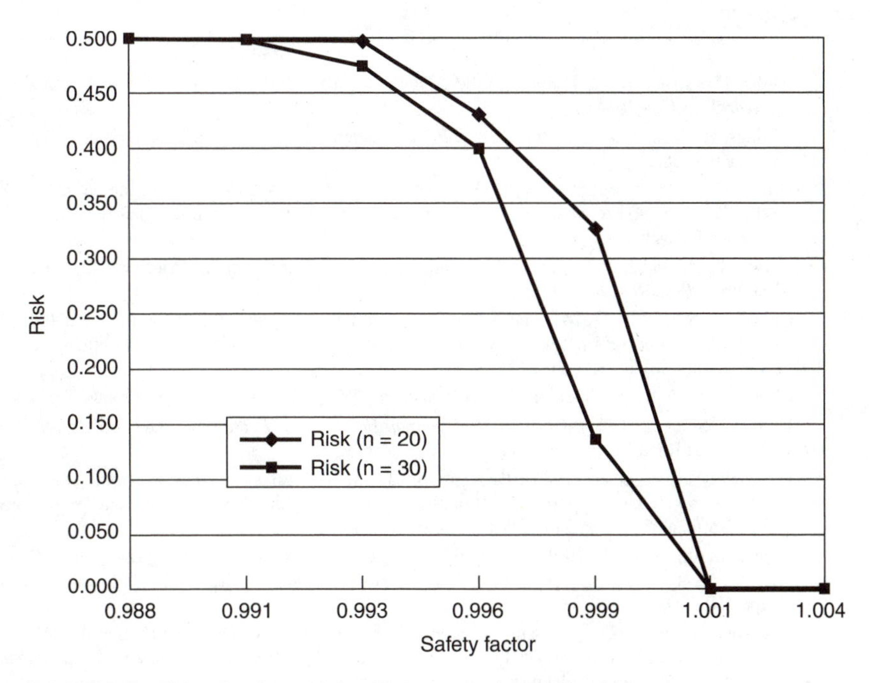

FIGURE 2.19 Risk–safety factor relationship for sample sizes of 20 and 30.

2.5 SUMMARY AND CONCLUSIONS

The damages caused due to lack of urban water supplies during drought conditions may result in adverse and undesirable effects on the general welfare. The efforts which may be taken basically include water conservation practices and/or new water development projects. In the case where the availability of this resource is limited due to the drought, water conservation is a viable target. Marginal pricing of water demand is found to be one of the best alternatives in urban water conservation efforts. This is achieved due to the fact that urban water demand is elastic to price changes. In other words, an increase in urban water price decreases its demand.

For urban water planning purposes during a drought condition, the inclusion of a third dimension termed the risk, the probability that the demand exceeds the supply, plays an important role in deciding, for a given drought condition, the tolerable risk for a given affordable price or vice versa. Thus the risk-price-return period relationship developed will help as one of the decision support systems for urban water planning under drought conditions.

The use of the risk–safety factor approach under sustained drought conditions may be also useful. It not only has the advantage of lowering the demand during such conditions, but it is also a more effective way of controlling water use. This is because the water-use control framework is enforced at the water meters of the individual customers.

Although water demand is believed to be elastic to price, it is apparently not significant for small price increases. The fact that the risk is highly sensitive to the return period and less sensitive to the price demands that there be a strong commitment to conserve water under adverse drought conditions. The price-elasticity study has been geographically limited, mainly to the United States. More research is needed in regions with a different socioeconomic status and hydrometeorological setup, that is, in regions where the incomes of the customers are relatively low and the weather conditions are more uncertain and/or the drought occurrences are more frequent. It is likely that the use of the concept of price elasticity will be more effective in such areas.

2.6 REFERENCES

Agthe, D. E., and R. B. Billings (1980). Dynamic Models of Residential Water Demand, *Water Resources Research,* 16(3):476–480.

Billings, R. B., and D. E. Agthe (1980). Price Elasticities for Water: A Case of Increasing Block Rates, *Land Economics,* 56(1):73–84.

Bruins, H. (1993). Drought Risk and Water Management in Israel: Planning for the Future, in Donald A. Wilhite (ed.), *Drought Assessment, Management and Planning: Theory and Case Studies,* Kluwer Academic Publishers, Boston.

Carver, P. H., and J. J. Boland (1980). Short- and Long-Run Effects of Price on Municipal Water Use, *Water Resources Research,* 16(4):609–616.

Changnon, S. (1993). Are We Doomed to Fail in Drought Management? in Martin Ruess (ed.), *Water Resources Administration in the United States: Policy, Practice, and Emerging Issues,* Michigan State University Press, East Lansing, Mich., 194–202.

Chow, V. T., D. R. Maidment, and L. W. Mays (1988). *Applied Hydrology,* McGraw-Hill, New York.

Dixon, L. S., et al. (1996). *Drought Management Policies and Economic Effects on Urban Areas of California, 1987–1992,* Rand Corporation, Santa Monica, Calif.

Dziegielewski, B., et al. (1986). Drought Management Options, in *Drought Management and Its Impact on Public Water Systems,* report on a colloquium sponsored by the Water Science and Technology Board, Sept. 5, 1985, National Academy Press, Washington, D.C.

Dziegielewski, B., et al. (1991). *National Study of Water Management During Drought: A Research Assessment for the US Army Corps of Engineers Water Resources Support Center,* Institute for Water Resources, IWR Report 91-NDS-3.

Easterling, W. E. (1993). Assessing the Regional Consequences of Drought: Putting the MINK Methodology to Work on Today's Problems, in Donald A. Wilhite (ed.), *Drought Assessment, Management and Planning: Theory and Case Studies,* Kluwer Academic Publishers, Boston.

Garen, D. C. (1993). Revised Surface-Water Supply Index for Western United States, *Journal of Water Resources Planning and Management,* ASCE, 119(4):437–454.

Gibbs, K. (1978). Price Variables in Residential Water Demand Models, *Water Resources Research,* 14(1):15–18.

Grebenstein, C. R., and B. C. Field (1979). Substituting for Water Inputs in U.S. Manufacturing, *Water Resources Research,* 15(2):228–232.

Griffin, R. C., and J. Stoll (1983). The Enhanced Role of Water Conservation in the Cost-Benefit Analysis of Water Projects, *Water Resources Bulletin,* 19(3):447–457.

Grunewald, O. C., et al. (1976). Rural Residential Water Demand: An Econometric and Simulation Analysis, *Water Resources Bulletin,* 12(5):951–961.

Hanke, S. H. (1970). Demand for Water under Dynamic Conditions, *Water Resources Research,* 6(5):1253–1261.

Hanke, S. H., and L. de Maré (1982). Residential Water Demand: A Pooled, Time Series, Cross-Section Study of Malmö, Sweden, *Water Resources Bulletin,* 18(4):621–625.

Hobbs, B. F. (1989). An Overview of Integrated Water Supply System Availability, in Larry W. Mays (ed.), *Reliability Analysis of Water Distribution Systems,* ASCE, New York.

Howe, C. W. (1982). The Impact of Price on Residential Water Demand: Some New Insights, *Water Resources Research,* 18(4):713–716.

Howe, C. W. (1993). Water Pricing: An Overview, *Water Resources Update,* 92:3–6.

Howe, C. W., and F. P. Linaweaver, Jr. (1967). The Impact of Price on Residential Water Demand and Its Relation to System Design and Price Structure, *Water Resources Research,* 3(1):13–32.

Hudson, Jr., H. E., and R. Hazen (1964). Drought and Low Stream Flows, in Ven Te Chow (ed.), *Handbook of Applied Hydrology,* McGraw-Hill, New York.

Jones, C. V., and J. R. Morris (1984). Instrumental Price Estimates and Residential Water Demand, *Water Resources Research,* 20(2):197–202.

Jordan, J. L. (1994). The Effectiveness of Pricing as a Stand-Alone Water Conservation Program, *Journal of the American Water Resources Association,* 30(5):871–877.

Kiefer, J. C. (1994). City of Phoenix Water Use Monitoring Program: A Multi-Objective Study of Single-Family Household Water Use (Unpublished Technical Report), Planning and Management Consultants, Ltd., Phoenix.

Lansey, K. E., et al. (1989). Methods to Analyze Replacement-Rehabilitation of Water Distribution System Components, in Larry W. Mays (ed.), *Reliability Analysis of Water Distribution Systems,* ASCE, New York.

Maddock, T. S., and W. G. Hines (1995). Meeting Future Public Water Supply Needs: A Southwest Perspective, *Journal of the American Water Resources Association,* 31(2):317–329.

Mays, L. W., and Y. K. Tung (1992). *Hydrosystems Engineering and Management,* McGraw-Hill, New York.

Moncur, J. E. T. (1987). Urban Water Pricing and Drought Management, *Water Resources Research,* 23(3):393–398.

Moncur, J. E. T. (1989). Drought Episodes Management: The Role of Price, *Water Resources Bulletin,* 25(3):499–505.

Montgomery, D. C., and G. C. Runger (1994). *Applied Statistics and Probability for Engineers,* John Wiley, New York.

National Research Council (1986). *Drought Management and Its Impact on Public Water Systems, Colloquium 1,* National Academy Press, Washington, D.C.

Palmer, W. C. (1965). Meteorological Drought, U.S. Weather Bureau, Research Paper No. 45.

Puckett, L. J. (1981). *Dendroclimatic Estimates of a Drought Index for Northern Virginia: Geological Survey Water Supply Paper 2080,* U.S. Government Printing Office, Washington, D.C.

Schneider, M. L., and E. E. Whitlatch (1991). User-Specific Water Demand Elasticities, *Water Resources Planning and Management,* 117(1):52–73.

Sheer, D. P. (1980). Analyzing the Risk of Drought: The Occoquan Experience, *Journal of American Water Works Association,* 72(12):246–253.

Steila, D. (1972). *Drought in Arizona: A Drought Identification Methodology and Analysis,* University of Arizona, Tucson.

Suter II, G. W. (1993). *Ecological Risk Assessment,* Lewis Publishers, Chelsea, Mich.

Tung, Y. K. (1996). Uncertainty and Reliability Analysis, in Larry W. Mays (ed.), *Water Resources Handbook,* McGraw-Hill, New York.

U.S. Soil Conservation Service Monthly Report (Jan. 1988). Colorado Water Supply Outlook, United States Department of Agriculture, U.S. Government Documents.

U.S. Water Resources Council (1979). Procedures for Evaluation of National Economic Development (NED) Benefits and Costs in Water Resources Planning (Level C), Final Rule, Federal Register, 44(242):72891–72976.

Wilhite, D. A. (1993). Planning for Drought: A Methodology, in Donald A. Wilhite (ed.), *Drought Assessment, Management and Planning: Theory and Case Studies,* Kluwer Academic Publishers, Norwell, Mass., pp. 87–108.

Wong, S. T. (1972). A Model on Municipal Water Demand: A Case Study of Northern Illinois, *Land Economics,* 48(1):34–44.

Young, R. A. (1973). Price Elasticity of Demand for Municipal Water: A Case Study of Tucson, Arizona, *Water Resources Research,* 9(4):1068–1072.

Young, R. A. (1996). Water Economics, in Larry W. Mays (ed.), *Water Resources Handbook,* McGraw-Hill, New York.

CHAPTER 3
COMPUTER PROGRAMS FOR INTEGRATED MANAGEMENT

Messele Z. Ejeta
California Department of Water Resources
Sacramento, California

Larry W. Mays
Department of Civil and Environmental Engineering
Arizona State University
Tempe, Arizona

3.1 INTRODUCTION

The facts that every living being depends on water to live and that water's availability is limited in terms of both quantity and quality make water a resource over which there is growing competition. The various competitors for water have made it often challenging in space and time to fully satisfy their needs. The viable solution under such conditions is balancing out. This may be achieved through integrated hydrosystems management.

The concept of integrated hydrosystems management and computer programs used for integrated hydrosystems management are discussed in this chapter. The term *integrated hydrosystems management* is used for various types of water systems. A review of computer programs for hydrosystems management developed prior to the 1960s up to the present day shows enormous evolution and revolution of computing applications to hydrosystems. These efforts, which started in the early days of computer programming for the simplification of calculations of analytical functions, have now reached the age of what is being referred to in computing technology as *artificial intelligence.* In essence, it has become possible to write computer programs that not only evaluate a hydrosystems problem, but also draw preliminary conclusions based on the results and recommend appropriate actions based on the conclusions.

Various definitions have been given in the past to integrated resource management in general and water resources management in particular by different authors and institutions involved in the study of water resources. In addition, various terms such as hydrosystems management, integrated water management, integrated regional water management, water resources management, river basin management, watershed management, and total water management have been used to refer to the management of water resources on a large scale. Herein, the term integrated hydrosystems management is consistently used unless otherwise specified.

The definitions of integrated hydrosystems management as used by various institutions and individuals are cited and an attempt is made to give a definition that considers its wide range of aspects. The evolution of simulation computer programs and the structure of optimization techniques for hydrosystems problems are revisited. Examples of a relatively new set of computer programs, generally termed *decision support systems* (DSSs), are reviewed. Some of the examples of DSSs given for integrated hydrosystems management manifest the possibility of incorporating or at least monitoring water policy issues in the process of allocating water to all the competing users.

Computer programs and computing techniques useful for hydrosystems management have been categorized into simulation programs, optimization techniques, and DSSs. The discussion of different optimization techniques ranging from mathematical programming to heuristic search techniques including genetic algorithms and simulated annealing shows the potential resources available for computer programming for integrated hydrosystems management. Incorporating established water resources operation policies in simulation and optimization computer programs may help develop DSSs that can be used for integrated hydrosystems management. The study of a few such computer programs manifests the relative importance of these computer programs for integrated hydrosystems management. Only a limited number of DSSs for this purpose have been developed and used in the past. However, the availability of technical resources including database management systems, simulation computer programs, optimization techniques, and advanced computing technology provide the opportunity for more exploration to develop DSS for integrated hydrosystems management.

3.2 INTEGRATED HYDROSYSTEMS MANAGEMENT

3.2.1 Definition

Mitchell (1990) noted that integrated water management may be contemplated in at least three ways. These include (1) the systematic consideration of the various dimensions of water: surface and groundwater, quality and quantity; (2) the implication that while water is a system it is also a component which interacts with other systems; and (3) the interrelationships between water and social and economic development. In the first thought, the concern is the acceptance that water comprises an ecological system, which is formed by a number of interdependent components. In the second one, the interactions between water, land, and the environment, which involve both terrestrial and aquatic issues, are addressed. Finally, the concern is with the relationships between water and social and economic development, since availability or lack of water may be viewed as an opportunity for or a barrier against economic development. Each aspect of integrated hydrosystems management depends on and is affected by other aspects. Loucks (1996) points out that "integrated water resources systems planning and management focuses not only on the performance of individual components, but also on the performance of the entire system of components."

Water policy issues, of which limited effort was made in the past to incorporate them into hydrosystems computer programs, are some of the major factors that affect integrated hydrosystems management. Grigg (1998) describes water policy as dealing with finding satisfactory ways to allocate resources to balance between diverse and competing objectives of society and the environment. He referred to *integrated water management* as blending together actions and objectives favored by different players to achieve the best total result. Mitchell (1990) states that integration in water management deals with "...problems that cut across elements of the hydrological cycle, that transcend the boundaries among water, land and environment, and that interrelate water with broader policy questions associated with regional economic development and environmental management." The policies that are needed for integrated water resources management require coordination and collaboration among governments and agencies engaged in water management (Viessman, 1998). Grigg (1998) notes that improving coordination is the most promising route to the conceptual and perhaps utopian vision of integrated water management.

AACM, a consulting company in Australia, and the Center for Water Policy Research, Australia, in 1995 defined integrated resource management, of which water resources is a part, as the coordinated management of land and water resources within the region, with the objectives of controlling and/or conserving the water resource, ensuring biodiversity, minimizing land degradation, and achieving specified and agreed-upon land and water management and social objectives (Hooper, 1995). The American Water Works Association Research Foundation (AWWARF) (1996) defined the concept of total water management, which comprehends wide aspects of integrated hydrosystems management, through the following statements.

> Total Water Management is the exercise of stewardship of water resources for the greatest good of society and the environment. A basic principle of Total Water Management is that the supply is renewable, but limited, and should be managed on a sustainable use basis. Taking into consideration local and regional variations, Total Water Management:
>
> - Encourages planning and management on a natural water systems basis through a dynamic process that adapts to changing conditions;
> - Balances competing uses of water through efficient allocation that addresses social values, cost effectiveness, and environmental benefits and costs;
> - Requires the participation of all units of government and stakeholders in decision-making through a process of coordination and conflict resolution;
> - Promotes water conservation, reuse, source protection, and supply development to enhance water quality and quantity; and
> - Fosters public health, safety, and community good will.

Table 3.1 shows an elaboration by Grigg (1998) of the definition of total water management as related to the concept of coordination. He emphasized what is implied by each of the important phrases used in the definition. These phrases which are apparently the central aspects of integrated hydrosystems management include society and environment, stakeholder, watershed and natural water systems, means of water management, time-wise, intergovernmental, water quality and quantity, local and regional concerns, and competing uses. Integrated hydrosystems management is as much challenging as compromising between these different aspects in making decisions.

The foregoing definitions and discussions indicate that integrated hydrosystems management is multiobjective. It is necessary both for economic efficiency (which is measured in monetary units) and for environmental quality (which may be measured in terms of pollutant concentration). Shortly, it balances between societal welfare and ecosystem sustainability. To summarize, integrated hydrosystems management in a watershed involves a multidisciplinary approach of developing and using water resources by making possible balances between all the competing water uses and through coordination between all parties without causing detrimental consequences to the ecosystem and/or future requirements.

3.2.2 History

Jamieson and Fedra (1996) report that practitioners have recognized the concept of integrated hydrosystems management since the early 1970s. The United Nations in the Dublin Statement endorsed this perception in 1992. Thus, integrated hydrosystems management on a regional scale is a newly emerging approach. The lack of a clear definition of a water resources region has, perhaps, contributed to the absence in the past of a regional approach to hydrosystems management.

River basin boundaries often differ from political boundaries. Groundwater flow has obviously never been dictated by political boundaries, and neither has the movement of atmospheric water. Furthermore, the question of the size of a region has been a challenge and will probably remain so in the near future. Viessman (1998) states that it is not clear that integrated regional water plans can be fit within the geographic limits of large river basins or watersheds. Vlachos (1998) poses a very important question: Can integrated planning and management work in the vast expanses of the Nile, the Amazon, the Paraná/La Plata, or should it be restricted to more regional, specific sociopolitical conflicts of rather well-defined geographic, cultural, environmental, physiographic, and economic boundaries?

Defining a water resources region now appears to be driven more by the watershed approach than the other factors mentioned above. A national forum convened in January 1994 by the Conservation Fund and the National Geographic Society clearly recognized the critical need for the watershed approach for integrated hydrosystems management rather than political jurisdiction or boundaries. Similarly, the Environmental Advisory Board (EAB) of the U.S. Army Corps of Engineers (USACE) recommended in 1994 that the watershed or ecosystem approach be used as the holistic, integrated concept on which to base (water resources) planning (Bulkley, 1995). Furthermore, the

TABLE 3.1 Types of Coordination from Total Water Management Definition

Type of coordination	Phrase from total water management definition	Discussion	Effective-ness ranking
Society and environment	The exercise of stewardship of water resources for the greatest good of society and the environment	This statement provides a general organizing framework for balancing. It is adequately understood, but needs more explanation.	1
Stakeholder	Requires the participation of all…stakeholders in decision making through a process of coordination and conflict resolution	Process is known as stakeholder and public involvement. Good and improving. A central issue of democratic government.	2
Watershed and natural water systems	Encourages planning and management on a natural water systems basis	It is recognized and currently popular that water management on a basin or watershed basis is desirable. Further progress will require more effort.	3
Means of water management	Promotes water conservation, reuse, source protection, and supply development	This means to coordinate different ways to meet needs and sustain the environment. A central planning and management issue.	4
Time-wise	Through a dynamic process that adapts to changing conditions	This requires valid planning methods to preserve institutional memory and keep processes on track and requires much improvement.	5
Intergovern-mental	Requires the participation of all units of government in decision making through a process of coordination and conflict resolution	Intergovernmental coordination is given as separate from stakeholders because of the different kinds of authorities that government has.	6
Water quality and quantity	To enhance water quality and quantity	This is handled through water quality law and regulation. Many problems still require solution.	7
Local and regional concerns	Taking into consideration local and regional variation	This is a difficult issue requiring intergovernmental cooperation in arenas which lack adequate incentives and often cannot be mandated. It is not working too well.	8
Competing uses	Balances competing uses of water through efficient allocation that addresses social values, cost-effectiveness, and environmental benefits and costs	This is handled through state and federal water law regulations, court decisions, and other institutions. A very difficult arena.	9

Source: Grigg (1998).

U.S. General Accounting Office (1994) listed the importance of the watershed approach for integrated management. Accordingly, watershed boundaries

1. Are relatively well defined
2. Can have major ecological importance
3. Are systematically related to one another hierarchically and thus include smaller ecosystems
4. Are already used in some water management efforts
5. Are easily understood by the public

Many water resources projects in the past lacked the integrated planning aspect. Hall (1998) states that throughout history, water management systems have been developed in a linear fashion; i.e., they had a piecemeal development in which the components of integrated water management were put into place as the need for each component arose. Similarly to the management aspect, the modeling aspect of water resources has also followed the same piecemeal fashion. Jamieson and Fedra (1996) state that "although the principle of integrated river basin management models has been aspired to in many countries, more often than not the problems have been considered in a piecemeal fashion, with experts from different disciplines using separate computer programs (water resources, surface-water pollution control, groundwater contamination, etc.), to tackle parts of the overall problem in a reactive way." Uncoordinated hydrosystems modeling efforts often result in incompatibilities. As a result, these systems have not been sufficient and effective enough, thus leading to the emergence of the regional management and modeling approach.

3.2.3 Importance

We are becoming increasingly aware, with time, of the fact that our water supplies are limited both in quantity and quality. Because water has multiple and often competing uses, hydrosystems are interrelated with other physical and socioeconomic systems. In some locations, when water supplies become extremely limited, their further use is based on the determination of which user has the oldest "right" to them, or on a judgment about which uses have the highest priority (Hall, 1998). Hall (1998) also warns that unless dealt with appropriately, the forces of population growth; urbanization; and increased water demands for home, industry, and agriculture, coupled with an increasingly global economy and culture, will in the future produce spreading, perilous degradation of water quality everywhere, and a continuously widening gap between water needs and the availability of useful water in all too many locations. As a solution to this problem, he suggested a different approach, which includes: (1) management across political boundaries; (2) the collective management of atmospheric water, surface waters, and groundwater; (3) the combined management of water quality and water quantity.

Schultz (1998) gives the criteria for water resources management projects at present and those criteria emerging as new ones in the future. Accordingly, the factors that have to be satisfied include: (1) economic benefits; (2) technical efficiency; and (3) performance reliability. The approach, which seems to become more and more dominant, includes

1. Principle of sustainable development
2. Ecological quality
3. Consideration of macroscale systems and effects
4. Planning in view of changes in natural and socioeconomic systems

It is evident from these comparisons that hydrosystems projects are geared toward integrated management. These new planning approaches for integrated hydrosystems management necessitate new ways of modeling. Schultz (1998) adds that new planning tools are required to plan and design water resources systems on the basis of the new criteria. He concludes "since no planning tools following the four new criteria are available, we are faced with a vacuum."

In a different argument, an integrated hydrosystems project needs to be evaluated on the following important factors: technical, economic, financial, environmental, and sociopolitical. Technically, it must be feasible to build; economically, it must be reasonably affordable; financially, it must have a source; environmentally, its effect must be mitigated with ease; and sociopolitically, it must be acceptable to the public. The project can be successful if effective coordination prevails between the parties involved and if such parties are mandated to monitor clearly defined scope and regional coverage. In this argument also, integrated hydrosystems management is perceived to be a viable approach in planning efficient and successful water resources projects. In England and Wales, for example, regional water authorities whose boundaries were defined by the watersheds of the country enabled the replacement of 1600 separate water service entities with 10 regional watersheds (Bulkley, 1995).

3.3 COMPUTER PROGRAMS FOR INTEGRATED MANAGEMENT

The implementation of the ideals of integrated hydrosystems management necessitates analytical tools to simplify or assist the balancing-out process. Water policies need to be transformed into forms that can be understood and interpreted using analytical tools such as computer programs. Consequently, robust computer programs that not only solve the problems with analytical structure or mathematical formula but also are capable of reducing and incorporating water policies into analytical structure are required. Furthermore, they may be required to interpret the result of the computations, give conclusions based on the result, and make appropriate recommendations based on the conclusions reached.

As we apply computing technology to water resources modeling, the lack of conventional terms is apparent. In general, the terms *computer programs* and *models* have been used interchangeably. Models are also used to refer to the applications developed for specific computer programs. Herein, we will refer to computer programs as the tools developed to solve any problem in a specific general area of problems. In contrast, we will refer to a model as a specific representation of a physical or real system that is prepared to be solved by a given computer program. Thus, a computer program is generic whereas a model is specific. For instance, we refer to the U.S. Environmental Protection Agency's EPANET as a computer program and a representation of a particular city's water distribution system problem that can be solved by EPANET as a model.

Although tremendous work has been done in the past to develop computer programs for integrated hydrosystems management, only a few exist that address the overall framework of problems associated with integrated hydrosystems management. A few of the reasons may be attributable, among others, to

1. the lack of clear definition and better understanding of integrated hydrosystems management
2. the variation of water needs with space and time
3. the evolution (revolution) of computer programming

Computer modeling approaches that at least partly tried to address some of the concepts of integrated hydrosystems management are highly based on interfacing simple computer programs written and used for the analysis of specific hydrosystems problems. At the core of some advanced computer programs used for integrated hydrosystems management lie simple simulation modules, rule-based simulation modules (also known as expert systems), and optimization modules of hydrosystems problems. While many simulation and optimization modules have been developed and interfaced over the years by different institutions and agencies, the incorporation of rule-based simulation modules in computer programs for integrated hydrosystems management appears to have emerged recently as a sound approach. By incorporating rule-based simulation modules, it has become easier to manage decisions that involve several factors and water policies.

Section 3.3.1 discusses the development of simulation computer programs that emerged in the United States over the past few decades for the simulation of various types of hydrosystems problems. Section 3.3.2 discusses the basic mathematical structure of optimization computer programs, which may be viewed as generic tools that can be customized to specific hydrosystems problems.

3.3.1 Development of Hydrosystems Simulation Computer Programs

Over the past few decades, water resources professionals have witnessed the development of quite a number of hydrosystems simulation computer programs. Wurbs (1995) points out that a tremendous amount of work has been accomplished during the past 3 decades in developing computer programs for use in water resources planning and management. The majority of these programs, perhaps most of the earliest computer programs to be developed for water resources problems, may be viewed as simulation computer programs.

As information technology advances, hydrosystems simulation computer programs have generally gone through an evolutionary process. Figure 3.1 depicts the evolution of hydrosystems computer programs as classified into five generations (derived from the explanation given by Jamieson and Fedra, 1996). The first-generation codes, which tremendously simplified calculation of analytical functions through generic computer codes, are but mediocre by today's standards. One may draw an analogy between the coming into being of these codes and the transition of computation methods from using the slide rule to scientific calculators. In both cases, similar jobs are done, but the new method highly reduced the time required for numerical computations. The succeeding generations of programs enhanced the robustness of the programs and/or the ease with which the programs can be used. The fifth generation of programs is embodied with artificial intelligence that not only performs analytical computations but also draws some preliminary conclusions and recommends appropriate actions.

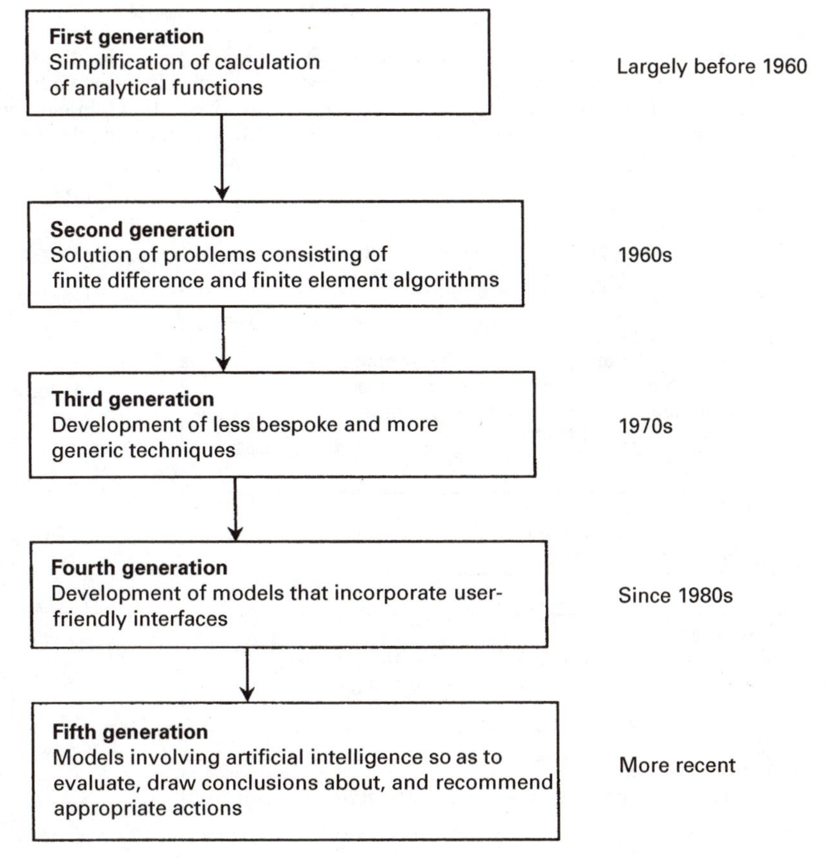

FIGURE 3.1 Schematic diagram showing the evolution of hydroinformatics. (After Jamieson and Fedra, 1996)

3.3.2 Optimization Formulations

Background and General Formulation. Various optimization techniques, in general, and their application to various hydrosystems problems, in particular, have shown remarkable progress over the past 3 decades. The progress of the application of these techniques has occurred alongside the revolution of computer programs, and as such, similar explanations can be given to the development of simulation computer programs and optimization techniques over the past 3 or more decades. Figure 3.2 gives the development of the application of optimization techniques to hydrosystems problems, which is analogous to that given for simulation computer programs in Fig. 3.1.

The general formulation for optimization problems in water resources, which are generally nonlinear programming (NLP) problems, can be expressed in terms of state (or dependent) variables ($\mathbf{x}$) and control (or independent) variables ($\mathbf{u}$) as (Mays, 1997; Mays and Tung, 1992)

$$\text{Optimize } f(\mathbf{x}, \mathbf{u}) \tag{3.1}$$

subject to process simulation equations

$$G(\mathbf{x}, \mathbf{u}) = 0 \tag{3.2}$$

and additional constraints for operation on the dependent ($\mathbf{u}$) and independent ($\mathbf{x}$) variables

$$\underline{\mathbf{w}} \leq w(\mathbf{x}, \mathbf{u}) \leq \overline{\mathbf{w}} \tag{3.3}$$

where $\underline{\mathbf{w}}$ and $\overline{\mathbf{w}}$ represent lower and upper bounds, respectively. The term *optimize* in Eq. (3.1) refers to either maximization or minimization, whereas the constraint equation [(Eq. (3.3)] dictates the feasibility of the objective with respect to each and every constraint. In other words, the solution to the simulation equation [Eq. (3.2)] must satisfy the constraints defined by Eq. (3.3). The

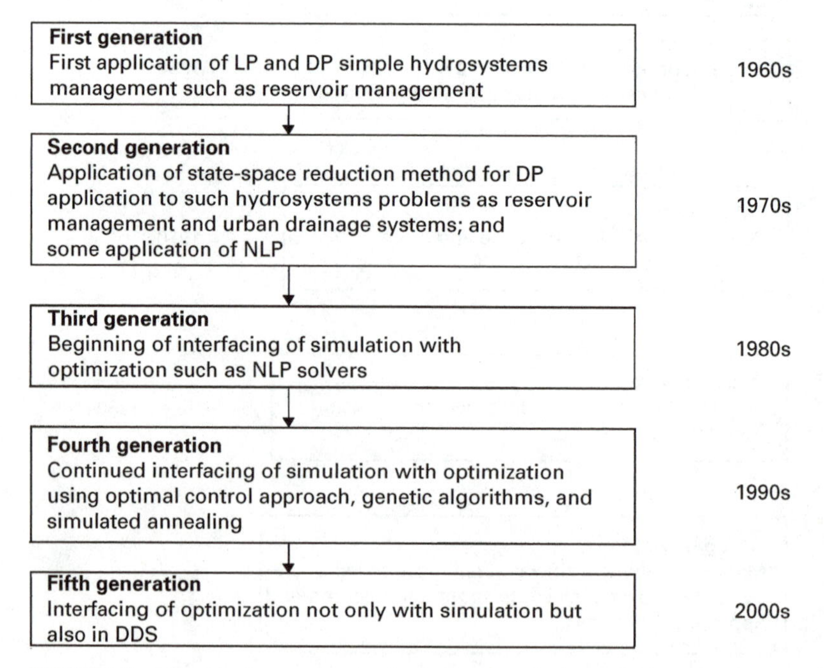

FIGURE 3.2 Schematic diagram showing the developments in the application of optimization techniques to hydrosystems problems. (Jamieson and Fedra, 1996)

process simulation equations basically consist of the governing physical equations of mass, energy, and momentum.

Many hydrosystems problems can be formulated as discrete-time optimal control problems. The basic mathematical definition of a discrete-time optimal control problem is stated as

$$\underset{\mathbf{u}}{\text{Min}}\ Z = \sum_{t=1}^{T} f_t(\mathbf{x}_t, \mathbf{u}_t, t) \tag{3.4}$$

subject to

$$\mathbf{x}_{t+1} = g_t(\mathbf{x}_t, \mathbf{u}_t, t) \qquad t = 1, 2, \ldots, T \tag{3.5}$$

$$\mathbf{x}_t \geq 0 \qquad \mathbf{u}_t \geq 0 \tag{3.6}$$

where $\mathbf{x}_t$ = vector of state variables at time t
$\mathbf{u}_t$ = vector of control variables at time t
T = number of decision times

A few possible optimization formulations for different hydrosystems problems are given below.

Water Distribution System Operation. Mays (1997) defines the optimization problem for water distribution system operation in terms of the nodal pressure heads, **H**; pipe flows, **Q**; tank water surface elevations, **E**; pump operating times, **D**; and water quality parameter, **C**, as follows.

$$\text{Minimize energy costs } = f(\mathbf{H}, \mathbf{Q}, \mathbf{D}) \tag{3.7}$$

subject to

$$\mathbf{G}(\mathbf{H}, \mathbf{Q}, \mathbf{D}, \mathbf{E}, \mathbf{C}) = \mathbf{0} \tag{3.8}$$

$$\mathbf{w}(\mathbf{E}) = \mathbf{0} \tag{3.9}$$

$$\underline{\mathbf{H}} \leq \mathbf{H} \leq \overline{\mathbf{H}} \tag{3.10}$$

$$\underline{\mathbf{D}} \leq \mathbf{D} \leq \overline{\mathbf{D}} \tag{3.11}$$

$$\underline{\mathbf{E}} \leq \mathbf{E} \leq \overline{\mathbf{E}} \tag{3.12}$$

$$\underline{\mathbf{C}} \leq \mathbf{C} \leq \overline{\mathbf{C}} \tag{3.13}$$

where Eqs. (3.8) and (3.9) express the energy and flow constraints and the pump operation constraints. Equations (3.10) to (3.13) express the bound constraints on the nodal pressure head, pump operating times, tank water surface elevations, and water quality, respectively.

Reservoir System Operation for Water Supply. The optimization for this kind of hydrosystems problem can be expressed as (Mays, 1997)

$$\text{Maximize benefits} = \sum_{0}^{T} f(\mathbf{S}_t, \mathbf{U}_t, t) \tag{3.14}$$

subject to

$$\mathbf{G}(\mathbf{S}_{t+1}, \mathbf{S}_t, \mathbf{U}_t, \mathbf{I}_t, \mathbf{L}_t) = \mathbf{0} \qquad t = 0, \ldots, T-1 \tag{3.15}$$

$$U_t \leq \mathbf{U}_t \leq \overline{\mathbf{U}}_t \qquad t = 1, \ldots, T \tag{3.16}$$

$$S_t \leq \mathbf{S}_t \leq \overline{\mathbf{S}}_t \qquad t = 1, \ldots, T \tag{3.17}$$

$$P[\mathbf{S}_t \geq S_t] \leq \alpha_t^{\min} \qquad t = 1, \ldots, T \tag{3.18}$$

$$P[\mathbf{S}_t \le S_t] \le \alpha_t^{\max} \qquad t = 1, \ldots, T \tag{3.19}$$

$$\mathbf{w}(\mathbf{S}_t, \mathbf{U}_t) = \mathbf{0} \tag{3.20}$$

where $\mathbf{S}_t$ and $\mathbf{U}_t$ are the vectors of reservoir storage and releases and t represents discrete time period. Equation (3.15) defines the system of equations of conservation of mass for the reservoirs and river reaches. $\mathbf{S}_{t+1}$ and $\mathbf{S}_t$ are, respectively, the vectors of reservoir storage at the beginning of time period $t + 1$ and t, $\mathbf{I}_t$ is the vector of hydrologic inputs, and $\mathbf{L}_t$ is the vector of reservoir losses. Equations (3.16) and (3.17) define the bound constraints on reservoir releases and storages, respectively. Equations (3.18) and (3.19) define the bound constraints on reservoir storage in probabilistic form where $P[\ ]$ denotes the probability and $\alpha_t^{\min}$ and $\alpha_t^{\max}$ represent the minimum and maximum reliability or tolerance levels, respectively. Equation (3.20) expresses the other constraints on reservoir operation.

Groundwater Management Subsystems. The general groundwater management problem can be expressed mathematically as (Mays, 1997)

$$\text{Optimize } Z = f(\mathbf{h}, \mathbf{q}) \tag{3.21}$$

subject to

$$\mathbf{G}(\mathbf{h}, \mathbf{q}, \mathbf{c}) = \mathbf{0} \tag{3.22}$$

$$\underline{\mathbf{q}} \le \mathbf{q} \le \overline{\mathbf{q}} \tag{3.23}$$

$$\underline{\mathbf{h}} \le \mathbf{h} \le \overline{\mathbf{h}} \tag{3.24}$$

$$\underline{\mathbf{C}} \le \mathbf{C} \le \overline{\mathbf{C}} \tag{3.25}$$

$$\underline{\mathbf{w}}(\mathbf{h}, \mathbf{u}) \le \mathbf{0} \tag{3.26}$$

where $\mathbf{h}$ and $\mathbf{q}$ in the objective function are vectors of heads and pumpages (or recharges), respectively. $\mathbf{C}$ is a parameter that measures quality such as chlorine content and so on. Equation (3.22) gives the general groundwater flow constraints, which represent a system of equations governing groundwater flow and transport. Equations (3.23) and (3.24) represent, respectively, the upper and the lower bounds on the pumpages (recharges) and on the heads. Equation (3.25) gives the groundwater quality constraints, whereas Eq. (3.26) may be taken as additional constraints which can be included to impose restrictions such as water demands, operating rules, and budgetary restrictions. It may be noted here that the lower and upper bounds on pumpages may or may not exist, whereas those on the head can be the bottom elevation of the aquifer and the groundwater surface elevations for the unconfined cells, respectively.

3.3.3 Interfacing Optimization and Simulation Computer Programs

The general form of the objective functions and the constraints in hydrosystems problems including the foregoing examples can be linear, nonlinear, or differential equations. The solution of each equation needs a different approach. Several computer codes have been written for each of these types of formulations.

For those hydrosystems optimization problems which involve solving general governing differential equations of mass, energy, and momentum, the approach used can be solving the optimization problem directly by embedding finite difference or finite element equations of the governing process equations (Mays, 1997). This approach is relatively tedious to apply to real-world problems. Alternatively, an appropriate process simulator can be used to solve the constraint process simulation equations when they need to be evaluated for the optimizer. Consequently, the following general and simpler optimization problem can be used.

$$\text{Minimize } F(\mathbf{u}) = f(\mathbf{x}(\mathbf{u}), \mathbf{u}) \tag{3.27}$$

subject to

$$\underline{\mathbf{w}} \leq \mathbf{w}(\mathbf{x}(\mathbf{u})) \leq \overline{\mathbf{w}} \tag{3.28}$$

Different techniques have been successfully applied to solve optimization problems that are formulated in the above form. The most common techniques are given below.

Mathematical Programming. *Mathematical programming* includes linear programming and nonlinear programming problems (Jeter, 1986). Herein we will refer to the mathematical programming approach as interfacing simulation computer programs with nonlinear programming codes such as GRG2. This programming technique has been found useful in several hydrosystems problems such as groundwater management systems (Wanakule et al., 1986), and water distribution systems operation (Brion and Mays, 1989; Sakarya and Mays, 1999).

Various computer programs are available that solve either linear programming problems, nonlinear programming problems, or both. Table 3.2 gives a summary of some of the more popular optimization computer programs in the United States.

TABLE 3.2 Summary of Some of the Most Popular Optimization Computer Programs in the United States

Model name	Developed by	Model purpose	Remarks
LINDO	Lindo Systems, Inc.	Solves linear, quadratic, and integer programming problems	A user-friendly linear interactive and discrete optimizer (hence, the name LINDO)
LINGO	Lingo Allegro USA, Inc.	Solves linear and nonlinear programming problems	A sophisticated matrix generator; helps the user create linear constraints and objective function terms by writing one-line code
GRG2	Univ. of Texas	Solves nonlinear programming problems	Uses the generalized reduced gradient algorithm to find the optimal solution
GINO		Solves nonlinear programming problems	This model is a microcomputer version of GRG2
GAMS	GAMS Development Corporation	Solves linear programming problems	
MINOS	Saunders and Murthagh	Solves linear and nonlinear programming problems	Uses different algorithms when the problem has linear objective function and constraints, nonlinear objective function and linear constraints, and nonlinear objective function and constraints
GAMS/ ZOOM		Solves mixed integer programming problems	Adapted ZOOM (Zero/One Optimization Method).
GAMS/ MINOS		Solves linear and nonlinear programming problems	Adapted MINOS (Modular In-Core Nonlinear Optimization System)

Differential Dynamic Programming. *Differential dynamic programming* (DDP) is a stagewise, nonlinear programming procedure that has been successfully applied to hydrosystems problems that are based on discrete-time optimal control, such as multireservoir operation and groundwater hydraulics (Mays, 1997).

A modified form of DDP, known as *successive approximation linear quadratic regulator* (SALQR), has been used for optimization problems in which nonlinear simulation equations are made linear in the optimization step (Culver and Shoemaker, 1992). For example, Carriaga and Mays (1995a,b) applied DDP to reservoir release optimization to control sedimentation. SALQR was applied to the operation of multiple reservoir systems to control sedimentation in alluvial river networks by Nicklow and Mays (2000, 2001); to the operation of soil aquifer treatment systems by Tang and Mays (1999); and to the determination of optimal freshwater inflows to bays and estuaries by Li and Mays (1995).

Genetic Algorithms and Simulated Annealing. *Genetic algorithms* (GAs) are nonconventional search techniques patterned after the biological processes of natural selection and evolution (Tang and Mays, 1999). Genetic algorithms can be useful for the selection of parameters to optimize the performance of a system and for testing and fitting quantitative models (Chambers, 1995). Every solution of the optimization problem is represented in the form of a string of bits (integers or characters) that consists of the same number of elements, say n. Each candidate solution represented as a string is known as an *organism* or a *chromosome*. The variable in a position on the chromosome and its value in the chromosome are called the *gene* and the *allele*, respectively. For example, if $n = 3$, a general chromosome is $x = (x_1, x_2, x_3)$ where x_1, x_2, and x_3 are the genes on this chromosome in the three positions (Murty, 1995).

Genetic algorithms for optimization problems are developed by first transforming the problem into an unconstrained optimization problem so that every string of length n can be looked upon as a solution vector for the problem (Murty, 1995). Five tasks are required in the performance of a GA to solve the optimization problem: encoding, initialization of the population, fitness evaluation, evolution performance, and working parameters (Adeli and Hung, 1995).

The decision variable vector is encoded as a chromosome using mostly a binary number coding method. Therefore if there are m decision variables and if each decision variable is encoded as an n-digit binary number, then a chromosome is a string of $n \times m$ binary digits as shown in Fig. 3.3.

A population of chromosomes is initialized which requires randomly generating the initial population in such a way that all values for each bit have equal probability of being selected. The fitness measure at every feasible solution is equal to the objective function value at that point. Thus, a fitness evaluation is used to determine the probability that a chromosome will be selected as a parent chromosome to generate new chromosomes. Evolution performance involves selection, crossover, and mutation. Selection chooses the chromosome to survive for a new generation. Crossover is used to recombine two chromosomes (parent strings) and generate two new chromosomes (offspring strings) based on a predefined crossover criterion. Mutation serves as an operator to reintroduce "lost alleles" into the population based on a predefined mutation criterion. Working parameters guide the genetic algorithm and include chromosome length, population size, crossover rate, mutation rate, and stopping criterion.

Simulated annealing (SA) stems from an algorithm that is used for the application of statistical thermodynamics concepts to combinatorial optimization problems. A solution to a combinatorial

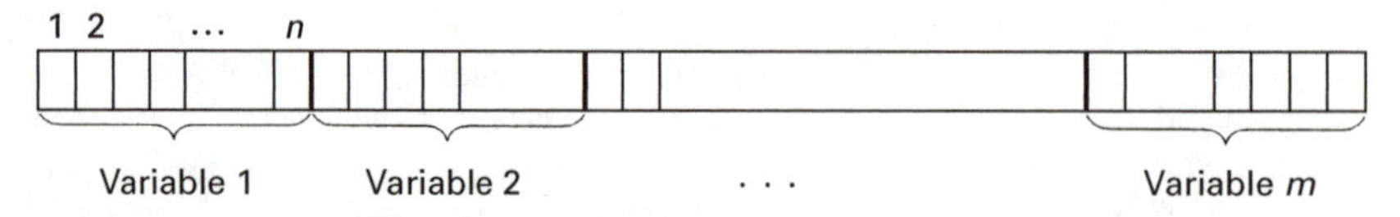

FIGURE 3.3 Typical representation of a chromosome of m decision variables of n bits each, used for encoding.

optimization problem is based on statistical mechanics in which the best solution is obtained from a large set of feasible solutions. In essence, it is a type of local search (descent method) heuristic that starts with an initial solution and has a mechanism for generating a neighbor of the current solution. For minimization problems, if the generated neighbor has a smaller objective value, it becomes the new current solution; otherwise the current solution is retained. The process is repeated until a solution is reached with no possibility of improvement in the neighborhood (Murty, 1995).

This algorithm has the disadvantage that the local search stops at a local minimum (see Fig. 3.4). This can be avoided by running the local search several times starting randomly from different initial solutions. By doing so, the global minimum can be taken as the best of the local minima found.

A better approach to finding the global minimum was introduced in 1953 by Metropolis et al. (Murty, 1995). In this attempt, annealing was applied to the search of the minimum energy configuration of a system after it is melted. At each iteration, the system is given a small displacement and the change in the energy of the system, δ, is calculated. If $\delta < 0$, the change in the system is accepted; otherwise, the change is accepted with probability $\exp(-\delta/T)$ where T is a constant times the temperature. This optimization technique has been applied to different problems in engineering, such as groundwater restoration (Skaggs and Mays, 1999), operation of water distribution systems (Sakarya et al., 1998; Goldman 1998; Goldman and Mays, 1999; Sakarya and Mays, 2000), for water quality purposes.

Comparison of Heuristic Search Methods to Other Optimization Techniques. Whereas the heuristic search methods involve trial solutions, mathematical programming and DDP/SALQR follow some given procedures. On the other hand, mathematical programming and DDP/SALQR require derivative information. The optimal solution found by the mathematical programming approach may result in a very short operating time during one time interval that cannot be followed for practical purposes. In the simulated annealing approach, this problem can be minimized by setting the minimum period of operation (Sakarya et al., 1998).

The mathematical programming approaches find the optimum solution in much shorter operating times than the heuristic search approaches. Tang and Mays (1998) have developed a new methodology for the operation of soil aquifer treatment systems, formulated as a discrete-time optimal control problem. This new methodology is based upon solving the operations problem using a genetic algorithm interfaced with the one-dimensional unsaturated flow model HYDRUS (Kool and van Genuchten, 1991). The same problem has been solved by Li et al. (2000) using SALQR interfaced with

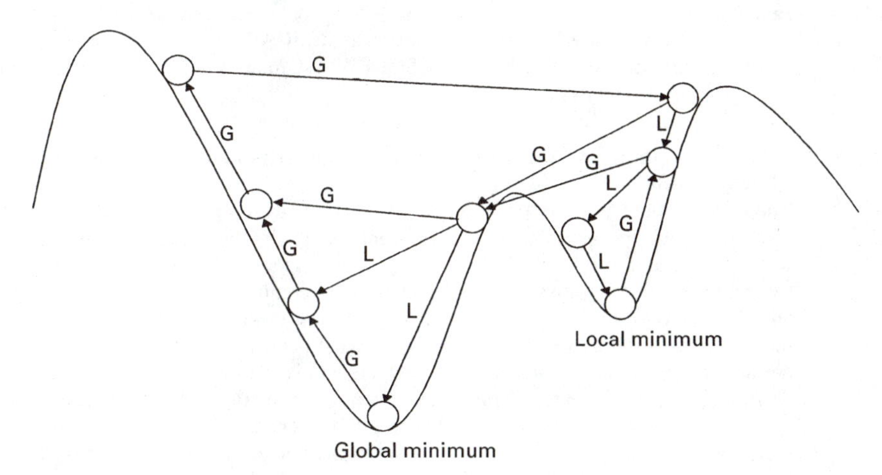

FIGURE 3.4 The objective function landscape. G = global, L = local. (After Topping et al., 1993)

the HYDRUS model. The computer time for a 10-cycle operation with the SALQR algorithm was reported as 654 CPU seconds, while with the genetic algorithm, it needed about 46,600 CPU seconds [about 13 hours (h)] on the same computer to obtain the optimal solution for a three-cycle operation.

Sakarya et al. (1998) have compared two newly developed methodologies, a mathematical programming approach and a simulated annealing approach, for determining the optimal operation of a water distribution system considering both quantity and quality aspects. Both methodologies formulate the problem as a discrete-time optimal control problem. The mathematical programming approach interfaces the GRG2 model (Lasdon and Warren, 1986), a generalized reduced gradient procedure, with the U.S. Environmental Protection Agency EPANET model (Rossman, 1994) for water distribution system analysis. The simulated annealing approach is also interfaced with the EPANET model. The study showed that while different results were obtained for total pump operation hours, the total 24-h energy costs were comparable.

3.3.4 Computer-Based Information Systems: Supervisory Control Automated Data Acquisition (SCADA)

SCADA is a computer-based system that can control and monitor several hydrosystems operations such as pumping, storage, distribution, and wastewater treatment. Several such systems have been developed in the past for different water supply agencies. For instance, the Metropolitan Sewer District of Cincinnati planned to integrate a SCADA system in the 1980s to monitor its wastewater treatment plants and pump stations. This system was planned for an area which consisted of 7 major treatment plants, 30 package wastewater plants serving individual subdivisions, and about 130 pump stations (Clement, 1996). A SCADA system developed in 1986 for the city of Honolulu had the capability of controlling and monitoring 57 source pumping stations, 126 storage reservoirs, and 73 booster pumping stations (Wada et al., 1986). In general, SCADA systems are designed to perform the following functions:

- Acquire data from remote pump stations and reservoirs and send supervisory controls
- Allow operators to monitor and control water systems from computer-controlled consoles at one central location
- Provide various types of displays of water system data using symbolic, bar graph, and trend formats
- Collect and tabulate data and generate reports
- Run water control software to reduce electrical power costs

Remote terminal units (RTUs) are used to process data from remote sensors at pump stations and reservoirs. The processed data are transmitted to the SCADA system also by the RTUs. Conversely, supervisory control commands from the SCADA system prompt the RTUs to turn pumps on and off and open and close valves.

3.3.5 Prospects of Computer Programs for Integrated Hydrosystems Management

There is no doubt that the first computer programs developed to solve hydrosystems problems targeted specific problems such as catchment runoff simulation, streamflow characterization, and water quality monitoring. With the enhancement of computing efficiency and speed over the past several years, more sophisticated and user-friendly computer programs for hydrosystems problems have been developed. However, the objective of most of the computer programs was not to address the problems of integrated hydrosystems management inasmuch as a consensus exists as to the definition of integrated hydrosystems management given in Sec. 3.2.

More recently, computer programs that attempt to provide support for decision makers have been brought into the picture. Such computer programs, generally termed DSSs, have manifested themselves at this time as promising computer programs for integrated hydrosystems management. Section 3.4 discusses the DSS applications for integrated hydrosystems management.

3.4 DSSs AS TOOLS FOR INTEGRATED HYDROSYSTEMS MANAGEMENT

3.4.1 Definition of DSSs

Decision support systems, as might be inferred from the name, do not refer to a specific area of specialty. It is not easy to connote a specific definition to DSSs based on their uses. Reitsma et al. (1996) point out that although some consensus exists as to the purpose of DSSs, "a single, clear, and unambiguous definition is lacking." Generally, however, DSSs give pieces of information, sometimes real-time information, that help make better decisions. Sprague and Carlson (1982) defined a DSS as an interactive computer-based support system that helps decision makers utilize data and computer programs to solve unstructured problems.

3.4.2 Basic Structure of DSSs

A DSS generally consists of three main components: (1) state representation, (2) state transition, and (3) plan evaluation (Reitsma et al., 1996). State representation consists of information about the system in such forms as databases. State transition takes place through modeling such as simulation. Plan evaluation consists of evaluation tools such as multicriteria evaluation, visualization, and status checking (Reitsma, 1996). The three main components comprise the database management subsystem, model base management subsystem, and dialog generation and management subsystem, respectively. Figure 3.5 depicts these subsystems including their specific purposes and functions. Some examples of DSSs for different integrated hydrosystems management are presented later in this section.

Jamieson and Fedra (1996) elaborated on the basic structure of the WaterWare DSS (Fig. 3.6). As shown in Fig. 3.6, each subsystem is made up of different components. The data management subsystem can use different tools such as geographic information systems (GIS) as well as other simplistic data. The model base subsystem consists of simple simulation computer programs, optimization techniques, and expert systems (also sometimes known as rule-based simulation computer programs). The dialog generation and management subsystem helps in visualization and making decisions through interactive user interface.

The structure of DSSs discussed above has, perhaps, made them the best structured and most promising computer programs for integrated resource management. These programs are believed to contribute largely to this objective. Reitsma et al. (1996) pointed out that "...the next few years will be most interesting" for DSSs. This stems from the fact that DSSs are promising computer programs for integrated hydrosystems management and the advance in the computing and information technology is remarkable.

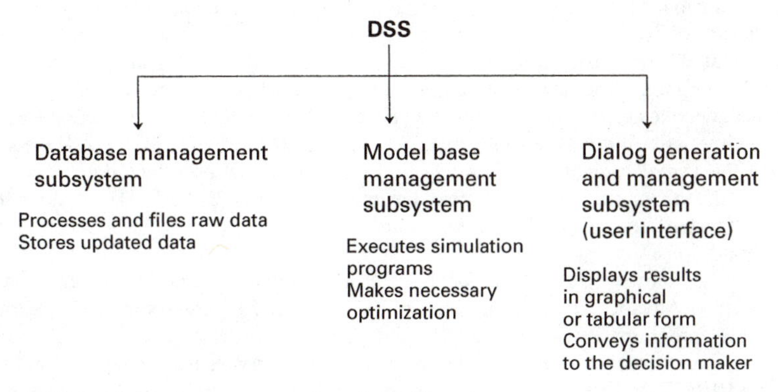

FIGURE 3.5 Basic components of a typical DSS.

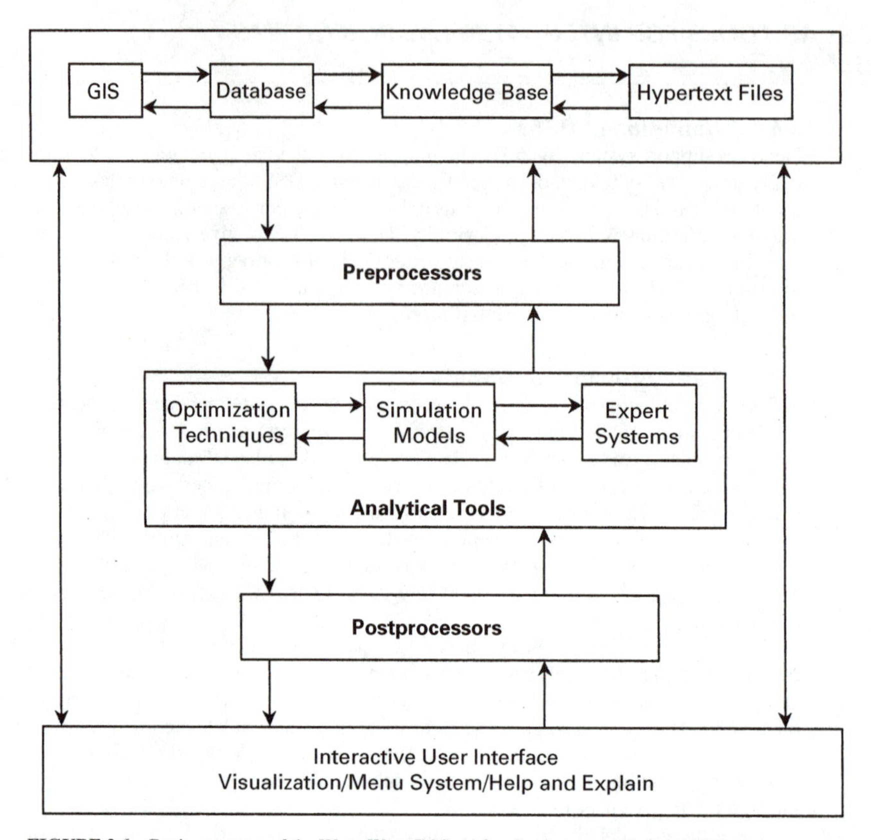

FIGURE 3.6 Basic structure of the WaterWare DSS. (After Jamieson and Fedra, 1996)

3.4.3 Examples of DSSs for Integrated Hydrosystems Management

DSS for Trinity River Basin, Texas. One of the integrated DSSs in regional hydrosystems management was developed for the Trinity River in Texas (Ford and Killen, 1995). This DSS has the capability of integrating three major hydrosystems problems. Accordingly, it has three components which perform the following tasks: (1) retrieve, process, and file rainfall and streamflow data; (2) estimate basin average rainfall and forecast runoff; and (3) simulate reservoir operation in order to forecast regulated flows basinwide. Each of the tasks is done by the DSS subsystems, which use existing computer programs. The first subsystem, the *data-retrieval, processing, and filing subsystem,* retrieves data that are collected from an existing precipitation and streamflow gauge network and stores the data using a time series *database management system* (DBMS) designated as HEC-DSS. The second subsystem, the *rainfall estimating and runoff forecasting subsystem,* uses the following computer programs: (1) PRECIP to compute catchment areal-average rainfall and (2) HEC-1F for forecasting runoff. The third subsystem, the *reservoir simulation subsystem,* uses HEC-5 that is customized and fitted to basin conditions.

Figure 3.7 shows different components of this DSS that are used for forecasting streamflow. The Trinity River Advanced Computing Environment (TRACE) is the forecaster's interface of the DSS. It executes programs PRECIP, HEC-1F, and HEC-5 with the proper input. It also serves as a file manager, input processor, and DBMS interface. Furthermore, it executes, behind the scenes, programs PREFOR and PREOP to complete the HEC-1F and HEC-5 files, respectively. The DBMS-interface component of TRACE executes program EXTRCT to create working copies of data

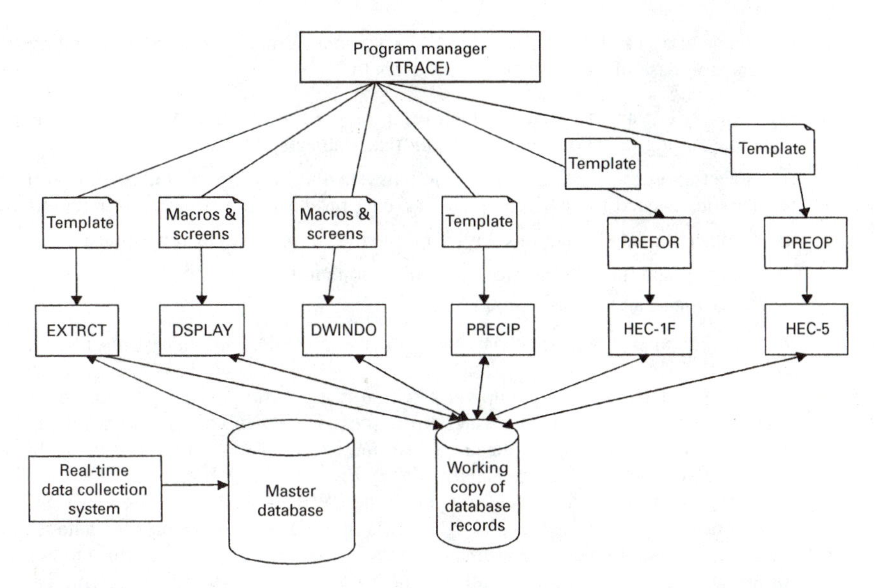

FIGURE 3.7 Components of forecasting software, Trinity River basin, Texas. (Ford and Killen, 1995)

records, program DISPLAY to graph data, and program DWINDO to tabulate and edit data (Ford and Killen, 1995).

The size of the Trinity River basin for which this DSS was developed is approximately 4.6 million hectares (17,800 mi^2). Seven multipurpose major reservoirs having a total capacity of approximately 13.63 billion m^3 (11,080,000 acre·ft) are found in the basin.

TERRA (TVA Environment and River Resource Aid). TERRA is a DSS developed for the Tennessee Valley Authority (TVA) and the Electric Power Research Institute (EPRI) (Reitsma et al., 1996). It was developed for the management of the TVA river, reservoir, and power resources. TERRA has the following characteristics:

1. Consists of georelational database
2. Serves as the central data-storage and retrieval system
3. Records the TERRA information flow
4. Supports interfacing specialized data management software
5. Has various visualization tools
6. Checks the data entering the database or data from both resident and nonresident computer programs against various sets of operational constraints (environmental, recreational, special/emergency, navigational, and so on)

TERRA consists of the three essential components of a DSS, namely, (1) management of the state information of the TVA river basin, (2) the computer programs for conducting simulations and optimizations, and (3) a comprehensive set of reporting and visualization tools for studying, analyzing, and evaluating current and forecast states of the river system.

PRSYM (Power and Reservoir System Model). This model is used for river, reservoir, and power systems. It provides a tool for scheduling, forecasting, and planning reservoir operations. It integrates the multiple purposes of reservoir systems such as flood control, navigation, recreation, water supply, and water quality, with power system economics by solving the problem based on pure simulation, rule-driven simulation, or a goal programming optimization (Zagona et al., 1995).

Shane et al. (1995) note that PRSYM represents a major advance in modeling flexibility, adaptability, and ease of use, which enables users to

1. Visually construct a model of their reservoir configuration using "icon programming" with icons representing reservoir objects, stream reach objects, diversions, etc.
2. Select appropriate engineering functions, standardized by the industry, to reflect object characteristics needed for schedule planning, e.g., reservoir and stream routing methods.
3. Replace outdated functions with improved versions developed by industry.
4. Develop and include functions that are unique to their systems.
5. Experiment with operating policies.
6. Use data display and analysis objects to customize data summary presentations.

CALSIM. The CALSIM computer program is a general-purpose planning program developed by the California Department of Water Resources (DWR) in collaboration with the U.S. Bureau of Reclamation (USBR). It was developed for the coordinated operation of the California State Water Project (SWP) and the Federal Central Valley Project (CVP). It replaces DWR's prior planning simulation program DWRSIM used to operate the SWP, as well as USBR's PROSIM and SANJASM programs used to operate the CVP. It utilizes optimization techniques to route water through a network of nodes and links. The user can formulate the operation of water resources systems as a linear programming (LP) or a mixed integer linear programming (MILP) problem. The operational constraints are formulated using a programming language known as Water Resources Engineering Simulation Language (WRESL). Data can be managed using a relational database management system and/or the United States Army Corps of Engineers data storage system (dss, not to be confused with DSS). CALSIM uses an LP/MILP solver, known as XA solver, to determine the optimal set of decisions given a set of user-defined priorities or weights and a set of system constraints. Figure 3.8 gives the CALSIM program components and structure.

Conjunctive Stream-Aquifer Management. This DSS is used for conjunctive management of surface water and groundwater under the prior appropriation water right (Fredericks et al., 1998). It has the three components which are typical of a DSS: database management subsystem, model base management subsystem, and a dialog generation and management subsystem or user interface. It is possible to prepare input data files for this DSS using GIS. The overlay of the GIS raster or grid database with other aquifer grid data enable the finite groundwater model MODFLOW to readily read these data.

RiverWare. Developed by the Center for Advanced Decision Support for Water and Environmental Systems (CADSWES) at the University of Colorado, this DSS was designed for general river basin modeling for a wide range of applications (Zagona, 1998). It has three fundamental solution methods: simple simulation, rule-based simulation, and optimization.

To abate the problems of complicated water policies, a different programming language called RiverWare Rule Language (RWRL) is used. Policy descriptions can be designed as a structured ruleset in RWRL. Once these policy descriptions are saved as ruleset files, a simulation may be guided by the ruleset. Furthermore, the policies can be modified between runs, without requiring the simulator to be changed or rebuilt (Wehrends and Reitsma, 1995).

Wehrends and Reitsma (1995) gave the following examples of how water policies can be formulated and interpreted.

```
IF Mead's elevation > 1229.0 THEN
   Mead's release = Mead's inflow
END IF
```

This approach gives a conditional water policy, which may be considered to be easy enough to be incorporated in a general simulation model.

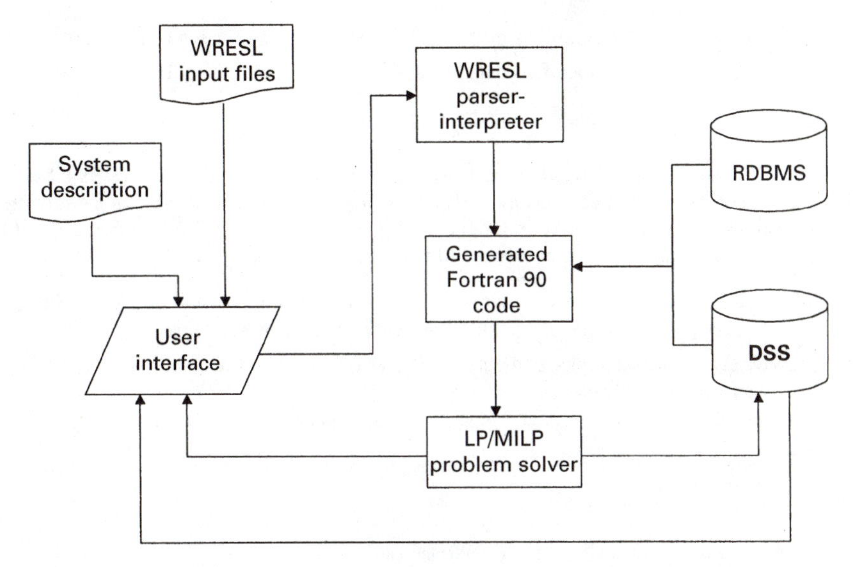

FIGURE 3.8 CALSIM model components and structure.

```
IF Mead's elevation > value THEN
   Mead's release = mead's inflow
END IF
```

In this approach, the user has the choice of changing ***value*** at run-time without rebuilding the program. However, the policies expressed in this fashion may be still very specific.

A more comprehensive approach is to allow policies to be completely modifiable without requiring the underlying system to be rebuilt. As such, policies can be written in a rule language that interprets the policies and can be interfaced with the simulation computer programs. The policies are interpreted during run-time, which makes the running time of the program longer.

The general architecture of the RiverWare program employs the representation of a river basin by *objects.* The objects that are included in RiverWare include the following (Zagona et al., 1998):

- *Storage Reservoir.* Mass balance, evaporation, bank storage, spill
- *Level Power Reservoir. Storage Reservoir* plus hydropower, energy, tailwater, operating head
- *Sloped Power Reservoir. Level Power Reservoir* plus wedge storage for very long reservoirs
- *Pumped Storage Reservoir. Level Power Reservoir* plus pumped inflow from another reservoir
- *Reach.* Routing in a river reach, diversion and return flows
- *Aggregate Reach.* Many *Reach* objects aggregated to save space on the workspace
- *Confluence.* Brings together two inflows to a single outflow as in a river confluence
- *Canal.* Bidirectional flow in a canal between two reservoirs
- *Diversion.* Diversion structure with gravity or pumped diversion
- *Water User.* Depletion and return flow from a user of water
- *Aggregate Water User.* Multiple *Water Users* supplied by a diversion from a *Reach* or *Reservoir*
- *Aggregate Delivery Canal.* Generates demand and models supplies to off-line water users
- *Groundwater Storage Object.* Stores water from return flows
- *River Gage.* Specified flows imposed at a river node

- *Thermal Object.* Economics of thermal power system and value of hydropower
- *Data Object.* User-specified data: expression slots or data for policy statements

Table 3.3 shows user methods for selected objects in RiverWare.

AQUATOOL. Developed at the Universidad Politécnica de Valencia (UPV), Spain, as a result of continuing research over a decade, AQUATOOL is a generalized DSS that has attracted several river basin agencies in Spain (Andreu et al., 1996). Andreu et al. (1996) also note that AQUATOOL has various capabilities that can be used in water resource systems to:

1. Screen design alternatives by means of an optimization module, obtaining criteria about the usefulness and performance of future water resource developments.
2. Screen operational management alternatives by means of the optimization module, obtaining criteria from the analysis of the results.
3. Check and refine the screened alternatives by means of a simulation module.
4. Perform sensitivity analysis by comparing the results after changes in the design or in the operating rules.

TABLE 3.3 Selected User Methods in RiverWare

Object type	User method category	User methods
Reservoirs	Evaporation and precipitation	No evaporation Pan and ice evaporation Daily evaporation Input evaporation CRSS evaporation
	Spill	Unregulated spill Regulated spill Unregulated plus regulated Regulated plus bypass Unregulated plus regulated plus bypass
Power reservoirs	Power	Plant power Unit generator power Peak base power LCR power
	Tailwater	Tailwater base value only Tailwater base value plus lookup table Tailwater storage flow lookup table Tailwater compare Hoover tailwater
Reaches	Routing	No routing Time lag routing Variable time lag routing SSARR Muskingum Kinematic wave Muskingum-Cunge MacCormack
Water user (on Agg-Diversion)	Return flow	Fraction return flow Proportional storage Variable efficiency

Source: After Zagona et al. (1998).

5. Use different computer programs, once an alternative is implemented, as an aid in the operation of the water resource system, mainly for water allocation among conflicting demands and to study impacts of changes in the system.
6. Perform risk analysis for short- and medium-term operational management to decide, for instance, the appropriate time to apply restrictions and their extent.

AQUATOOL has been accepted by the Sagura and Tagus river basin agencies in Spain as a standard tool to develop their basin hydrologic plan and to manage the resource efficiently in the short to medium term (Andreu et al., 1996).

WaterWare. This DSS is a comprehensive model for integrated river basin planning. It has the capabilities of combining geographic information systems, database technology, modeling techniques, optimization procedures, and expert systems (Jamieson and Fedra, 1996). The aspects of integrated river basin management that this DSS incorporates are briefly as follows (Fedra and Jamieson, 1996):

1. *Groundwater pollution control.* Simulation of flow and contaminant transport, and reduction of the level of contaminant in the aquifer and/or protecting groundwater resources.
2. *Surface-water pollution control.* Estimation of the level of effluent treatment required to meet the river water quality objectives.
3. *Hydrologic processes.* Estimation of ungauged tributary for use in the water resources planning component (see no. 5); assessment of daily water balance for ungauged subcatchments, and the impact of land-use changes on runoff; and evaluation of the effects of conjunctive use of surface and groundwater.
4. *Demand forecasting.* Use of rule-based inference computer programs which use generic expert systems.
5. *Water resources planning.* Consists of
   a. A model capable of simulating the dynamics of demand, supply, reservoir operations, and routing through the channel system
   b. A module for reservoir site selection which assesses 10 problem classes:
      (1) Landscape and archaeological or historical sites
      (2) Land-use restrictions
      (3) Drainage, soil, and microclimate
      (4) Natural habitats and associated communities
      (5) Water quality, aquatic biology and ecology
      (6) Water resources and cost implications
      (7) Reservoir construction
      (8) Reservoir operations
      (9) Socioeconomic effects of reservoir operations
      (10) Recreational provisions

3.5 STATE OF PRACTICE AND PROSPECTS OF HYDROSYSTEMS COMPUTER PROGRAMS

3.5.1 State of Practice

The concept of integrated water resources management is a comprehensive representation of several components, each of which requires sufficient representation or modeling within the whole system. Modeling needs to be driven by coverage of all aspects of integrated hydrosystems management, not by the convenience or simplicity of the modeling of each aspect of the problem. Loucks (1996) clearly states that "an integrated view of water-resource systems can not be compartmentalized into either surface water or groundwater and either water quantity or water quality just because the respective time and space scales make the modeling or study of such divisions convenient."

On the contrary, as mentioned previously, computer programming generally started out with the simplification of calculations of analytical functions that required a very long time to solve by hand. Through time, the capability enhanced to the level of tackling complex hydrosystems problems. It is through improvements of the programming methodologies and new technological discoveries that more sophisticated hydrosystems computer programs have been developed. Therefore, hydrosystems computer programs have been approaching the essence of integrated hydrosystems management from the bottom up.

The important aspects of integrated hydrosystems problems which have been tackled using computer programs include simulation, database management, data collection, and storage. These efforts have reached a level of promising prospect and have diminished the gap between the concept of computer programs for integrated hydrosystems management and the actual programs. For instance, GIS generally provides facilities for storage and management of very large geo-information. It has been possible to represent the terrain of the entire United States as a database of the Digital Elevation Model (DEM). Automatic data collection systems such as SCADA and radar provide readily available input data for real-time analysis of integrated hydrosystems problems. By integrating together different computer programs, it has been possible to develop DSSs that have manifested to address these issues. A few of these systems have been designed not only to solve the problem, but also to attempt to interpret the result. Jamieson and Fedra (1996) point out that DSSs have the capabilities of predicting what may happen under a particular set of planning assumptions and of providing expert advice on the appropriate course of action.

Most of the available computer programs for hydrosystems problems, however, address only limited issues of the general concept of integrated hydrosystems management. While they may be satisfactory tools to solve the particular problem they are designed for, only a few DSSs currently available such as TERRA, RiverWare, AQUATOOL, and WaterWare are useful as stand-alone computer programs for integrated hydrosystems management. Therefore, it can be inferred that because of the availability of only a limited number of DSSs for integrated hydrosystems management, the state of practice of DSSs for integrated hydrosystems management is premature, yet evolving.

3.5.2 Prospects

Advances in software engineering appear to be promising for integrated hydrosystems management computer programs. They have enabled the development of computer programs that not only incorporate easy-to-use analytical capabilities, but also offer expert advice and intelligent interrogation facilities. With these types of computer programs, the artificial intelligence involved can be provided by a mixture of optimization techniques and expert systems that can evaluate, draw preliminary conclusions, and recommend appropriate actions. This stage of development of hydrosystems computer programs is the emergence of what has been referred to as the fifth generation of the hydroinformatics system (Jamieson and Fedra, 1996).

Even though the efforts made in the past to develop simulation computer programs have been tremendous, many hydrosystems computer programs were written to address specific hydrosystems problems such as reservoir operation, water distribution, urban drainage, and streamflow. The focus of the objectives of many of these computer programs is limited. The parts are out, but we are faced with the painstaking task of integrating these simple computer programs to create a more integrated hydrosystems computer program.

Some promising efforts in this regard have already been undertaken. The successful developments of TERRA, WaterWare, RiverWare, AQUATOOL, and so on are very good examples. The efforts made at the USACE Hydrologic Engineering Center to enhance the old computer programs to the new ones, generally known as the Next Generation (NexGen) computer programs, may form one of the strong cores of DSSs, and simulation computer programs.

Decision support systems in general are, perhaps, the most promising tool to integrate the simple computer programs to be used for integrated hydrosystems management. The three subsystems of

DSSs—database management subsystem, model base management subsystem, and dialog generation and management subsystem—constitute a logical construct of the concept of integrated hydrosystems management. Figure 3.9 shows a representation of most of the possible components of a typical DSS. The dotted lines in the figure show the components that can be included in the DSS in the future or used to enhance its current proposed structure.

The database management subsystem provides the opportunity for easy collection, storage, and alteration of data, including on a real-time basis. GIS and SCADA, among others, are important systems for this purpose. The proliferation of simulation computer programs and the availability of some advanced optimization techniques provide valuable resources in dealing with different aspects of hydrosystems problems. The graphics-supported user-friendly interface environment also helps to draw appropriate conclusions and make necessary decisions that agree with predefined integrated hydrosystems management policies.

If there are challenges to overcome to use DSSs for integrated hydrosystems management problems, one of the most difficult challenges, perhaps, will be not having appropriate integrated hydrosystems management policies clearly defined. It may be noted that while it is possible to code policies in a computer program, no code may be written for a policy that does not exist. Likewise, it cannot be easy to write a clear computer code for an ambiguous or ill-defined policy.

3.6 SUMMARY AND CONCLUSIONS

Water as a limited resource poses the challenge of allocating a sufficient amount to all the competing users efficiently and effectively. An integrated hydrosystems management approach enables us to have knowledge in space and time of the purposes and amounts of water requirements, thereby allowing for balancing out between the competing needs. As discussed herein, an integrated hydrosystems management may be the most promising means to provide the water requirements of all the competing users, requiring the involvement of all parties concerned. This approach helps the formulation of

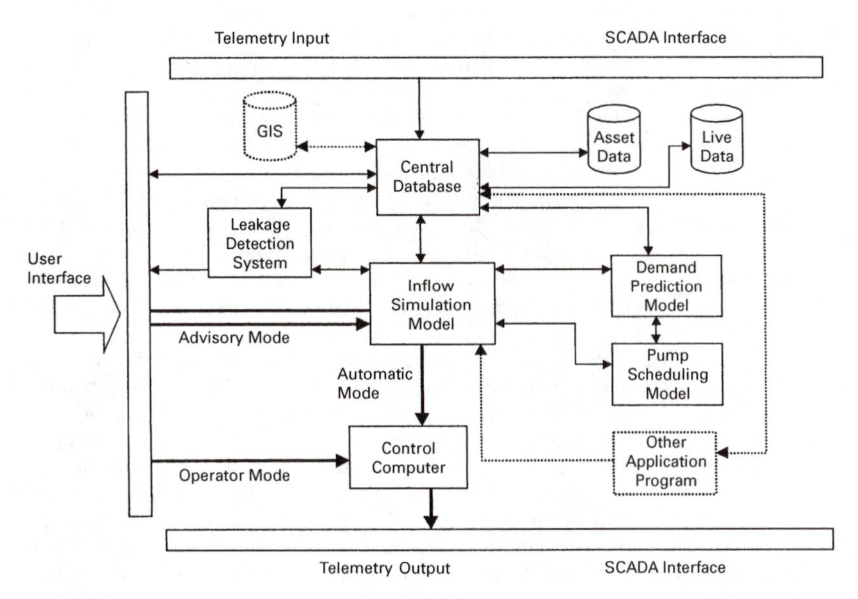

FIGURE 3.9 Proposed water management system for the Great Man-Made River Scheme, Libya. (Khalil, 1998)

viable water policies acceptable to most parties or satisfying most objectives. Nonetheless, although this concept is a strategy that is increasingly advocated in the literature, it is still relatively new. The challenge in this regard is yet to be fully overcome.

Because the concepts of integrated hydrosystems management can be best explained in terms of hydrosystems policies or rules and because such policies can be interpreted and coded in computer programs, it is very important to have these policies clearly defined for a given watershed. It may be noted that it is these policies that we begin with to deal with integrated hydrosystems management. Furthermore, the scope and areal coverage of integrated hydrosystems management that is mandated to an institution or water agency should be unambiguously defined. Implementing its concept would be more complete if the ideals of integrated hydrosystems management are clearly defined and understood, and if the policies can be easily interpreted and coded into computer programs. The authors agree with the watershed approach strategy already recommended by different institutions. This approach entails hydrosystems policies that transcend political boundaries for the purpose of integrated hydrosystems management and, therefore, it is necessary that this approach be acceptable by different parties so that the best overall result is obtained. To this effect, a river basin or watershed approach for regional coverage is a sound strategy.

Lack of efficient techniques in the past that could be used to code hydrosystems policies in computer programs might have had a negative impact on their development for integrated hydrosystems management. The advance in computing technology appears to be at a stage where it is capable of overcoming such problems. Today, a computer programming language specifically used for rulesets (a set of simulation rules) has been developed at CADSWES and therefore can be used for modeling integrated hydrosystems problems, should such languages become the requirement of the state of the art for this purpose.

Computer programs for integrated hydrosystems management can be very important tools for fast computations, easy data management, and drawing conclusions about certain water policies. Such computer programs, generally termed DSSs, have been introduced recently by different institutions. As computing speed and ease become more powerful, more complex yet more comprehensive computer programs are being developed. Such computer programs as CALSIM, TERRA, RiverWare, AQUATOOL, and WaterWare are examples of DSSs that can be used for integrated hydrosystems management. These DSSs are embodied with water policies in the form of rulesets (to use the term used in RiverWare) or expert systems (to use the term used in WaterWare). These computer programs have become successful as programs of integrated hydrosystems management by the incorporation of water policies that are formulated in a form understandable in the computation processes.

At the center of DSSs are found simulation and/or optimization computer programs. A tremendous amount of work has been done in the past to develop simulation and optimization computer programs that solve problems in the areas of hydrology, hydraulics, and water resources. An effort was also made to interface simulation and optimization computer programs to solve optimal control problems in water resources. In the process, water policy issues such as water rights, ecosystem sustainability, and amenity are easily incorporated in the form of rulesets or expert systems. In this regard, more effort is needed because rulesets or expert systems are relatively new approaches in modeling, and the concept of integrated hydrosystems management approach is yet to come to fruition.

In conclusion, some useful computer programs in the form of DSSs, which address integrated hydrosystems management problems, have been written. Several of these programs have proved the importance of the use of DSSs in integrated hydrosystems management problems. The availability of various hydrosystems computer programs that address specific hydrosystems problems and different optimization techniques, in conjunction with the advance in the information technology, provide a wealth of resources that are useful in designing DSSs. Thus, we may conclude that not enough work has been done to develop DSSs for integrated hydrosystems management. However, we have the technical resources—database management systems, simulation computer programs, optimization techniques, and advanced computing technology—and we are faced with the decision to make use of these resources to bring out more DSSs for integrated hydrosystems management.

3.7 REFERENCES

Adeli, H., and S. L. Hung (1995). *Machine Learning—Neural Networks, Genetic Algorithms, and Fuzzy Systems,* John Wiley & Sons, New York.

American Water Works Association Research Foundation (1996). Minutes of Seattle Workshop on Total Water Management, Denver, Colo.

Andreu, J., et al. (1996). AQUATOOL: A Generalized Decision Support System for Water-Resources Planning and Operational Management, *Journal of Hydrology* **177:**269–291.

Brion, L. M., and L. W. Mays (1989). Methodology for Optimal Operation of Pumping Stations in Water Distribution Systems, *Journal of Hydraulic Engineering,* ASCE, **117**(11):1551–1569.

Bulkley, J. W. (1995). Integrated Watershed Management: Past, Present and Future, *Water Resources Update,* Issue No. 100, Universities Council on Water Resources, Carbondale, Ill.

Carriaga, C. C., and L. W. Mays (1995a). Optimization Modeling for Simulation in Alluvial Rivers, *Journal of Water Resources Planning and Management,* ASCE, **121**(3):251–259.

Carriaga, C. C., and L. W. Mays (1995b). Optimal Control Approach for Sedimentation Control in Alluvial Rivers, *Journal of Water Resources Planning and Management,* ASCE, **121**(6):408–417.

Chambers, L. (1995). *Practical Handbook of Genetic Algorithms Applications,* Vol. 1, CRC Press.

Clement, D. P. (1996). SCADA System Using Packet Radios Helps to Lower Cincinnati's Telemetry Costs, *Water Engineering and Management* **134**(8):18–20.

Culver, T. B., and C. A. Shoemaker (1992). Dynamic Optimal Control for Groundwater Remediation with Flexible Management Periods, *Water Resources Research* **28**(3):629–641.

Fedra, K., and D. G. Jamieson (1996). The "WaterWare" Decision Support System for River-Basin Planning. 2. Planning Capability, *Journal of Hydrology* **177**:177–198.

Ford, D. T., and J. R. Killen (1995). PC-Based Decision-Support System for Trinity River, Texas, *Journal of Water Resources Planning and Management* **121**(5):375–381.

Fredericks, J. W., et al. (1998). Decision Support System for Conjunctive Stream-Aquifer Management, *Journal of Water Resources Planning and Management* **124**(2):69–78.

Goldman, F. E. (1998). The Application of Simulated Annealing for Optimal Operation of Water Distribution Systems, Ph.D. dissertation, Arizona State University, Tempe.

Goldman, F. E., and L. W. Mays (2002). Simulated Annealing Approach for Operation of Water Distribution Systems Considering Water Quality, ASCE (in review).

Grigg, N. S. (1998). Coordination: The Key to Integrated Water Management, *Water Resources Update,* Issue No. 111, Universities Council on Water Resources, Carbondale, Ill.

Hall, M. W. (1998). Extending the Resources: Integrating Water Quality Considerations into Water Resources Management, *Water Resources Update,* Issue No. 111, Universities Council on Water Resources, Carbondale, Ill.

Hooper, B. (1995). Towards More Effective Integrated Watershed Management in Australia: Results of a National Survey, and Integrated Implications for Urban Catchment Management, *Water Resources Update,* Issue No. 100, Universities Council on Water Resources, Carbondale, Ill.

Jamieson, D. G., and K. Fedra (1996). The "WaterWare" Decision Support System for River-Basin Planning. 1. Conceptual Design, *Journal of Hydrology* **177**:163–175.

Jeter, M. W. (1986). *Mathematical Programming: An Introduction to Optimization,* Marcel Dekker, New York.

Khalil, H. M. (1998). Proposed Water Management System for the Great Man-Made River Project, in W. R. Blain (ed.), *Hydraulic Engineering Software VII,* Wessex Institute of Technology Press, Boston, pp. 361–379.

Kool, J. B., and M. T. van Genuchten (1991). *HYDRUS: One-Dimensional Variable Saturated Flow and Transport Model, Including Hysteresis and Root Uptake*, U.S. Department of Agriculture, Agriculture Service, Riverside, Calif.

Lasdon, L. S., and A. D. Waren (1986). *GRG2 User's Guide,* Department of General Business, The University of Texas at Austin.

Li, G. L., and L. W. Mays (1995). Differential Dynamic Programming for Estuarine Management, *Water Resources Planning and Management* **121**(6):455–462.

Li, G., Z. Tang, L. W. Mays, and P. Fox (2000). New Methodology for Optimal Operation of Soil Aquifer Treatment Systems, *Water Resources Management* **14**(1):13–33.

Loucks, D. P. (1996). Surface Water Resource Systems, in L. W. Mays (ed.), *Water Resources Handbook,* McGraw-Hill, New York.

Mays, L. W. (1997). *Optimal Control of Hydrosystems,* Marcel Dekker, New York.

Mays, L. W., and Y. K. Tung (1992). *Hydrosystems Engineering and Management,* McGraw-Hill, New York.

Mitchell, B. (ed.) (1990). *Integrated Water Management: International Experiences and Perspectives,* Belhaven Press, London.

Murty, K. G. (1995). *Operations Research: Deterministic Optimization Models,* Prentice Hall, Englewood Cliffs, N.J.

Nicklow, J. W., and L. W. Mays (2000). Optimization of Multiple Reservoir Networks for Sedimentation Control, *Journal of Hydraulic Engineering* **126**(4):232–243.

Nicklow, J. W., and L. W. Mays (2001). Optimal Control of Reservoir Releases to Minimize Sedimentation in Rivers and Reservoirs, *Journal of the American Water Resources Association* **37**(1):197–211.

Reitsma, R. F., et al. (1996). Decision Support Systems (DSS) for Water-Resources Management, in L. W. Mays (ed.), *Water Resources Handbook,* McGraw-Hill, New York.

Rossman, L. A. (1994). EPANET Users Manual, Project Summary Report, Risk Reduction Engineering Laboratory, U. S. Environmental Protection Agency (EPA), Cincinnati, Ohio.

Sakarya, A. B., and L. W. Mays (2000). Optimal Operation of Water Distribution Pumps Considering Water Quality, *Journal of Water Resources Planning and Management* **126**(4):210–220.

Sakarya, A. B., et al. (1998). New Methodologies for Optimal Operation of Water Distribution Systems for Water Quality Purposes, in W. R. Blain (ed.), *Hydraulic Engineering Software VII,* Wessex Institute of Technology Press, Boston, pp. 101–110.

Schultz, G. A. (1998). A Change of Paradigm in Water Sciences at the Turn of the Century? *Water International,* **23**(1):37–44.

Shane, R. M., et al. (1995). The INTEGRAL Project: Overview in Computing in Civil Engineering, in *Proceedings of the Second Congress,* ASCE, June 5–8, Atlanta, Vol. 1, pp. 203–205.

Sprague, R. H., and E. D. Carlson (1982). *Building Effective Decision Support Systems,* Prentice-Hall, Englewood Cliffs, N.J.

Tang, A., and L. W. Mays (1998). Genetic Algorithms for Optimal Operation of Soil Aquifer Treatment Systems, *Water Resources Management,* **12**:375–396.

Topping, B. H. V, et al. (1993). Topological Design of Truss Structures Using Simulated Annealing, in B. H. V. Topping and A. I. Khan (eds.), *Neural Networks and Combinatorial Optimization in Civil and Structural Engineering,* Civil-Comp Press, Edinburgh, Scotland, pp. 151–165.

U.S. General Accounting Office (1994), Ecosystem Management Additional Actions Needed to Adequately Test a Promising Approach, GAO/RCED-94-111.

Viessman, W., Jr. (1998). Water Policies for the Future: Bringing It All Together, *Water Resources Update,* Issue No. 111, Universities Council on Water Resources, Carbondale, Ill.

Vlachos, E. C. (1998). Practicing Hydrodiplomacy in the 21st Century, *Water Resources Update,* Issue No. 111, Universities Council on Water Resources, Carbondale, Ill.

Wada, R. N., et al. (1986). Honolulu's New SCADA System, *Journal of the American Water Works Association* **78**(8):43–48.

Wanakule, N., et al. (1986). Optimal Management of Large Scale Aquifers: Methodology and Applications, *Water Resources Research* **22**(4):447–465.

Wehrends, S. C., and R. F. Reitsma (1995). A Rule Language to Express Policy in a River Basin Simulator in Computing in Civil Engineering, in *Proceedings of the Second Congress,* ASCE, June 5–8, Atlanta, Vol. 1, pp. 392–395.

Wurbs, R. A. (1995). *Water Management Models: A Guide to Software,* Prentice-Hall PRT, Englewood Cliffs, N.J.

Zagona, E. A. (1995). The INTEGRAL Project: The PRYSM Reservoir Scheduling and Planning Tool in Computing in Civil Engineering, in *Proceedings of the Second Congress,* ASCE, June 5–8, Atlanta, Vol. 1.

Zagona, E. A. (1998). RiverWare: A General River and Reservoir Modeling Environment, in *Proceedings of the First Federal Interagency Hydrologic Modeling Conference,* April 19–23, Las Vegas.

CHAPTER 4
OPTIMAL DESIGN OF WATER DISTRIBUTION SYSTEMS

Kevin E. Lansey
Department of Civil Engineering and Engineering Mechanics
University of Arizona
Tucson, AZ

4.1 OVERVIEW

A common design process for a water distribution system is trial and error. An engineer selects alternative designs and simulates these designs using a network solver. Because of the complex interactions between components, identifying changes to improve a design can be difficult even for mid-sized systems. In addition, this approach does not provide assurance that a good, let alone optimal, solution has been determined. Thus, significant research effort has been placed in developing approaches to solve for optimal designs of distribution systems with some success. However, given the complexity of the problem and the limitations of mathematical programming tools, the complete problem has not been resolved. This chapter is a brief overview of the general directions of this research, and it highlights representative work in the area. Other reviews can be found in Goulter (1987, 1992), Walski (1985), and Walters (1988).

4.2 PROBLEM DEFINITION

When designing or rehabilitating a water distribution system using trial-and-error methods or with formal optimization tools, a broad range of concerns can be considered. Cost is likely to be the primary emphasis and includes the costs for construction, operation, and maintenance. The initial capital investment is for system components: pipes, pumps, tanks, and valves. Energy consumption occurs over time to operate the system. The main constraints are that the desired demands are supplied with adequate pressure head being maintained at withdrawal locations. Also, the flow of water in a distribution network and the nodal pressure heads must satisfy the governing laws of conservation of energy and mass.

In summary, the problem can be verbally stated as:
Minimize capital investment plus energy costs, subject to:
Meeting hydraulic constraints,
Fulfilling water demands, and
Satisfying pressure requirements

In practice, additional complexities beyond the concise formulation presented above can have a significant impact on both optimal design methods and trial-and-error approaches. For example, the number of demand conditions and the demand distributions to be satisfied must be defined prior to any analysis. It has been shown that an optimal design for a single demand pattern will be a branched pipe network; thus, to introduce reliability and redundancy, multiple demands must be considered. No general guidance is available to select the set of loading conditions to examine during the design process.

In addition to demands, other simple constraints can be added, such as type, size, or material or allowing different rehabilitation alternatives (cleaning, relining, or both).

Further complexity may be added if the system layout is not defined. In most distribution systems, pipes are restricted to be placed beneath roadways or in right-of-ways. In some cases, most notably major supply systems, the layout may be more flexible and must be determined. Another significant factor is budgetary constraints that may require staging of construction over time.

System operations have been included in the formulation above. This alone is a significant problem in both defining the demands and in addressing the large number of constraints and decisions associated with operations. To operate the system with tanks, a time series of demands is considered that multiplies the number of constraints by the number of demand conditions. Decisions are made during each period to define pump operations. From an optimization perspective, these decisions are discrete (integer) and are extremely difficult to determine efficiently. Determining operations for existing networks also has been widely studied (Ormsbee and Lansey, 1994).

Additional constraints may be added to represent simple limits or complex constraints that are related to both the component sizes and the flow and pressure-head distributions. For example, a common engineering rule of thumb is to limit the range of flow velocities, which are functions of the flow rate and pipe diameter, between 2 and 5 ft/s for normal operating conditions. Water-quality requirements may also be added to the problem. Complex relationships exist to describe changes in water quality in a system. Since water quality decays with time in the network after disinfection, smaller pipes and tanks that would be introduced to improve water quality would conflict with meeting pressure requirements and reducing operation costs.

As noted above, an engineer typically defines one or more demand patterns that are considered during design. The intent in this selection is to produce a design that will operate effectively for the full range of conditions applied to the system. Indirectly, this implies some degree of reliability. The reliability measures are related to the number of components and their sizes. Improving reliability, however, increases system costs and may adversely affect water quality under normal operating conditions. The addition of reliability in optimal design is beyond the scope of this discussion. The interested reader is referred to Mays (1989).

Location of valves for the purposes of reliability and their operation to reduce pressures and leakage have also been considered in the literature (Jowitt and Xu, 1990; Reis et al., 1997). This specialized problem is not discussed further here.

4.3 MATHEMATICAL FORMULATION

Including general relationships to account for the constraints noted above, the overall optimization problem for the design of water distribution systems can be stated mathematically in terms of the nodal pressure heads H and the various design operational parameters D as follows

Objective: *Minimize* $Cost = f(H, D)$ (4.1)

subject to:

$G(H, D) = 0$ Conservation of mass and energy equations, (4.2)

$\underline{H} \leq H < \overline{H}$ Nodal pressure head bounds, (4.3)

$\underline{u} \leq u(D) \leq \overline{u}$ Constraints related to design/operational parameters, and (4.4)

$\underline{w} \leq w(H, D) < \overline{w}$ Constraints related to design parameters and pressure heads (4.5) where the decision variables D define the dimensions for each component in the system, such as the pipe diameter, pump size, valve setting, and tank volume or elevation. The objective function (Eq. 4.1) can be linear or nonlinear. Since each component may have a term associated with it in the objective function, variability in costs due to installation location or period can be considered. In this formulation, system expansions or new systems can be designed.

Upper and lower limits on nodal pressure heads are given as strict bounds in this formulation (Eq. 4.3). Other reliability-based optimization extensions recognize that slight deviations from absolute ranges may be acceptable, depending on the change in system cost (Cullinane et al. 1992; Halhal et al., 1997; Wagner et al., 1988).

Bounds on the design variables (u) are written in general form (Eq. 4.4). However, most often they are simple bounds defining the size of the component. Decisions, as presented, can be continuous or discrete. Many decisions, such as commercially available pipe diameters, are discrete and make the problem especially difficult to solve. Templeman (1982) showed that solving for discrete pipe sizes is NP-hard. NP-hardness implies that the computation time for an N-pipe system is an exponential function of N; therefore, solution time increases rapidly with the number of pipes. Many researchers consider pipe diameters to be continuous and assume that rounded solutions are nearly optimal, or they split pipes into two sections with varying diameter in the final design. Neither approach is appealing to practitioners.

The general constraint set w includes limits on terms that are functions of both the nodal pressure heads and the design variables (Eq. 4.5). As noted above, pipe velocity, water-quality constraints, budgetary limits, and reliability measures are examples of these terms.

The conservation equations (Eq. 4.2) are the set of nonlinear equations relating the flow and pressure-head distribution to the nodal demands and selected design variables. These equality constraints are written for each loading condition examined. The external nodal demands are included in these equations. These equations can be written for branched or looped networks for one or more demands, including an extended-period simulation. The conservation equations pose a significant difficulty in solving this optimization problem. Most approaches attempt to simplify these equations or to solve them outside the optimization model to avoid embedding them as constraints. Many methods can theoretically consider multiple demand conditions, but the computational expense may be high.

At this point, a general problem has been presented that can incorporate most concerns of water distribution engineers: that is, cost with budgetary limits and construction staging, water quality, reliability, operations, and discrete solutions. Because of the complexity of the general problem, however, it has not been solved. The majority of effort has been on cost minimization subject to simple bounds (Eqs. 4.3 and 4.4) without operations. This review considers this problem, with minor discussion of the other problems. Clearly, research should be expanded to incorporate realistic and practical concerns that make the results more applicable to engineering practice (Walski, 1996).

4.4 OPTIMIZATION METHODS

Virtually every optimization method has been applied to the problem of water distribution optimization. A progression can be seen over time. Branched networks were considered first, followed by looped networks that considered only pipe sizing. Both were solved using linear programming (LP). Nonlinear programming (NLP) was used later to solve a more general problem. Finally, to overcome continuous NLP decisions, more recent work has focused on stochastic search techniques, such as genetic algorithms and simulated annealing. The following sections describe representative work in each area. Dynamic, integer, and geometric programming methods also have been applied to this problem with limited general applicability and aye not discussed further (Kim and Mays, 1991; Schaake and Lai, 1969; Yang et al., 1975).

4.4.1 Branched Systems

To size pipes in branched networks, LP was applied using split-pipe formulations. Here, a pipe link is broken into segments of different diameters and the pipe lengths of each diameter are optimized

(Altinbilek, 1981; Calhoun, 1971; Gupta, 1969; Karmeli et al., 1968). Branched systems can thus be posed without further simplification to linear optimization models. Later, to overcome the limitation of split-pipe lengths, Walters and Lohbeck (1993) applied a genetic algorithm (GA) for the layout and design of a tree network.

The LP formulation is presented for introduction. Branched or tree systems, such as irrigation systems, have a single path of flow to each node (Fig. 4.1). Thus, by conservation of mass, the flow rate in each pipe can be computed for a defined set of nodal demands by summing the total demand downstream of the pipe. The energy loss in each pipe is a function of the pipe diameter and the selected pipe material. In addition, the energy loss and pipe cost increase linearly with the length of pipe. Thus, the LP model can be stated as follows:

$$\textit{Minimize } Z = \sum_{(i,j)\varepsilon I} \sum_{m\varepsilon M_{i,j}} c_{i,j,m} X_{i,j,m} \tag{4.6}$$

subject to

$$\sum_{m\varepsilon M_{i,j}} X_{i,j,m} = L_{i,j} \quad (i,j)\varepsilon I \tag{4.7}$$

$$H_{\min,n} \leq H_s + E_P - \sum_{(i,j)\varepsilon I_n} \sum_{m\varepsilon M_{i,j}} J_{i,j,m} X_{i,j,m} \leq H_{\max,n} = 1,\ldots, N \tag{4.8}$$

$$X_{i,j,m} \geq 0 \tag{4.9}$$

where $c_{i,j,m}$, $X_{i,j,m}$, and $J_{i,j,m}$, are the cost per unit length, length, and hydraulic gradient of pipe diameter m connecting nodes i and j, respectively. The objective function sums the cost of each pipe segment making up the total pipe length $L_{i,j}$ between nodes i and j, and sums over all pipes I. Constraint Eq. (4.7) requires that the total length of the segments of different diameters in the set of admissible M equals the distance between nodes i and j for all pipes. Equation (4.9) ensures that each pipe segment has a non-negative length.

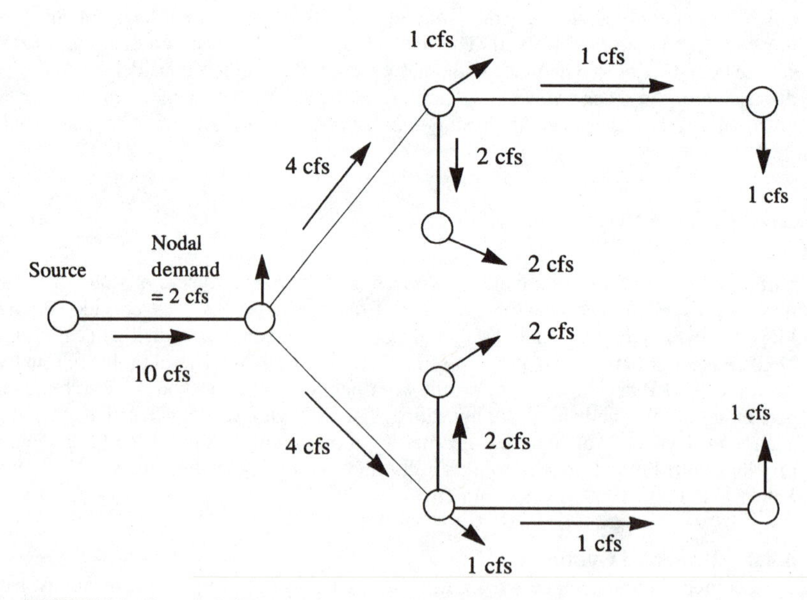

FIGURE 4.1 Branched pipe network.

As noted, the nodal conservation of mass equation defines the flow rate in each pipe. The hydraulic gradients can then be computed using the desired flow equation: for example, for the Darcy-Weisbach equation

$$J = \frac{8fQ^2}{\pi^2 g D^5} \tag{4.10}$$

where f is the pipe friction factor, Q is the flow rate (ft^3/s), and D is the pipe diameter (ft). The hydraulic equation is the headloss along the path from the source with total head H_s to each node N. This energy loss is bounded to ensure the nodal pressure head does not exceed minimum and maximum pressures H_{min} and H_{max}, respectively, in Eq. (4.8). To account for pumps, a pump-head added term E_P is included in the path equation.

In this form, the model is linear with respect to the pipe segment lengths and can be solved by standard LP methods. This single source, single load model can be extended to multiple sources and to more than one loading condition. Finally, if the cost of added pump head is linear, it can be included in the objective function (Altinbilek, 1981).

4.4.2 Looped Pipe Systems via Linearization

Some of the most successful looped-system optimization methods linearize the looped network hydraulics so the design problem can be solved quickly by LP (Alperovits and Shamir 1977; Featherstone and El-Jumaily, 1983; Fujiwara et al., 1987; Morgan and Goulter, (1985a,b; Quindry et al., 1981). Alperovits and Shamir first developed this linearization approach described as the linear programming gradient method. All these methods apply iterative processes by fixing pipe flow rates or pressure heads, optimizing the pipe sizes for the given flow or pressure distribution, updating the flow/pressure head distribution, and reoptimizing. The search for an updated flow/pressure head distribution varies. This process continues until an overall convergence criterion is met. Both discrete diameter and continuous pipe length models have been formulated. Some can handle decisions beyond pipe sizing. Most of these models are limited to considering only a single loading condition during the optimization process.

Morgan and Goulter (1985b) considered multiple demand patterns for pipe sizing. Their work is described below for a single loading condition, and the extension to multiple demands is straightforward. Unlike other linearization schemes and branched network models that determine pipe lengths directly, the decisions here are to determine the length of the pipe segment that will change the present solution's diameter m to an adjacent commercially available pipe diameter $m + 1$ or $m - 1$ The result, however, is still a split pipe solution.

$$\textit{Minimize } Z = \sum_{i,j} [(c_{i,j,m+1} - c_{i,j,m})\, X_{i,j,m+1} + (c_{i,j,m-1} - c_{i,j,m})\, X_{i,j,m-1}] \tag{4.11}$$

subject to

$$\sum_{(i,j)\varepsilon I_n} [(J_{i,j,m+1} - J_{i,j,m})\, X_{i,j,m+1} + (J_{i,j,m-1} - J_{i,j,m})\, X_{i,j,m-1}] \le H_{\min,n} - H_n \tag{4.12}$$

$$L_{i,j} \ge X_{i,j,m+1} \ge 0 \tag{4.13}$$

$$L_{i,j} \ge X_{i,j,m-1} \ge 0 \tag{4.14}$$

where $X_{i,j,m+1}$ and $X_{i,j,m-1}$ are the pipe segment replacement lengths of the adjacent commercially available pipe diameters to diameter m that make up the majority of the pipe length $L_{i,j}$ between nodes i and j. The lengths of these segments are restricted to be nonnegative and less than the entire distance (Eqs. 4.13 and 4.14).

For the present set of pipe segment lengths, the flow and nodal pressure-head distribution can be computed by the network solver and the hydraulic gradients can be determined (e.g., Eq. 4.10). Equation (4.12) represents the change in nodal pressure head from H_n toward its limiting value $H_{\min,n}$ along path I_n. If slack exists between H and $H_{\min,n}$, the next smaller-diameter pipe may replace some segments of larger pipe, reducing H and resulting in a lower objective function value. Since changing the pipe lengths alters the flow and pressure-head distributions, the new flow distribution is computed using the network solver after the LP is solved. New J's are inserted in Eq. (4.12) and another problem is solved. This process is repeated until the final solution has been determined, when no pipe segments have been changed in a linear program. In this method, the choice of paths is a concern since only pipes along the selected path can be modified (Goulter et al., 1986).

All linearization approaches noted above follow similar procedures. Although a linear program is solved at each iteration, the problem is nonlinear and a global optimal solution cannot be guaranteed. Fujiwara and Khang (1990), Eiger et al. (1994), and Loganathan et al. (1995) have developed approaches to move toward global optimal solutions by developing two-phase searches. In their first phase, Fujiwara and Khang solved a linear program for changing the headlosses (and pipe-segment lengths) for a fixed flow pattern. The flow pattern is then changed to find a local optimum using a nonlinear search until a local optimum is determined (Fujiwara et al., 1987). In the second phase for the optimal solution with fixed headlosses and pipe lengths, a concave linear optimization problem is solved to determine a new flow distribution. This new flow pattern will have a lower cost than the previous local optimum and is used as the starting point for another iteration of the first-phase problem. The process continues until no better local optimal solution is found. Loganathan et al. (1995) also used the modified LP gradient method to solve for pipe size in an inner problem while using simulated annealing to solve the inner problem of modifying the pipe flows.

Eiger et al. (1994) considered the problem to be a nonsmooth optimization and used duality theory to converge to a global optimum using a similar two-phase approach. The so-called inner or primal problem is solved given a flow distribution using LP for the optimum pipe segment lengths and an upper bound on the global optimal solution. An outer nonconvex, nonsmooth dual problem is solved that provides a lower bound on the global optimal solution. The difference between the upper and lower bound is known as a *duality gap*. A branch and bound algorithm, in which the inner and dual problems are solved numerous times, is applied to reduce the duality gap to an acceptable level by changing the flow rates in each loop. Ostfeld and Shamir (1996) later modified and applied this approach to consider reliability and water quality in the optimization process. Both of these methods have been shown to provide good results.

4.4.3 General System Design via Nonlinear Programming

Typically, NLP has also been applied to more general problems (Lansey and Mays, 1989; Ormsbee and Contractor, 1981; Shamir, 1974). In these models, a network simulation model was linked with the optimization model so that the hydraulic constraints (Eq. 4.2) could be removed from the optimization problem. The NLP models generally can consider multiple demand conditions and a broad range of design variables. Similar methods also were applied to the optimal operation problem. In addition, the methods can be extended to consider reliability measures—at a heavy computational expense, however.

The general method for solving the design problem efficiently uses an optimal control framework that links a network simulation model with a nonlinear optimizer. In this case, a set of decision variables D, known as the control variables in this formulation, is passed from the optimizer to the network solver. The simulation model solves the hydraulic equations and determines the values of the nodal pressure heads H known as the state variables. The set of hydraulic equations is therefore implicitly satisfied and the other constraints containing the state variables (Eqs. 4.2 and 4.4) can be evaluated. This information is then passed back to the optimizer. D is then modified to move to a lower-cost solution and the process continues until a stopping criterion is met.

In terms of the optimization problem, when the simulator is linked with an optimizer (Fig. 4.2) and assuming that H can be determined for all D, the optimization problem (Eq. 4.15) transformed to a reduced problem:

Objective: *Minimize Cost* $= f(H(D), D) = F(D)$ (4.15)

subject to

$\underline{H} \leq H(D) < \overline{H}$ Nodal pressure head bounds, (4.16)

$\underline{u} \leq u(D) \leq \overline{u}$ Constraints related to design/operational parameters, and (4.17)

$\underline{w} \leq w[H(D), D] \leq \overline{w}$ Constraints related to design parameters and pressure heads. (4.18)

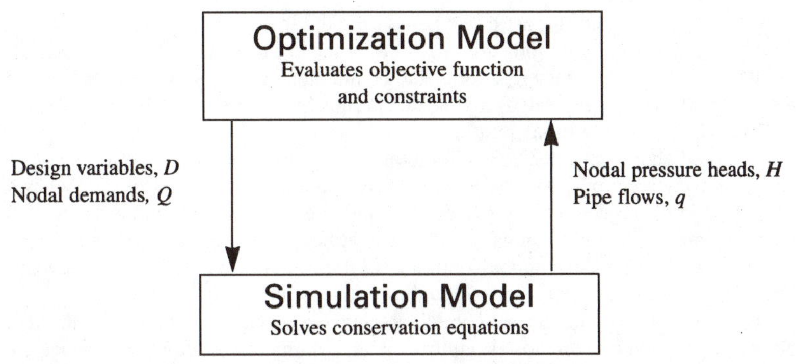

FIGURE 4.2 Optimization-simulation model link for nonlinear programming methodology.

The problem is reduced since Eq. (4.2) is no longer included because the network simulator satisfies it.

The means to modify D depend on the optimizer. Direct search methods use information from a few points to determine the next D. More efficiently, gradients of the objective function with respect to changes in the design variables can be computed using the reduced gradient equations to provide information and define a search direction. The movement in the search direction is then guided by one of several possible optimization techniques (e.g., Newton type methods, conjugate gradients).

Difficulties remain in handling constraints containing the state variables (Eq. 4.2 and 4.4). Most typically, they are included in the objective function in the form of a penalty function, such as an augmented Lagrangian term. Violations of those constraints are multiplied by some factor and increase the cost of the solution. Gradients of the objective with the penalty term are then used to change D, and the constraints are directly considered when determining changes in the design parameters.

Since a network simulator is used in this formulation, any decision that can be formulated as a continuous variable in terms of cost can be considered in this model. Also, the number of demand conditions to be evaluated is not limited, with the exception of increased computational effort. Thus, multiple demands for fire conditions or a sequence of demands describing daily operations can be considered. In addition, more complex constraints, such as reliability constraints, can be addressed (Duan et al., 1990; Su et al., 1987).

4.4.4 Stochastic Search Techniques

More recently, evolutionary optimization, genetic algorithms (Dandy et al., 1996; Savic and Walters, 1997; Simpson et al., 1994), and simulated annealing (Loganathan et al., 1995) have been used to solve the pipe-sizing problem. A logical extension is to move to other decision variables (Savic and Walters, 1997). Details on the optimization algorithms are not presented here. For reference, genetic algorithms were proposed by Holland (1975) and further developed by Goldberg (1989) and others. A good description of simulated annealing is found in Loganathan et al. (1995).

In general, when searching for an optimal solution with a stochastic search algorithm, the objective function (Eq. 4.1) is evaluated for a set of solutions (D). Based only on the objective function of the previous solutions, new decision vectors are generated. The generation process varies by approach but always contains a random element. The new solution vectors are evaluated and another set of solutions is generated using that information. The process of evaluating and generating solutions continues until a defined stopping criterion is met.

Stochastic search techniques can directly solve unconstrained optimization problems: that is, without constraints containing the control variables H. Pipe diameters are only selected within their acceptable ranges. In the evaluation step of the network design problem, the hydraulic relationships (Eq. 4.2) and the remaining constraints (Eq. 4.4) are typically also computed by executing a network solver. If a solution does not satisfy Eq. (4.2) or Eq. (4.4), the objective function can be assigned a large value so that the search will not continue near this point or the violations of the constraints can be included in a penalty term in the objective function (e.g., Savic and Walters 1997).

Because many solutions are examined during the search process, the optimal solution and a number of nearly optimal solutions are provided by these techniques. A penalty term has the advantage of allowing slightly unfeasible solutions to be maintained during the search. The decision-maker can then judge the cost of not satisfying the constraints by comparing the cost of the unfeasible solutions with feasible, nearly optimal solutions.

As noted above, stochastic search methods only use objective function values to move to better solutions compared to LP and NLP, which use gradient information. Since stochastic search methods only depend on objective function evaluations, they can easily handle discrete decisions inherent with pipe sizing. Also, in comparison to NLP, although not proved, experience has shown that those methods tend to produce global optimal solutions that cannot be guaranteed by NLP methods. Promising results have been seen in finding low-cost pipe network designs.

The advantage of not requiring gradients, however, is also the methods' shortcoming. Since only objective function information is required, a large number of simulations is necessary to move to the optimal solution. This is time consuming and limits the size of the problem that can be solved. However, this complaint applies to all the optimization techniques discussed. Faster computers and parallel processing provide hope for these general techniques.

4.5 APPLICATIONS

The previous section provided a general view of the present use of and the need for moving the theoretical field of optimal design to practical use. It also is interesting to compare the abilities of the various approaches. Table 4.1, from Savic and Walters (1997), presents solutions for a well-posed problem of adding parallel tunnels to the New York City water supply system. This system, shown in Fig. 4.3, was first examined in 1969 by Schaake and Lai and subsequently was studied by many researchers to test their approaches. The purpose is to identify pipes to install in parallel with the existing 21 pipes to ensure that adequate pressure head is supplied as the demands are increased to a defined level. The network has two loops, two dead-ends and a known constant source of energy. As seen in the table, most methods converged to solutions that had similar costs and introduced many of the same parallel pipes. Thus, most methods work reasonably well for small networks but may be affected by local optima.

A few notes of clarification regarding the table are warranted. First, two GA solutions are provided from Savic and Walters (1997). These solutions account for the range of solutions that can be developed through rounding of constant terms in the Hazen-Williams headloss equation. Differences in solutions can be partially attributed to the coefficients used in the models. Second, Gessler's solution (1982) is not truly optimization; it is an enumeration scheme. Third, the solution by Fujiwara and Khang, the only NLP method and an extension of Alperovits and Shamir (1977), is slightly unfeasible with violations of less than 0.2 ft of pressure head. This result shows the sensitivity of optimal cost to the defined nodal pressure requirements. It also points to the importance of setting those values and the potential impact of the uncertainty in forecasting desired service levels and pipe roughness coefficients. An unpublished NLP methodology from the author's work at the University of Texas also arrived at several solutions with similar optimal costs ($37.9 million and $39.3 million). The approach embedded the hydraulic equations as constraints in the NLP model.

The robustness of the solutions in terms of cost and new pipe locations is reassuring and provides confidence in the usefulness of the methods. Savic and Walters (1997) presented similar results for two other pipe networks in the literature containing 8 and 34 pipes under a single loading condition.

A more difficult problem was posed to researchers in the Battle of the Network Models (Walski et al., 1987). Expansion and rehabilitation decisions were necessary for a realistic network in Anytown, USA. Pipe costs were varied by location. Additional tanks and pumps could be added over time to meet demands at three time horizons. Energy costs were computed for daily demands, and a large set of peak and fire conditions were to be satisfied.

Solution approaches varied between modelers. Most often, a single demand condition was used to optimize the network, and other decisions were made via judgment. Energy consumption and

TABLE 4.1 Comparison of Results from Various Methods for New York City Water Supply Problem

	Schaake and Lai (1969)	Quindry et al. (1981)	Gessler (1982)	Bhave (1985)	Morgan and Goulter (1985)	Fujiwara and Khang (1990)	Murphy et al. (1993)	Loganathan et al. (1995)	Savic and Walters (Solution 1) (1997)	Savic and Walters (Solution 2) (1997)
Decisions	C	C	D	C	S	C	D	S	D	D
Pipe	D (in)	D (in)	D (in)	D (in)	D (in)	D (in)	D (in)	D (in)	D (in)	D (in)
1	52.02	0	0	0	0	0	0	0	0	0
2	49.90	0	0	0	0	0	0	0	0	0
3	63.41	0	0	0	0	0	0	0	0	0
4	55.59	0	0	0	0	0	0	0	0	0
5	57.25	0	0	0	0	0	0	0	0	0
6	59.19	0	0	0	0	0	0	0	0	0
7	59.06	0	100	0	144	73.62	0	120–132	108	0
8	54.95	0	100	0	0	0	0	0	0	0
9	0	0	0	0	0	0	0	0	0	0
10	0	0	0	0	0	0	0	0	0	0
11	116.21	119.02	0	0	0	0	0	0	0	0
12	125.25	134.39	0	0	0	0	0	0	0	0
13	126.87	132.49	0	0	0	0	0	0	0	0
14	133.07	132.87	0	0	0	0	0	0	0	0
15	126.52	131.37	0	136.43	0	0	120	0	0	144
16	19.52	19.26	100	87.37	96	99.01	84	96–108	96	84
17	91.83	91.71	100	99.23	96	98.75	96	96–108	96	96
18	72.76	72.76	80	78.17	84	78.97	84	84–96	84	84
19	72.61	72.64	60	54.40	60	83.82	72	72–84	72	72
20	0	0	0	0	0	0	0	0	0	0
21	54.82	54.97	80	81.50	84	66.59	72	72–84	72	72
Cost ($ million)	78.09	63.58	41.80	40.18	39.20	36.10	38.80	38.0	37.13	40.42

Note: Decisions are defined as D – Discrete pipe diameters, C – Continuous diameters, S – Split pipe lengths.

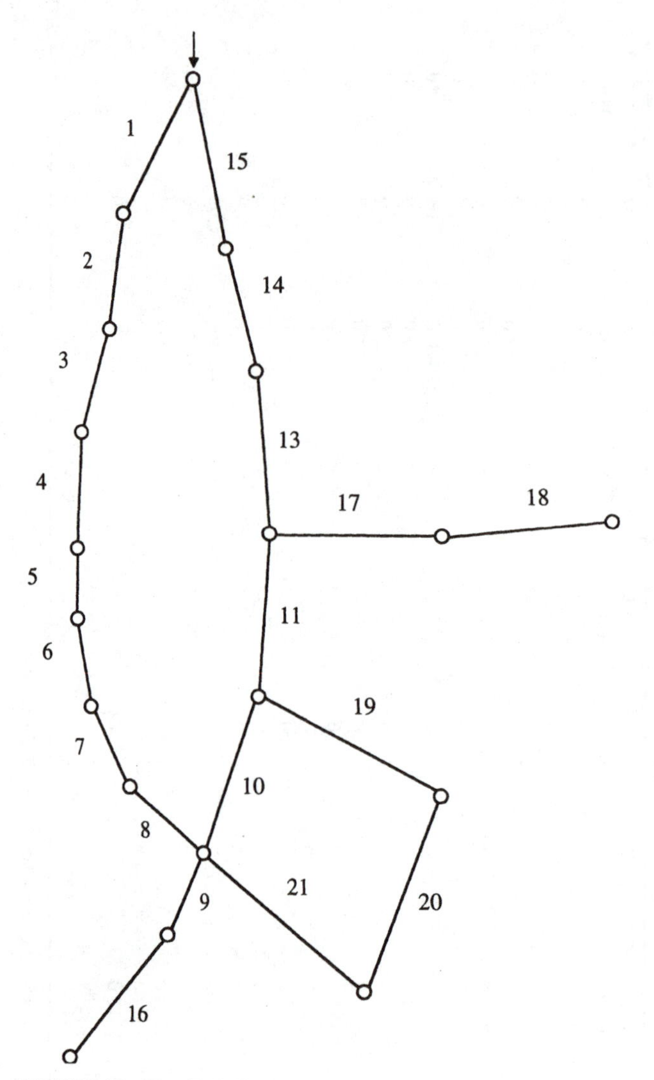

FIGURE 4.3 New York City water supply tunnels.

pumping decisions also were made using judgment. This approach is a valid use of optimization tools. The tools are intended to provide guidance with regard to reducing costs but are not used without detailed evaluation of the solutions. As noted by Goulter (1992), the end results for Anytown, however, were surprisingly similar in terms of cost (within 12 percent).

A significant difficulty at the time of the Battle was that all the computer programs but one were research oriented and required detailed knowledge of their workings. The gap between development and application resulted from a perceived lack of a market by researchers and other software developers and a lack of interest among practicing engineers. Recently, several commercially available optimization programs based on genetic algorithms have been produced. The models are more intuitive and hopefully begin a broader application of technology that has been developing over the last 30 years.

4.6 SUMMARY

Although much effort has been placed in developing methods, optimization models have seen limited usage in practice. Criticism has followed the use of analytical methods to determine pipe sizes since Camp (1939) presented an approach for economic pipeline design. The criticism of those methods parallels that of optimization methods, as expressed by Lischer (1979), Walski (1985), and Walski (1996). These criticisms focus on the uncertainty in defining design parameters, including cost coefficients, future roughness coefficients, and, most significantly, water demands. Solutions providing continuous diameters or split pipe also cause concern. Optimization models also have been correctly criticized for removing redundancy in design because they focus solely on minimizing costs.

In light of these complaints, Lischer (1979) stated that optimization and mathematical modeling should not be used without experienced engineering judgment. In other words, optimization models are another tool for engineers to determine efficient low-cost solutions for a set of design conditions. The choice of these conditions will have an impact on trial-and-error procedures and optimization models. A sensitivity analysis of the resulting solution can be completed in either case, and modifications can be made to improve it. If viewed from that perspective, continuous pipe sizes are also not a critical problem.

Reliability of least-cost optimal water networks is a critical issue. When designing by judgment, engineers explicitly or implicitly introduce reliability and redundancy. However, this reliability is typically based on their judgment rather than on a systematic and quantifiable method. Guidelines for demands and criteria for acceptable design, such as pressure heads, are lacking and, as Templeman suggested in 1982, a first step is to establish a firm set of design criteria for water distribution systems. Designing for multiple demand conditions will introduce redundancy to a network. Quantifying reliability and including it as a constraint within optimization models is a second way to force redundancy. However, no universal reliability measure or acceptable reliability levels have been developed, and inclusion of such measures comes at a high computation expense. Overcoming these weaknesses of present optimization methods will improve practicing engineers' view of the models. In turn, recognition by these engineers of how optimization models that are presently available can be used as part of a design process should lead to wider application of the tools.

4.7 REFERENCES

Alperovits, E., and U. Shamir, "Design of optimal water distribution systems," *Water Resources Research*, 13(6):885–900, 1977.

Altinbilek, H., "Optimum design of branched water distribution networks by linear programming," *Proceedings of the International Symposium on Urban Hydrology, Hydraulics and Sediment Control*, Lexington, KY, 249–254, 1981.

Bhave, P., "Optimal expansion of water distribution systems," *Journal of Environmental Engineering, ASCE*, 111(2):177–197, 1985.

Calhoun, C., "Optimization of pipe systems by linear programming," in J. E Tullis (ed.), *Control of Flow in Closed Conduits*, Colorado State University, Ft. Collins, 175–192, 1971.

Camp, T., "Economic pipe sizes for water distribution systems," *Transactions of ASCE*, 104:190–213, 1939.

Cullinane, M., K. Lansey, and L. Mays, "Optimization of availability-based design of water distribution networks," *ASCE, Journal of Hydraulic Engineering*, 118(3):420–441, 1992.

Dandy, G., A. Simpson, and L. Murphy, "An improved genetic algorithm for pipe network optimization," *Water Resources Research*, 32(2):449–458, 1996.

Duan, N., L. Mays, and K. Lansey, "Optimal reliability-based design of pumping and distribution systems," *ASCE, Journal of Hydraulic Engineering*, 116(2):249–268, 1990.

Eiger, G., U. Shamir, and A. Bent Tal, "Optimal design of water distribution networks," *Water Resources Research*, 30(9):2637–2646, 1994.

Featherstone, R., and K. El-Jumaily, "Optimal diameter selection for pipe networks," *ASCE Journal of Hydraulic Engineering*, 109(2):221–233, 1983.

Fujiwara, O., and D. Khang, "A two-phase decomposition method for optimal design of looped water distribution networks," *Water Resources Research*, 26(4):539–549, 1990.

Fujiwara, O., B. Jenchaimahakoon, and N. Edirisinghe, "A modified linear programming gradient method for optimal design of looped water distribution networks," *Water Resources Research*, 23(6):977–982, 1987.

Gessler, J., "Optimization of pipe networks," *International Symposium on Urban Hydrology, Hydraulics and Sediment Control, University of Kentucky*, Lexington, KY, 1982.

Gessler, J., "Pipe network optimization by enumeration," *Proceedings of Specialty Conference on Computer Applications in Water Resources, American Society of Civil Engineers (ASCE)*, New York, 572–581, 1985.

Goldberg, D., *Genetic Algorithms in Search, Optimization, and Machine Learning*, Addison-Wesley, Reading, MA, 1989.

Goulter, I., "Current and future use of systems analysis in water distribution network design," *Civil Engineering Systems*, 4(4):175–184, 1987.

Goulter, I., "Systems analysis in water-distribution network design: From theory to practice," *ASCE Journal of Water Resources Planning and Management*, 118(3):238–248, 1992.

Goulter, I., B. Lussier, and D. Morgan, "Implications of head loss path choice in the optimization of water distribution networks," *Water Resources Research*, 22(5):819–822, 1986.

Gupta, I., "Linear programming analysis of water supply system," *AIIE Transactions*, 1(1):56–61, 1969.

Halhal, D., G. Walters, D. Ouazar, and D. Savic, "Water network rehabilitation with structured messy genetic algorithm," *ASCE Journal of Water Resources Planning and Management*, 123(3):137–146, 1997.

Holland, J., *Adaptation in Natural and Artificial Systems*, University of Michigan Press, Ann Arbor, MI, 1975.

Jowitt, P., and C. Xu, "Optimal valve control in water-distribution networks," *ASCE Journal of Water Resources Planning and Management*, 116(4):455–472, 1990.

Karmeli, D., Y Gadish, and S. Meyers, "Design of optimal water distribution networks," *Journal of Pipeline Division, ASCE*, 94(1):1–9, 1968.

Kim, H., and L. Mays, "Optimal rehabilitation model for water distribution systems," *ASCE Journal of Water Resources Planning and Management*, 120(5):674–692, 1994.

Lansey, K. and L. Mays, "Optimal design of water distribution system design," *ASCE Journal of Hydraulic Engineering*, 115(10):1401–1418, 1989.

Lischer, V., "Discussion of 'Optimal design of water distribution networks' by Cedenese and Mele," *ASCE Journal of Hydraulic Engineering*, 105(1):113–114, 1979.

Loganathan, G., J. Greene, and T. Ahn, "Design heuristic for globally minimum cost water distribution systems," *ASCE Journal of Water Resources Planning and Management*, 121(2):182–192, 1995.

Mays, L. ed., *Reliability Analysis of Water Distribution Systems, ASCE*, New York, 1989.

Morgan, D., and I. Goulter, "Optimal urban water distribution design," *Water Resources Research*, 21(5):642–652, 1985a.

Morgan, D., and I. Goulter, "Water distribution design with multiple demands," *Proceedings of Specialty Conference on Computer Applications in Water Resources, ASCE*, New York, 582–590, 1985b.

Murphy, L., A. Simpson, and G. Dandy, *Pipe Network Optimization Using an Improved Genetic Algorithm*, Research Report No. R109, Department of Civil and Environmental Engineering, University of Adelaide, Australia, 1993.

Ormsbee, L., and D. Contractor, "Optimization of hydraulic networks," *International Symposium on Urban Hydrology, Hydraulics, and Sediment Control*, Lexington, KY, 255–261, 1981.

Ormsbee, L., and K. Lansey, "Optimal control of water supply pumping systems," *ASCE Journal of Water Resources Planning and Management*, 120(2):237–252, 1991.

Ostfeld, A., and U. Shamir, "Design of optimal reliable multi-quality water-supply systems," *ASCE Journal of Water Resources Planning and Management*, 122(5):322–333, 1996.

Quindry G., E. Brill, and J. Liebman, "Optimization of looped water distribution systems," *ASCE Journal of Environmental Engineering*, 107(4):665–679, 1981.

Reis, L., R. Porto, and F. Chaudhry, "Optimal location of control valves in pipe networks by genetic algorithm," *ASCE Journal of Water Resources Planning and Management*, 123(6):317–326, 1997.

Savic, D., and G. Walters, "Genetic algorithms for least-cost design of water distribution networks," *ASCE Journal of Water Resources Planning and Management*, 123(2):67–77, 1997.

Schaake, J., and D. Lai, Linear Programming and Dynamic Programming Applications to Water Distribution Network Design, Report 116, Department of Civil Engineering, Massachusetts Institute of Technology, Cambridge, MA, 1969.

Shamir, U., "Optimal design and operation of water distribution systems," *Water Resources Research*, 10(1):27–35, 1974.

Simpson, A., G. Dandy, and L. Murphy, "Genetic algorithms compared to other techniques for pipe optimization," *ASCE Journal of Water Resources Planning and Management*, 120(4):423–443, 1994.

Su, Y, L. Mays, N. Duan, and K. Lansey, "Reliability-based optimization for water distribution systems," *ASCE, Journal of Hydraulic Engineering*, 113(12):589–596, 1987.

Templeman, A., "Discussion of 'Optimization of looped water distribution systems,' by Quindry et al," *ASCE Journal of Environmental Engineering*, 108(3):589–596, 1982.

Wagner, J., U. Shamir, and D. Marks, "Water distribution reliability: Simulation methods," *ASCE Journal of Water Resources Planning and Management*, 114(3):276–294, 1988.

Walski, T., "State of the art pipe network optimization," *Proceedings of Specialty Conference on Computer Applications in Water Resources, ASCE*, New York, 559–568, 1985.

Walski, T, "Discussion of 'Design heuristic for globally minimum cost water-distribution systems' by Loganathan et al." *ASCE Journal of Water Resources Planning and Management*, 122(4):313–314, 1996.

Walski, T., E. Brill, J. Gessler, I. Goulter, R. Jeppson, K. Lansey, H-L. Lee, J. Liebman, L. Mays, D. Morgan, and L. Ormsbee, (1987), "Battle of the network models: Epilogue," *ASCE Journal of Water Resources Planning and Management*, 113(2):191–203, 1987.

Walters, G., "Optimal design of pipe networks: A review," *Proceedings of the 1st International Conference on Computer Methods and Water Resources in Africa, Vol. 2; Computational Hydraulics*," Computational Mechanics Publications and Springer, Verlag, Southampton, England, 21–32, 1998.

Walters, G., and T. Lohbeck, "Optimal layout of tree networks using genetic algorithms," *Engineering Optimization*, 22(1):27–48, 1993.

Yang, K., T. Liang, and I. Wu, "Design of conduit systems with diverging branches," *ASCE Journal of Hydraulic Engineering*, 101(1):167–188, 1975.

CHAPTER 5

OPTIMAL OPERATION OF WATER SYSTEMS

Fred E. Goldman
GTA Engineering, Inc.
Phoenix, Arizona

A. Burcu Altan Sakarya
Department of Civil Engineering
Middle East Technical University
Ankara, Turkey

Larry W. Mays
Department of Civil and Environmental Engineering
Arizona State University
Tempe, Arizona

5.1 INTRODUCTION

5.1.1 Background

Until 1974, government regulators and water system operators concentrated their efforts on meeting water quality *maximum contaminant levels* (MCLs) at the water source or water treatment plant. However, with the implementation of the Safe Water Drinking Act in 1974 and amendments in 1986 and 1996, water system operators are being required to meet standards throughout their distribution systems and at the points of delivery. The *Surface Water Drinking Rule* requires that a detectable level of chlorine be maintained in the system. The *Total Coliform Rule* requires that total coliform standards be met at the customer's tap. The *Lead and Copper Rule* requires testing at the point of delivery (Clark 1994b; Clark et al. 1994, 1995; Pontius, 1996).

Water quality can undergo significant changes as the water travels through the water distribution system from the point of supply and/or treatment to the point of delivery. For example, chlorine concentration will decrease with time in pipes and tanks through bulk decay and through reaction with the pipe wall. Also, chlorine mass leaves the system at demand nodes or can be added to the system at chlorine booster nodes. Mixing of water from different sources can also affect water quality (Clark et al., 1995).

Pumps are usually operated according to an operating policy, which includes the scheduling of pump operation, which can affect water quality if the system pumps draw water from sources with differing water quality. Pump operation will affect the turnover rate of storage tanks (Ormsbee et al., 1989). Chlorine dosage may differ at each pump depending on the water source quality. For example, consider a pump that operates for 6 h divided into 1-h periods. The pump can either be on or off for any period. The number of combinations is $2^6 = 64$. A pump which operates for 24 h divided into 1-h periods has $2^{24} = 16,777,216$ combinations. The 6-h example can be solved by trial and error, but the 24-h example is prohibitively large to solve by trial and error. For large combinatorial optimization problems, simulated annealing provides a manageable solution strategy. These considerations have generated a need for computer models and optimization techniques that consider water quality in the water distribution system as well as system hydraulics.

Computer models that simulate the hydraulic behavior of water distribution systems have been available for many years (Wood, 1980). More recently these models have been extended to analyze the quality of water as well as the hydraulic behavior (Grayman et al., 1996; Rossman, 1993; Rossman et al., 1993; Elton et al., 1995; Boulos et al., 1995). These models are capable of simulating the transport and fate of dissolved substances in water distribution systems. The reason behind this is the governmental regulations and consumer-oriented expectations (Rossman and Boulos, 1996). As mentioned, the passage of the Safe Drinking Water Act in 1974 and its amendments in 1986 (SDWAA) changed the manner in which water is treated and delivered in the United States. The U.S. Environmental Protection Agency (USEPA) is required to establish maximum contaminant level goals for each contaminant that may have an adverse effect on the health of persons. These goals are set to the values at which no known or expected adverse effects on health can occur, by allowing a margin of safety (Clark et al., 1987; Clark, 1994).

The objective of this chapter is to describe optimization models that can be used to determine the optimal operation times of the pumps in a water distribution system for a predefined time horizon for water quality purposes while satisfying the hydraulic constraints, the water quality constraints, and the bound constraints of the system. Two solution methodologies are discussed. The first (described in Sec. 5.2) is based upon a mathematical programming approach (Sakarya, 1998; Sakarya and Mays, 2000) that links the nonlinear optimization code GRG2 by Lasdon and Waren (1986) with the water distribution simulation code EPANET by Rossman (1994). The second approach (described in Sec. 5.3) for optimal pump operation is based upon a simulated annealing approach interfaced with the EPANET model (Goldman, 1998). The simulated annealing approach can also be used to determine the optimal operation of chlorine booster stations as described in Sec. 5.4.

5.1.2 Previous Models for Water Distribution System Optimization

Ormsbee and Lansey (1994), and more recently, Goldman et al. (2000), reviewed methods used to optimize the operation of water supply pumping systems to minimize operation costs. Ostfeld and Shamir (1993) developed a water quality optimization model that optimized pumping costs and water quality with both steady-state and dynamic conditions that was solved using GAMS/MINOS [general algebraic modeling system (Brooke et al., 1988)/mathematical in-core nonlinear optimization systems (Murtagh and Saunders, 1982)]. Ormsbee (1991) suggested that on-off operation of pumps poses a difficulty for optimization techniques that require continuous functions. Computer-based water quality models exhibit the same difficulties. Most modeling methods include numerical methods rather than continuous functions. Linear superposition was used by Boccelli et al. (1998) for the optimal scheduling of booster disinfection stations by developing system-dependent discretized impulse response coefficients using EPANET and linking with the linear programming solver (MINOS by Murtagh and Saunders, 1987).

Dynamic programming has been used for the optimal scheduling of pumps in a water distribution system by Coulbeck and Orr (1982) and Coulbeck, Brdys, and Orr (1987). The method is sensitive to the number of reservoir states and the number of pumps. In general, the increase in number of discretizations and number of state variables increases the size of the problem dramatically, which is known as the curse of dimensionality (Mays and Tung, 1992). This has restricted the application of this method to small systems. Other optimization approaches include Coulbeck et al. (1988a,b), Chase and Ormsbee (1989, 1991), Lansey and Zhong (1990), Lansey and Awumah (1994, Nitivattananon et al. (1996), Ormsbee and Reddy (1995), Ostfeld and Shamir (1993a,b), Sadowski et al. (1995), and Zessler and Shamir (1989).

The nonlinear programming approach by Brion and Mays (1991) evaluates gradients using finite differences or a mathematical approach to provide derivatives required by GRG2, which solves nonlinear optimization problems by using the generalized reduced gradient method (Lasdon et al., 1978; Mays, 1997). The method has been used for scheduling pump operations to minimize energy and improve water quality by Sakarya (1998) and Sakarya and Mays (2000), who considered three objective functions: (1) the minimization of the deviations of actual concentrations from a desired concentration, (2) minimization of total pump duration times, and (3) the minimization of total energy while meeting water quality constraints. The model by Sakarya and Mays (2000) is discussed in Sec. 5.2.

5.2 OPTIMAL PUMP OPERATION—MATHEMATICAL PROGRAMMING APPROACH

5.2.1 Model Formulation

Different objective functions can be considered to solve the optimal operation of water distribution systems for water quality purposes. In this model three objective functions are considered: case I is the minimization of deviations of the actual concentrations of a constituent from the desired concentration values; case II is the minimization of the total pump operation times; and case III is the minimization of the total energy cost. The constraints that have to be considered at all time periods for all cases are basically the hydraulic constraints; the water quality constraints; and the bound constraints on pump operation times, pressures, and tank water storage heights. In addition to the constraint set which case I considers, the bound constraints on the substance concentrations have to be satisfied for cases II and III.

Consider a water distribution system with M pipes, K junction nodes, S storage nodes (tank or reservoir), and P pumps, which are operated for T time periods. The objective function for case I is

$$\text{Min } Z_{\text{I}} = \text{minimize} \sum_{t=1}^{T} \sum_{n=1}^{N} \min\left[0, \min\left(C_{nt} - \underline{C}_{nt}, \overline{C}_{nt} - C_{nt}\right)\right]^2 \tag{5.1}$$

where N = total number of nodes (junction and storage)

C_{nt} = substance concentration at node n at time t, mass/ft^3

$\underline{C}_{nt}$, $\overline{C}_{nt}$ = lower and upper bounds, respectively, on substance concentration at node n at time t, mass/ft^3

The objective function for case II considers the minimization of the total pump durations expressed as

$$\text{Min } Z_{\text{II}} = \text{minimize} \sum_{p=1}^{P} \sum_{t=1}^{T} D_{pt} \tag{5.2}$$

where D_{pt} is the length of time pump p operates during time period t in hours.

The objective function for case III considers the minimization of the total energy cost expressed as

$$\text{Min } Z_{\text{III}} = \text{minimize} \sum_{p=1}^{P} \sum_{t=1}^{T} \frac{\text{UC}_t 0.746\, \text{PP}_{pt}}{\text{EFF}_{pt}} D_{pt} \tag{5.3}$$

where UC_t = unit energy or pumping cost during time period t, \$/kWh

PP_{pt} = power of pump p during time period t, horsepower (hp)

D_{pt} = length of time pump p operates during time period t, h

EFF_{pt} = efficiency of pump p in time period t

The efficiency of the pump and the unit energy cost are considered to be constant for all time periods.

The distribution of flow throughout the network must satisfy the conservation of mass and the conservation of energy, which are defined as the hydraulic constraints. The conservation of mass at each junction node, assuming water is an incompressible fluid, is

$$\sum_{i} (q_{ik})_t - \sum_{j} (q_{kj})_t - Q_{kt} = 0 \qquad k = 1, \ldots, K;\ t = 1, \ldots, T \tag{5.4}$$

where $(q_{ik})_t$ is the flow in pipe m connecting nodes i and k at time t in cubic feet per second and Q_{kt} is the flow consumed (+) or supplied (−) at node k at time t in cubic feet per second. The conservation of energy for each pipe m connecting nodes i and j in the set of all pipes M is,

$$h_{it} - h_{jt} = f(q_{ij})_t \qquad \forall\, i, j, \in M;\ t = 1, \ldots, T \tag{5.5}$$

where h_{it} is the hydraulic grade line elevation in feet at node i (equal to the elevation head E_i plus the pressure head H_{it}) at time t, and $f(q_{ij})_t$ is the functional relation in feet between head loss and flow in a pipe connecting nodes i and j at time t.

The total number of hydraulic constraints is $(K + M)T$, and the total number of unknowns is also $(K + M)T$, which are the discharges in M pipes and the hydraulic grade line elevations at K nodes. The pump operation problem is an extended period simulation problem. The height of water stored at a storage node for the current time period y_{st} is a function of the height of water stored from the previous time period which can be expressed as

$$y_{st} = f(y_{s,t-1}) \qquad s = 1, \ldots, S;\ t = 1, \ldots, T \tag{5.6}$$

The water quality constraint which is the conservation of mass of the substance within each pipe m connecting nodes i and j in the set of all pipes M is

$$\frac{\partial(C_{ij})_t}{\partial t} = -\frac{(q_{ij})_t}{A_{ij}} \frac{(\partial C_{ij})_t}{\partial x_{ij}} + \theta(C_{ij})_t \qquad \forall\, i, j \in M;\ t = 1, \ldots, T \tag{5.7}$$

where $(C_{ij})_t$ = concentration of substance in pipe m connecting nodes i and j as a function of distance and time, mass/ft^3

x_{ij} = distance along pipe, ft

A_{ij} = cross-sectional area of pipe connecting nodes i and j, ft^2

$\theta(C_{ij})_t$ = rate of reaction of constituent within the pipe connecting nodes i and j at time t, mass/ft^3/day

All the constraints considered up to now are equality constraints. In addition, there also exist inequality constraints, which are the bound constraints. The lower and upper bounds on pump operation time are given as

$$\Delta t_{\min} \leq D_{pt} \leq \Delta t_{\max} \qquad p = 1, \ldots, P;\ t = 1, \ldots, T \tag{5.8}$$

where D_{pt} is the length of the operation time of pump p at time t, and $\Delta t_{\min}$ and $\Delta t_{\max}$ are the lower and upper bounds on D_{pt}, respectively. $\Delta t_{\min}$ can be zero in order to simulate the pump closing, and $\Delta t_{\max}$ is equal to the length of one time period.

The nodal pressure head bounds are

$$\underline{H}_{kt} \leq H_{kt} \leq \overline{H}_{kt} \qquad k = 1, \ldots, K;\ t = 1, \ldots, T \tag{5.9}$$

where $\underline{H}_{kt}$ and $\overline{H}_{kt}$ are the lower and upper bounds, respectively, on the pressure head at node k at time t, H_{kt}. There are no universally accepted values for the lower and upper bound values. The range of 20 to 40 psi is acceptable for minimum pressure for average loading conditions but may be lowered during emergency situations such as a fire.

The bounds on the height of water storage are

$$\underline{y}_{st} \leq y_{st} \leq \overline{y}_{st} \qquad s = 1, \ldots, S;\ t = 1, \ldots, T \tag{5.10}$$

where $\underline{y}_{st}$ and $\overline{y}_{st}$ are the lower and upper bounds, respectively, on the height of water stored at node s at time t, y_{st}. These limits are due to physical limitations of the storage tank.

Cohen (1982) stated that, if there is not a requirement for the periodicity in operation of a network, then the optimization of the operation has no meaning. Hence, to achieve this, in the previous studies made on the pump operation problem, final storage bounds are tightened so that the storage in the tanks will be more or less the same as the initial states. However, a system reaches (approaches) steady state if the daily pump operation schedules repeat themselves for a certain period of time. A water distribution never actually reaches these steady-state conditions in reality. The tank water levels at the beginning and at the end of the day are equal to each other, when the steady-state conditions are reached. Hence, the constraint that forces the tank water levels at the end of the day to be more or less equal to the initial state is not considered. Instead of this, a constraint, which forces the water levels in the tanks to be within the desired limits, is considered.

Cases II and III consider the same constraint set defined by Eqs. (5.4) to (5.10) as case I, with an additional constraint for the substance concentration values to be within their desired limits. The bounds on substance concentrations are

$$\underline{C}_{nt} \leq C_{nt} \leq \overline{C}_{nt} \qquad n = 1, \ldots, N;\ t = 1, \ldots, T \tag{5.11}$$

where $\underline{C}_{nt}$ and $\overline{C}_{nt}$ are the lower and upper bounds, respectively, on substance concentration at node n at time t, C_{nt}.

The above formulation results in a large-scale nonlinear programming problem with decision variables $(q_{ij})_t$, H_{kt}, y_{st}, D_{pt}, and $(C_{ij})_t$. In the proposed solution methodology, the decision variables are partitioned into two sets; control (independent) and state (dependent) variables. The pump operation times are the control variables. The problem is formulated above as a discrete time optimal control problem. Equations (5.1) to (5.3) are the objective functions for cases I, II, and III, respectively, and the hydraulic and water quality constraints, Eqs. (5.4) to (5.7) define the simulator equations. Equation (5.8) is the control variable bound constraint. Similarly, Eqs. (5.9) to (5.11) are the state variable bound constraints.

5.2.2 Solution Methodology—Mathematical Programming Approach

The solution algorithm that is used in this study is a reduction technique, which is similar to the algorithms used for a groundwater management model by Wanakule et al. (1986), for water distribution system design by Lansey and Mays (1989), for operation of pumping stations in water distribution systems by Brion and Mays (1991), for optimal flood control operation by Unver and Mays (1990), and for optimal determination of freshwater inflows to bays and estuaries by Bao and Mays (1994a, b). In all these studies, the optimal solution of the problem is obtained by using a hydraulic simulation code together with a nonlinear optimization code.

Mathematical formulation of the pump operation problem is a large-scale nonlinear programming problem. The problem is reformulated in an optimal control framework where an optimal solution to the problem is arrived at by linking a simulation code, EPANET (Rossman, 1994), with an optimization code, GRG2 (Lasdon and Waren, 1986). The decision variables are partitioned into control variables and state variables in the formulation of the reduced problem. The control variables are determined by the optimizer and used as input to the simulator, which solves for the state variables. Hence, the state variables are obtained as implicit functions of the control variables. This results in a large reduction in the number of constraints as both the hydraulic constraints and the water quality constraints are solved by the simulator. Only the bound constraints remain to be solved by the optimizer. Figure 5.1 shows the linkage between the optimization and the simulation codes.

Improvements in the objective function of the nonlinear programming problems are obtained by changing the control variables of the reduced problem. NLP codes restrict the step size by which the control variables change so that the control variable bounds are not violated. The state variables, which are implicit functions of control variables, are not taken into consideration in the determination of step size. If the bounds of the state variables are violated, more iterations would be needed to obtain a feasible solution. The *penalty function method* is used to overcome this problem. The state variable bound constraints are included in the objective function as penalty terms. The application of this technique is also beneficial from the size point of view of the problem. Including the state bounds into the objective function results in a significant reduction in the size of the problem, such that the number of the constraints is also reduced.

There are many kinds of penalty functions that can be used to incorporate the bound constraints into the objective function. In this research, two different penalty functions, the bracket and the augmented lagrangian, are used. The bracket penalty function method (Reklaitis et al., 1983; Li and Mays, 1995) uses a very simple penalty function expressed as

$$\text{PB}_j\,(V_{j,i}, R_j) = R_j \sum_i \left[\min\,(0, V_{j,i})\right]^2 \tag{5.12}$$

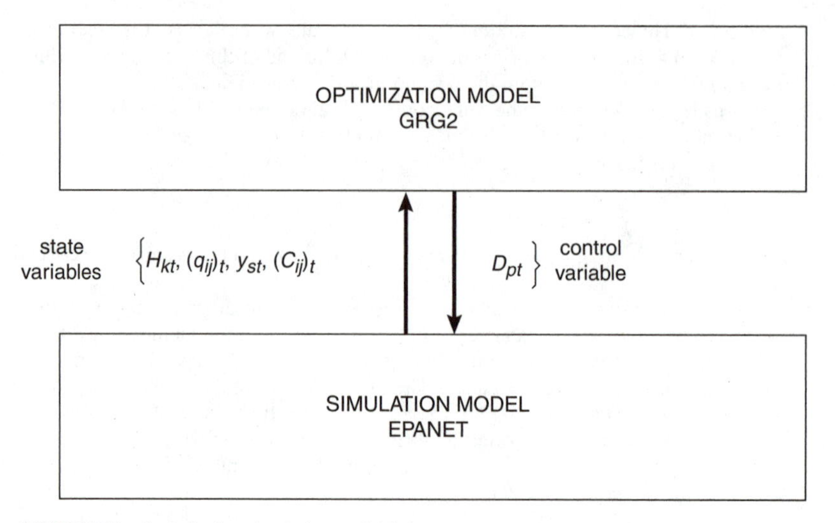

FIGURE 5.1 Optimization-simulation model linkage.

The augmented lagrangian method (Fletcher, 1975) uses the following penalty function

$$\mathrm{PA}_j(V_{j,i}, \mu_{j,i}, \sigma_{j,i}) = \frac{1}{2}\sum_i \sigma_{j,i} \min\left[0, \left(V_{j,i} - \frac{\mu_{j,i}}{\sigma_{j,i}}\right)\right]^2 - \frac{1}{2}\sum_i \frac{\mu_{j,i}^{\ 2}}{\sigma_{j,i}} \tag{5.13}$$

where the index j is the representation of H for pressure head, C for concentration, and y for water storage height bound constraints. The index i is a one-dimensional representation of the double index (k, t) for the pressure head penalty term, (n, t) for the concentration penalty term, and (s, t) for the storage bound penalty term. PB_j and PA_j define the bracket and the augmented lagrangian penalty functions for bound constraint j, respectively. $V_{j,i}$ is the violation of the bound constraint j, R_j is a penalty parameter used in the bracket penalty method; and $\sigma_{j,i}$ and $\mu_{j,i}$ are, respectively, the penalty weights and lagrangian multipliers used in the augmented lagrangian method. A detailed description of the bracket and the augmented lagrangian penalty function methods and the methods to update the penalty function parameters can be found in Sakarya (1998).

The violation of the pressure head constraint is defined as

$$V_{H,kt} = \min\left[(H_{kt} - \underline{H}_{kt}), (\overline{H}_{kt} - H_{kt})\right] \tag{5.14}$$

Similarly the violations of the substance concentration and the water storage height bound constraints can be defined.

The reduced problem for case I is

$$\mathrm{Min}\ L_1 = P_C(V_{C,nt}, F_C) + P_H(V_{H,kt}, F_H) + P_y(V_{y,st}, F_y) \tag{5.15}$$

subject to

$$0 \le D_{pt} \le \Delta t \qquad p = 1, \ldots, P;\ t = 1, \ldots, T \tag{5.16}$$

where P_C, P_H, and P_y define the bracket or the augmented penalty terms associated with the concentration, pressure head, and storage bound constraints, respectively, depending on the penalty function method used. Similarly, F_C, F_H, and F_y define the penalty function parameters which are the penalty parameters for the bracket or the penalty weights and the lagrangian multipliers for the augmented lagrangian penalty method, associated with the concentration, pressure head, and storage bound constraints, respectively.

For case II, the reduced objective function is subjected to the same constraint defined by Eq. (5.16) as case I, and has the following form

$$\text{Min } L_{\text{II}} = \sum_{p=1}^{P} \sum_{t=1}^{T} D_{pt} + P_C(V_{C,nt}, F_C) + P_H(V_{H,kt}, F_H) + P_y(V_{y,st}, F_y) \tag{5.17}$$

The reduced objective function for case III is defined as

$$\text{Min } L_{\text{III}} = \sum_{p=1}^{P} \sum_{t=1}^{T} \frac{\text{UC}_t 0.746\text{PP}_{pt}}{\text{EFF}_{pt}} D_{pt} + P_C(V_{C,nt}, F_C)$$

$$+ P_H(V_{H,kt}, F_H) + P_y(V_{y,st}, F_y) \tag{5.18}$$

subject to Eq. (5.16).

The solution of the final form of the problem is obtained by the two-step optimization procedure described in Wanakule et al. (1986), Lansey and Mays (1989), Brion and Mays (1991), and Mays (1997). The finite-difference approximations were used to calculate the derivatives of the objective function with respect to pump operation times, which are the reduced gradients that the optimization code needs to find the optimal solution. Initially, the original objective function and the penalty method are chosen depending on the case being considered. The penalty function parameters, which are lagrangian multipliers and penalty weights for the augmented lagrangian penalty method, and penalty parameters for the bracket penalty method associated with three terms in the objective function are fixed, and the inner level optimization is made. The control variable, D_{pt}, is solved by the optimizer. After the optimal values of D_{pt} for the fixed penalty function parameters are obtained, the outer-level loop is carried out to update the penalty function parameters, if necessary. A new problem is formed with these updated penalty function parameters, and the inner-level optimization is carried out again to solve for the control variables. The procedure is repeated until the penalty function parameters are not updated (i.e., overall optimum solution is found), the iteration limit of the outer loop is reached, or no improvements are achieved for some predefined consecutive iterations. The flowchart of the optimization model is given in Fig. 5.2.

During the solution procedure, GRG2 needs the values of reduced objective function and the reduced gradients while searching for the optimum solution. The calls made to calculate the values of the reduced gradients are named PARSH calls, and the calls made to calculate the reduced objective function are named FUNCTION calls. For each iteration step in the inner loop, GRG2 changes the control variables, depending on the values of the reduced objective function and the reduced gradients of the objective function.

A simplified method is used to reduce the number of EPANET calls. The main idea of this simplified method is that, if the maximum change in the control variables between consecutive iterations is small, the change that occurs in the state variables is also small. Hence, if the maximum change in the control variables between consecutive iterations is smaller than a specified limit, as the change in the values of the state variables will also be small, EPANET will not be called at that iteration to calculate the state variables; instead the previous values will be used.

5.2.3 North Marin Water District

The North Marin Water District (NMWD) serves a suburban population of approximately 53,000 people who live in the northern portion of Marin County, California. The NMWD system is divided into a series of zones. In this research, only the primary (zone I) zone of the NMWD is used, after making some changes. Figure 5.3 shows the water distribution system of the North Marin Water District zone I, used in this study. The network has 91 junction nodes, 115 links, 2 pumps, 3 storage tanks, and 2 reservoirs. Sixty of the junction nodes are demand nodes. The diameters of the pipes in the NMWD system range between 8 to 30 in. The minimum and maximum pressures at demand nodes were set at 20 and 100 psi, respectively. The desired minimum water storage heights at all tanks were 5 ft. The minimum and maximum allowable concentration limits at all demand nodes were set at 50 μg/L and 500 μg/L, respectively. The bulk and the wall rate coefficients used in the

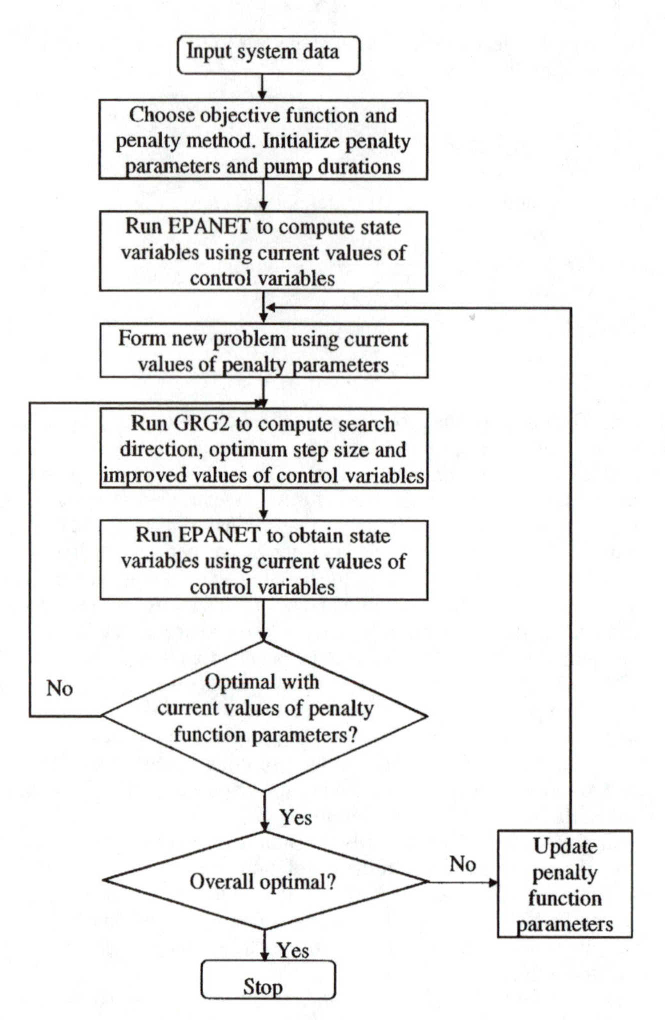

FIGURE 5.2 Flowchart of the optimization model.

simulation were −0.1/day and −1 ft/day, respectively. The simulation was conducted for a total of 12 days with 2-h time intervals where the values at the last day were used. The unit cost of energy was assumed to be constant at 0.07 $/kWh for all time periods, and the efficiency of both pumps was 0.75.

In this example, two different objective functions were used. The hypothetical example has shown that using a substance concentration violation term as a bound constraint gives better results than using it as the original objective function; therefore, case I which considers the minimization of substance concentration from its desired limits was not considered in this example.

The optimum pump operation schedules were determined using different concentration values at reservoirs 4 and 5. Table 5.1 summarizes the results obtained for cases II and III. Run number 1 in these tables shows the results obtained when the concentration values at both reservoirs are 500 μg/L. Run number 2 was based upon 200-μg/L and 400-μg/L concentrations at reservoirs 4 and 5, respectively. For run number 3, a 300-μg/L concentration was used at both reservoirs. Initially, both pumps were considered to be on for all time periods. Hence, the total pump operation time was 48 h and the total 24-h energy cost was $555.83 for all runs for both cases. The initial pump operation schedule resulted in a pressure violation of 5582.15 and there existed no violation of tank water sur-

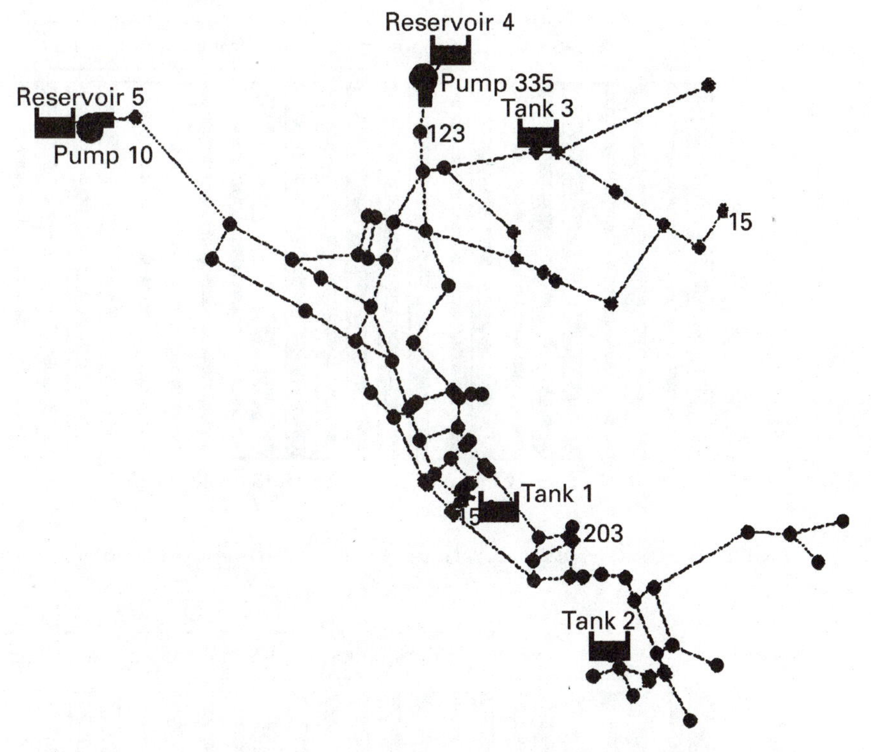

FIGURE 5.3 Water distribution system of North Marin Water District zone 1. (Rossman, 1994)

TABLE 5.1 Optimized Solutions Obtained by Using Different Solution Approaches for 500 μg/L of Concentration at Both Reservoirs

	Case II: minimize pump operation time		Case III: minimize pump operation cost	
	Mathematical programming	Simulated annealing	Mathematical programming	Simulated annealing
Original objective function	34.47	24.00	399.95	429.53
Concentration violation	0.00	0.00	0.00	0.00
Pressure violation	0.00	0.00	0.00	0.00
Tank water level violation	0.00	0.00	0.00	0.00
Operation time of pump 10, h	19.97	5.00	24.00	14.00
Operation time of pump 335, h	14.50	19.00	13.62	17.00
Total pump operation time, h	34.47	24.00	37.62	31.00
Total 24-h energy cost, $	401.06	433.63	399.95	429.53

face height level for all runs for both cases. Run numbers 1 and 3 resulted in no concentration violation, whereas a violation of 10.92 existed for run number 2 for both cases.

Figures 5.4 and 5.5 show the optimal pump operation times of pumps 10 and 335 for runs 1, 2, and 3 for case II, respectively. Similarly, the optimum pump operation times of pumps 10 and 335 obtained for all runs for case III are shown in Figs. 5.6 and 5.7. For case III the optimized solutions resulted in pump 10 operating for all time periods for all runs.

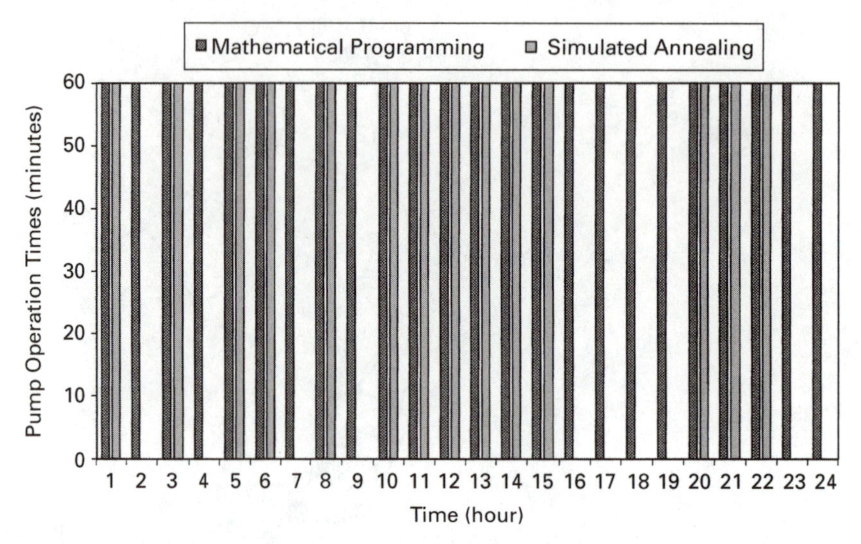

FIGURE 5.4 Optimal operation times of pump 10 used in NMWD for minimizing pump operation time.

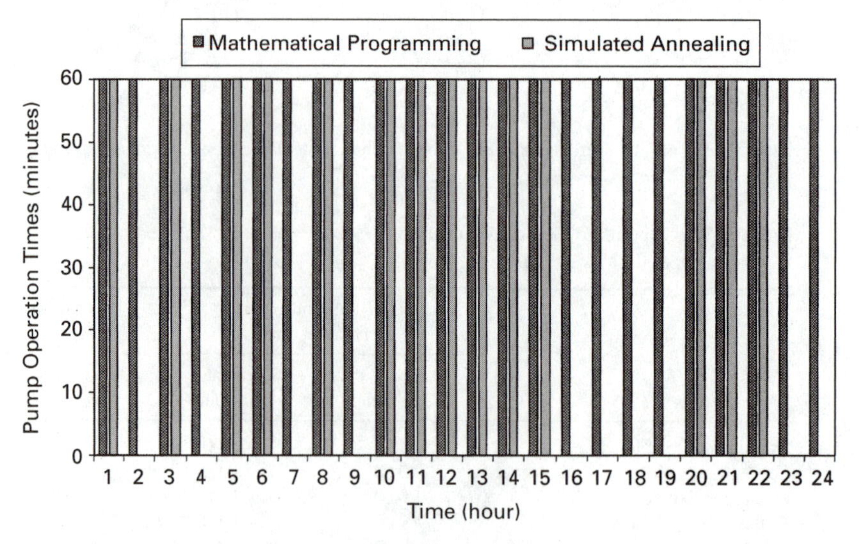

FIGURE 5.5 Optimal operation times of pump 335 used in NMWD for minimizing pump operation time.

Run number 1 gave the minimum pump operation time, which is the original objective function for case II. No concentration, pressure, or tank water level violation existed for this run. Note that this run used a 500-μg/L of concentration at both reservoirs. When the concentration values were reduced to 300-μg/L for run 3, the total pump operation time increased to 37.45 h, but still no violation of any penalty term existed. It should be noted that run 3 gave the minimum total 24-h energy cost of all three runs for case II. For run 2, concentrations of 200 μg/L and 400 μg/L were used at reservoirs 4 and 5, respectively. Concentration violations existed for this run.

For case III, when the total energy is considered as the original objective function, run 1 gave the best result. For the other two runs, there existed violations of the concentration and the total 24-h

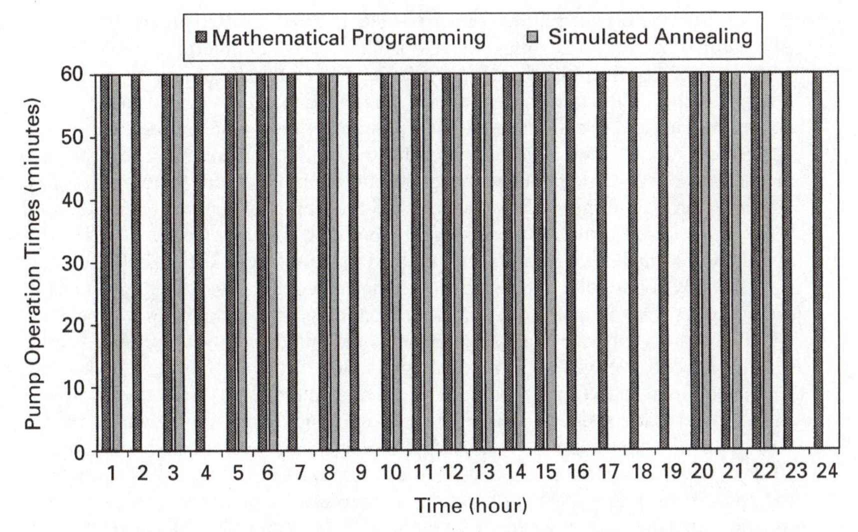

FIGURE 5.6 Optimal operation times of pump 10 used in NMWD for minimizing pump operation cost.

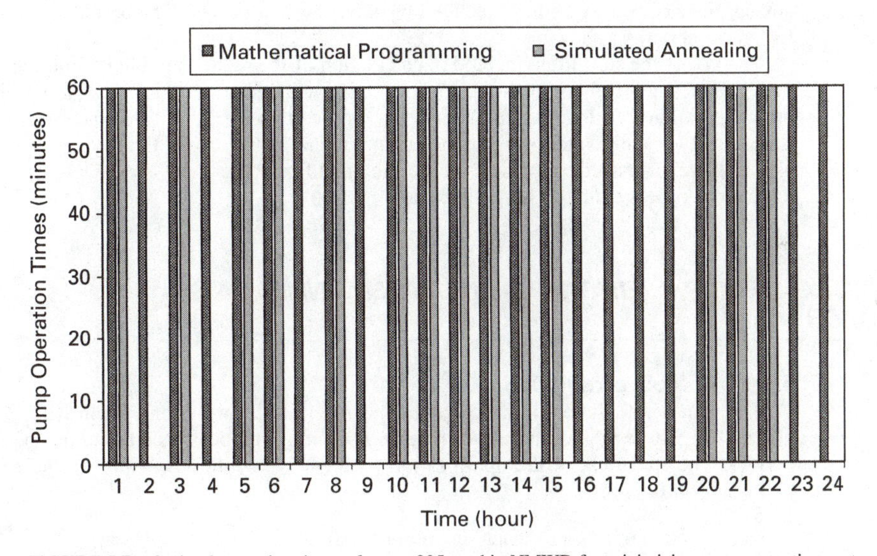

FIGURE 5.7 Optimal operation times of pump 335 used in NMWD for minimizing pump operation cost.

energy costs were higher. Even though the minimization of the total energy cost was the objective function for this case, the best result obtained from case III (run 1) had a higher value than the result obtained for case II (run 3), which considered the total pump operation as the original objective function.

At the optimized solutions, pump 335 was closed and pump 10 operated for all time periods, 24 h, except run number 1 for case II. The reason for this is that pump 335 is larger than pump 10; hence, closing pump 335 for a certain time period has more effect than closing pump 10. The reduced gradients obtained for pump 335 are greater than the reduced gradients of pump 10, which forces GRG2 to close pump 335 rather than pump 10.

5.2.4 Summary and Conclusions—Mathematical Programming Approach

The optimal control methodology developed herein, which interfaces the EPANET simulation code with the GRG2 optimization code, provides an efficient methodology to determine the optimal operation schedules of the pumps to minimize the deviations of actual substance concentrations from the desired values, and to minimize the total pump operation times or the total energy cost while satisfying the hydraulic, the water quality, and the bound constraints. The resulting optimum control problem is a large-scale nonlinear programming problem, which cannot be solved by using existing nonlinear programming codes. The discrete time optimal control approach used in this research makes it possible to solve the resulting problem. The simulation code EPANET solves the hydraulic and the water quality constraints, and the optimization code GRG2 finds the optimum pump operation times by solving the unconstrained nonlinear problem. The development of the optimal control methodology and computer program to solve the optimal pump operation schedules in a water distribution system for water quality purposes is the unique contribution of this research.

This model development is the first application of the discrete time optimal control approach to find the optimal operations of pumps considering water quality. The use of different objective functions proved that considering the minimization of the total pump operation time or the total energy cost as the objective function and using the substance concentration violation term as a bound constraint gives better results than using the minimization of the substance concentration from the desired values as the objective function. This research has shown that using two different penalty function methods, the augmented lagrangian method and the bracket penalty method, resulted in similar results. The solutions obtained depend on the values of the initial penalty function parameters, which makes finding the global optimum more difficult. Global optimum solutions cannot be guaranteed, since convexity of the objective function cannot be proven. The best local optima can be used for practical purposes.

The use of the simplified method in calculation of the reduced gradients and the objective function provides reduced computational effort and gives similar results. It was found that if the daily pump operation schedules repeat themselves for a certain period of time, the system reaches steady state where the tank water levels at the beginning and at the end of the day are equal to each other. Hence, there is no need to consider a constraint that forces the tank water levels at the end of the day to be more or less equal to the initial state.

5.3 OPTIMAL PUMP OPERATION—SIMULATED ANNEALING APPROACH

5.3.1 Model Formulation

First we can define the optimization problem for a water distribution system with M pipes, K junction nodes, S storage nodes (tanks or reservoirs), and P pumps, which are to be operated for T time periods. The statement of the optimization problem is as follows: Minimize the energy cost [Eq. (5.3)] subject to:

Conservation of mass for each junction node [Eq. (5.4)]

Conservation of energy for each pipe [Eq. (5.5)]

Storage tank continuity [Eq. (5.6)]

Water quality constraint [Eq. (5.7)]

Nodal pressure head bounds [Eq. (5.9)]

Water storage tank height bounds [Eq. (5.10)]

Bounds on substance concentrations [Eq. (5.11)]

The above formulation results in a large-scale nonlinear programming problem with decision variables $(q_{ij})_t$, H_{kt}, y_{st}, D_{pt}, and $(C_{ij})_t$. In the proposed solution methodology, the decision variables are partitioned into two sets: (1) D_{pt}, the control (independent) variable, and (2) the remaining state (dependent) variables. The operation of a pump during a time period (on or off) is the control variable. The problem is formulated above as a discrete time optimal control problem.

5.3.2 Solution Methodology—Simulated Annealing Approach

The Method of Simulated Annealing. Simulated annealing is a combinatorial optimization method that uses the Metropolis algorithm to evaluate the acceptability of alternate arrangements and slowly converges to an optimum solution. The method does not require derivatives and has the flexibility to consider many different objective functions and constraints. Simulated annealing uses concepts from statistical thermodynamics and applies them to combinatorial optimization problems. Kirkpatrick et al. (1983) explains how the Metropolis algorithm was developed to provide the simulation of a system of atoms at a high temperature that slowly cools to its ground energy state. If an atom is given a small random change, there will be a change in system energy, ΔE. If $\Delta E \leq 0$, the new configuration is accepted. If $\Delta E > 0$, then the decision to change the system configuration is treated probabilistically. The probability that the new system is accepted is calculated by:

$$P(\Delta E) = \exp - \frac{\Delta E}{k_B T} \tag{5.19}$$

A random number evenly distributed between 0 and 1 is chosen. If the number is smaller than $P(\Delta E)$, then the new configuration is accepted; otherwise it is discarded and the old configuration is used to generate the next arrangement. The Metropolis algorithm simulates the random movement of atoms in a water bath at temperature T. By using Eq. (5.19), the system becomes a Boltzmann distribution.

Entropy is a measure of the variation of energy at a given temperature. At higher temperatures there is significant energy variation which reduces dramatically as the temperatures are lowered. This implies that the annealing process should not get "stuck" since transitions out of an energy state are always possible. Also, the process is a form of "adaptive divide-and-conquer." Gross features of the eventual solution appear at the higher temperatures with fine details developing at low temperatures (Kirkpatrick et al. 1983).

The requirements for applying simulated annealing to an engineering problem are (Dougherty and Marryott, 1991).

1. A concise representation of the configuration of the decision variables
2. A scalar cost function
3. A procedure for generating rearrangements of the system
4. A control parameter (T) and an annealing schedule
5. A criteria for termination

Configuration of Decision Variables. The decision variables consist of the operational schedule of each pump during discrete time periods. For this application, the 24-h day was divided into 1-h periods. The program finds the optimum pump schedule where each pump is either on or off during each time period. A configuration is a given schedule of pump operation, which determines the 1-h periods each pump is operating. EPANET is then run using that pump configuration. The program results are used to rate the performance of the simulation. The decision on whether to retain or change the pump operation configuration is based on the Metropolis algorithm as shown in Fig. 5.8, a pseudocode description of the adaptation of the simulated annealing algorithm to pump operation optimization.

Development of Cost Function. The cost function $g(C_i)$ for a given configuration C_i is used in place of energy of a system of atoms. The temperature T can be considered a control parameter, which has the same units as the cost function. The cost function should include the cost of pumping and penalties for violations of the storage tank level, pressure, and water quality bounds. The pumping cost is calculated for each period for each pump running and is a function of the pump flow rate, head, efficiency,

```
INITIALIZE;
T = T0;
C = Ck0;
g = g(Ck0);
DO WHILE (stopping criterion not satisfied);
    DO WHILE (equilibrium not satisfied);
        PERTURB:
        Ck+1 = rearrangement of Ck;
        ·g = g(Ck+1) − g(Ck);
        IF · g · 0 THEN
            ACCEPT:
            C = Ck+1;
        ELSE
            IF random(0,1) < exp(−·g/T) THEN
                ACCEPT:
                C = Ck+1;
            ENDIF
        ENDIF
    ENDDO
    T = · T
ENDDO
```

FIGURE 5.8 Pseudocode description of the simulated annealing algorithm. T = psuedo temperature; C_k = pump operation schedule configuration; g = cost of C_k including pseudocosts for constraint violations. (After Dougherty and Marryott, 1991)

and electricity rate during that period. The violations of the pressure and water quality bounds are calculated using penalty functions. By adjusting the penalty functions, the optimization problem can be adjusted to bias one constraint over another constraint. The cost of pumping is added to the "pseudocost" penalty functions:

$$\text{COST}_{\text{pumps}} = K \sum_{ij} \frac{Q_{ij}\, \text{TDH}_{ij}\, P_j t_j}{\eta_{ij}} \tag{5.20}$$

where K = unit conversion factor
Q_{ij} = flow from pump i during period j
TDH_{ij} = operating point of total dynamic head for pump I during period j
P_j = power rate per kilowatt hour during period j
t_j = length of time that pump operates during period j
η_{ij} = wire to water efficiency of pump i during period j

Storage tanks have a minimum level to provide emergency fire flow storage. If the tanks are depleted below this level, then fire protection is compromised. A penalty term τ_1 has been developed to account for this constraint which is based on the constraint penalty terms developed by Brion (1990) and given by

$$\tau_1 = \sum_j \beta_s\,[\min\,(0, E_{sj} - E_{\min,\,s})]^2 \tag{5.21}$$

where τ_1 = penalty for violations of tank low water level bound
E_{sj} = water level for tank s during period j
$E_{\min,\,s}$ = lower bound or minimum level in tank s
β_s = penalty term for tank low water level constraint

Cohen (1982) stated that "optimization of a network over a limited horizon of, say, 24 hours has no meaning without the requirement of some periodicity in operation. A simple way to do that is to constrain the final states to be the same as the initial ones." A constraint τ_2 is developed to generate a cost if the tank levels do not return to their starting elevation.

$$\tau_2 = \sum_s \beta_{s2} \, [\min (0, 1 - |E_{s,1} - E_{s,j}|)]^2 \tag{5.22}$$

where $E_{s,1}$ = water level for tank s at beginning of simulation, during period 1
$E_{s,j}$ = water level for tank s at end of simulation, during final period j
β_{s2} = penalty term for beginning and ending tank level constraint for tank s

For a 24-h simulation the concept of returning tanks to their starting level is somewhat addressed by using a starting configuration where the pumps supply a volume of water equal to the sum of the nodal demands. Providing a total volume of pumped water equal to the total volume of the system demands will return the tanks to their original level exactly if there is only one tank. Even if pumping equals demand exactly, the tanks may not return to their original level if there are several tanks unless the tanks all start full. This is because one tank may "supply" water to another tank during the simulation and, hence, the total volume stored will be the same but shifted from one tank to another.

The Cohen condition may be unnecessarily restrictive. The second example involved modifying pump operation for a 24-h period that was repeated for 12 days to allow the water quality variations to overcome the initial conditions. It was observed that the tanks exhibited periodic behavior, adjusting themselves until the pumps were supplying a quantity equal to the demands (see Fig. 5.9). If the pumps operated for longer periods, the tanks remained closer to full and the pump heads moved to the left of the system curve reducing the flow. If the pumps operated for shorter periods, the tank levels lowered until the flows increased to meet the demands. By running the simulations over several days, the pump operation can be scheduled to optimize efficiency and perhaps reduce cost.

A water distribution system needs to deliver water to its customers at sufficient pressure to service the water system customers but at a pressure that will not damage water systems or customer's facilities. The Uniform Plumbing Code sets the normal pressure range as 15 to 80 psi (IAPMO, 1994). While a city may have a pressure range in a zone from 40 to 80 psi with a 20 psi residual during a fire flow (Malcolm Pirnie, 1996), the system operation needs to operate between two extreme pressures, $p_{\min}$ and $p_{\max}$.

The penalty functions for minimum and maximum pressure bounds at each node are

$$\tau_3 = \sum_{k,j} \gamma_1 \, [\min (0, p_{k,j} - p_{\min})]^2 \tag{5.23}$$

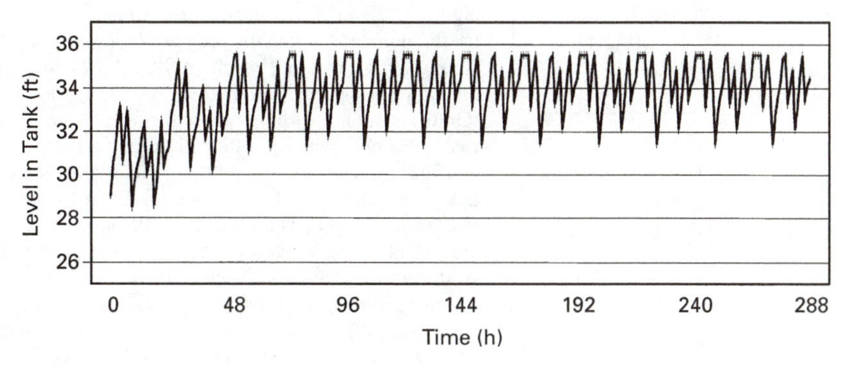

FIGURE 5.9 Periodic behavior.

$$\tau_4 = \sum_{k,j} \gamma_2 \, [\min\,(0, p_{\max} - p_{k,j})]^2 \tag{5.24}$$

where $p_{\min}$ = minimum system pressure bound
$p_{\max}$ = maximum system pressure bound
p_{kj} = pressure at node k during time period j
γ_1, γ_2 = penalty terms for minimum and maximum pressure violations

The penalty terms for minimum and maximum free chlorine concentration are

$$\tau_5 = \sum_{k,j} \gamma_3 \, [\min\,(0, c_{k,j} - c_{\min})]^n \tag{5.25}$$

$$\tau_6 = \sum_{k,j} \gamma_4 \, [\min\,(0, c_{\max} - c_{k,j})]^n \tag{5.26}$$

where $c_{\min}$ = minimum free chlorine concentration
$c_{\max}$ = maximum free chlorine concentration
c_{kj} = chlorine residual concentration at node k during time period j
β_3 = penalty term for minimum chlorine residual pressure bound violations
γ_4 = penalty term for maximum chlorine residual pressure bound violations

The value of n will usually be 2. In cases where the lower concentration bound is more important, a value of $n < 1$ will place a higher penalty on minimum free chlorine violations. There will also be a term for the amount of chlorine used. The goal will be to meet the chlorine residual bounds while using the smallest amount of chlorine. Not only will the operational cost be decreased, but the creation of total trihalomethane (TTHM) will also be reduced.

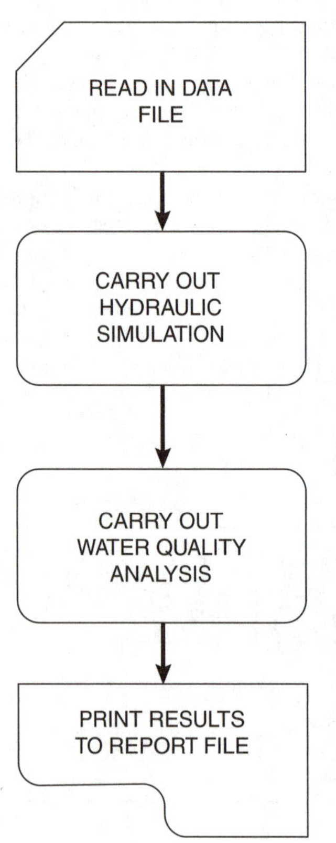

FIGURE 5.10 EPANET Flowchart.

Development of Software. The basic strategy was to use EPANET as the simulator and to develop new functions around EPANET that calculate penalties or costs, generate pump operation configurations, and evaluate the acceptance of potential configurations using the Metropolis algorithm. The EPANET program was developed for a single simulation as described by the flowchart in Fig. 5.10. In order to carry out the numerous simulations required for annealing, several modifications were required. New functions were written in the C programming language to carry out penalty calculations, generate pump operation configurations, modify the temperature according to the annealing schedule, and analyze each result using the Metropolis algorithm. Special routines were developed to collect information useful in understanding the progress of the annealing method.

Figure 5.11 is a flowchart that shows the operation of the annealing program. The chart shows that the number of loops and iterations per loop are set before the process begins. Each loop carries out a number of iterations at a given temperature. If a pump configuration is held for 50 times, it is considered to be the optimal configuration. If the program finishes normally, the final pump configuration is considered to be optimal. Frustration between constraints in a water distribution system provides many good configurations to recommend to the system operators. There may be pump configurations that provide better solutions than the final solution. A routine was added to save the 20 best configurations for examination by the user.

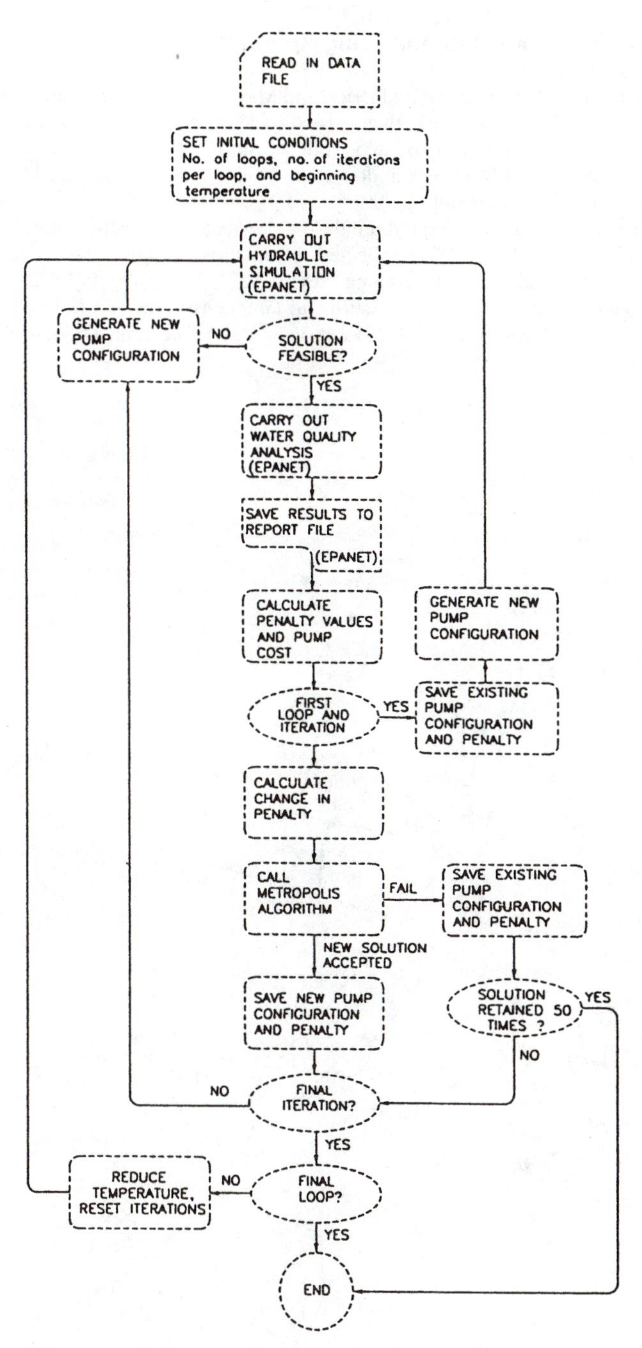

FIGURE 5.11 Simulated annealing flowchart.

5.3.3 Applications—Simulated Annealing Approach

Austin, Texas, Example. Brion (1990) and Brion and Mays (1991) applied their model to the City of Austin Northwest B Pressure Zone, which consisted of 3 pumps, 126 pipes of approximately 38.5 mi in total length, 98 junction nodes of which 5 were pressure "watch points," and one storage tank (Fig. 5.12). The 24-h simulation was broken down into twelve 2-h time periods. The system served about 31,000 residents located in about 32,000 acres.

The pipes, junctions, pumps, tanks, and diurnal flow distribution can be found in Goldman (1998) or Brion (1990). Brion and Mays (1991) collected information on the actual pumping times used on September 29, 1988, resulting in a pumping cost of $231 during the day. Using NLP with an augmented lagrangian description of the pressure and tank constraints and linking the GRG2 program with KYPIPE, the cost was reduced to $219, a savings of 5.2 percent. The resulting pump operation schedule is shown in Fig. 5.13.

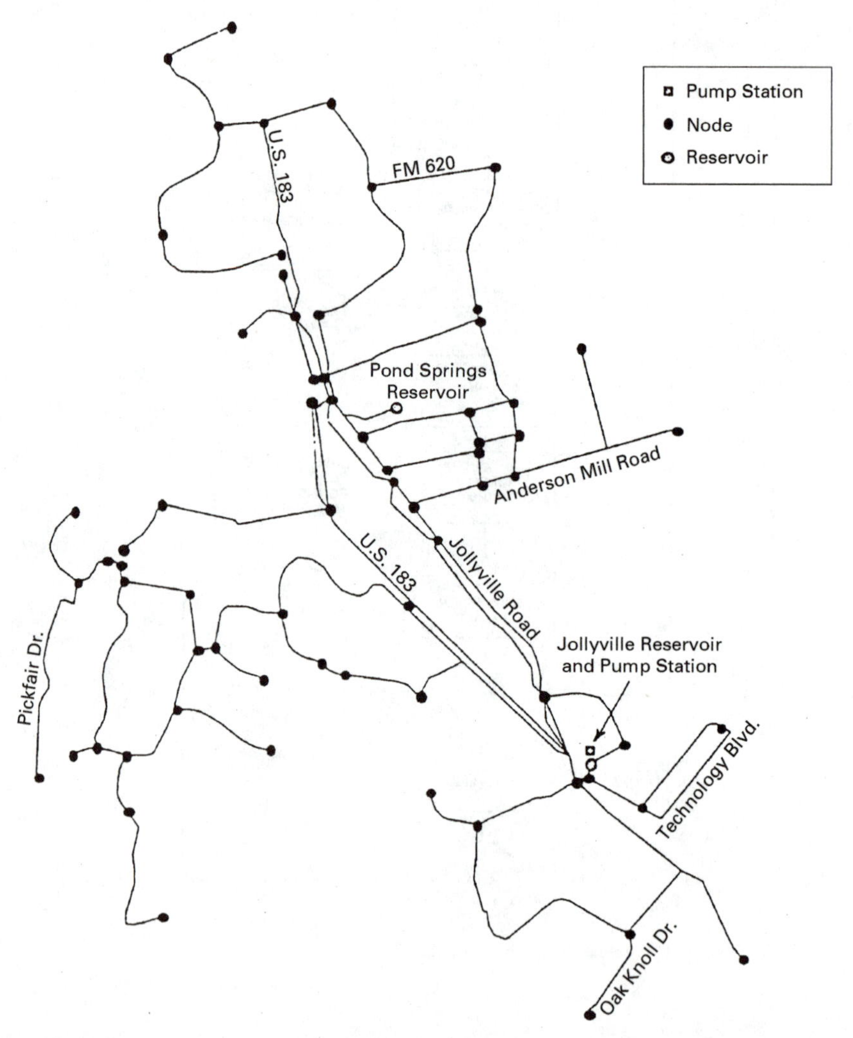

FIGURE 5.12 Water distribution system for city of Austin Northwest B Pressure Zone. (Brion and Mays, 1991)

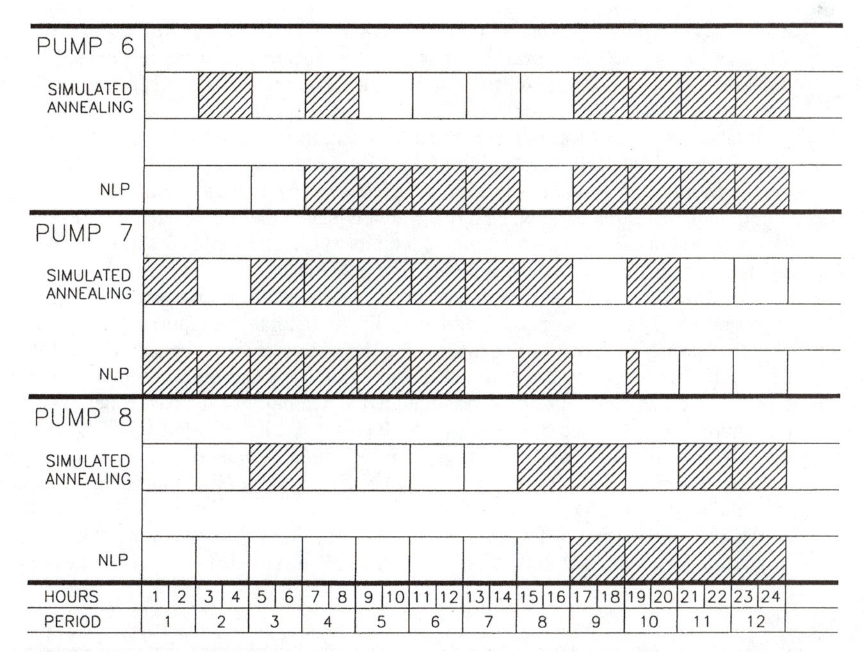

FIGURE 5.13 Austin example: simulated annealing and NLP pump operation schedules.

The series of trial pump configurations was generated by starting with the last trial configuration, and then using a random number generator to choose a pump and a period, and switching the operation. If the chosen pump was operating during the chosen period in the previous trial, it is turned off. If it was off during the chosen period, it is changed to operating. The stopping condition was having a pump schedule withstand 20 challenges from new pump schedules. Because of the high level of frustration, with many pump schedules having very similar costs, the ending condition was not met.

The best result was $221.38, which occurred in trial 67 of loop 3. This resulted in a savings of 4.1 percent. The annealing method did not isolate one solution because of the high level of frustration. The program found 173 solutions, which met the pressure and tank constraints and resulted in a cost less than $235; 50 pump schedules had a cost less than $230. The feasible solutions were reviewed to find the least costly pump operation schedule that met the pressure and tank constraints. The optimum pumping schedule found using simulated annealing and NLP is shown in Fig. 5.13.

A review of Fig. 5.13 shows the NLP solution had a 15-min sliver of operation occurring for pump 7 during the period between hours 18 and 20, which is undesirable for operating pumps and most likely would need to be adjusted by the operator. Short-period results can be expected from the NLP method, which is optimizing the duration of pump operation for a 2-h time period. Simulated annealing resulted in a minimum 2-h pumping period since it is a combinatorial optimization method where the pump has to be on or off during the entire period; thus the minimum operating time is 2 h.

The tank trajectory for the simulated annealing optimization pump schedule is shown in Fig. 5.14. The minimum height in the tank was 1 ft. A penalty term was used to return the tank level to the original tank level, within a tolerance of 1 ft.

North Marin Example. The piping system consists of 125 pipes, 102 nodes, 3 tanks, and 2 pumps as shown in Fig. 5.3. Twenty-four 1-h periods were considered with the pumping schedule repeated every 24 h. The goal of the optimization is to find the optimal operation of two pumps with differing capacity characteristics pumping from two different sources of water. Two strategies were studied. In the first, both sources had free chlorine concentrations of 0.5 mg/L, and the minimum free chlorine concentration at any demand node was taken as 0.05 mg/L. Pressures at demand nodes were to be kept between 20 and 100 psi. Global bulk decay was taken as 0.1/day, and global wall decay was taken as 1.0 ft/day. The second strategy had two different chlorine residuals. The free chlorine for the larger pump was 0.4 mg/L and for the smaller pump was 0.2 mg/L. The minimum free chlorine concentration was taken as 0.05 mg/L.

It takes many days for initial chemical concentrations to be overcome and repeatable steady-state behavior to be observed. This periodic behavior was discussed previously and was reported for water quality by Boccelli et al. (1998). Figure 5.15 shows the periodic behavior of the chlorine concentration in tank 1 of this example. To reach steady-state conditions, each iteration was run for 288 h. This resulted in each iteration taking 10 s on average. The results for the last 24 h were used to calculate pumping cost and constraint violation penalties.

The two optimization techniques, simulated annealing and the mathematical programming approach (NLP using reduced gradients with lagrangian multipliers for constraints and GRG2), obtained similar results. For the case where the source chlorine concentration for both sources was 0.5 mg/L, there were no pressure or chlorine violations and the minimum cost found by NLP was $399.95 per day versus $429.53 for simulated annealing. For the case where water qualities for the two sources were 0.2 and 0.4 mg/L, the NLP cost was $408.39 versus $407.66 for simulated annealing. There were no pressure violations, but the total chlorine violations measured in total chlorine excursions for all nodes during the 24 h was 69.76 μg/L for NLP solution as compared to 5.89 μg/L for simulated annealing solution.

There were distinct differences in computation times to arrive at a solution. The mathematical programming model was solved on a PC with a Pentium processor, while the simulated annealing

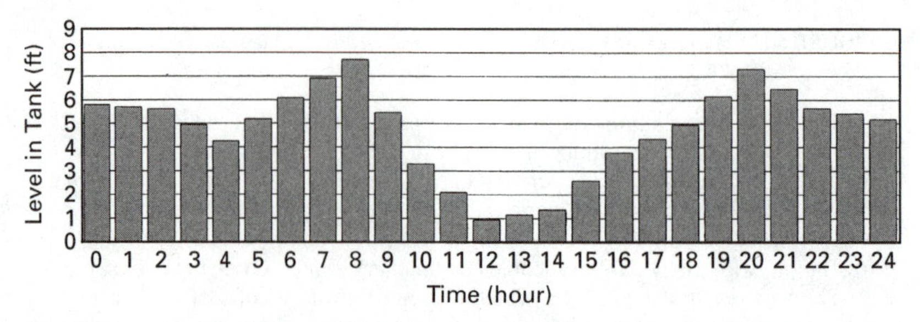

FIGURE 5.14 Austin example: simulated annealing and tank levels.

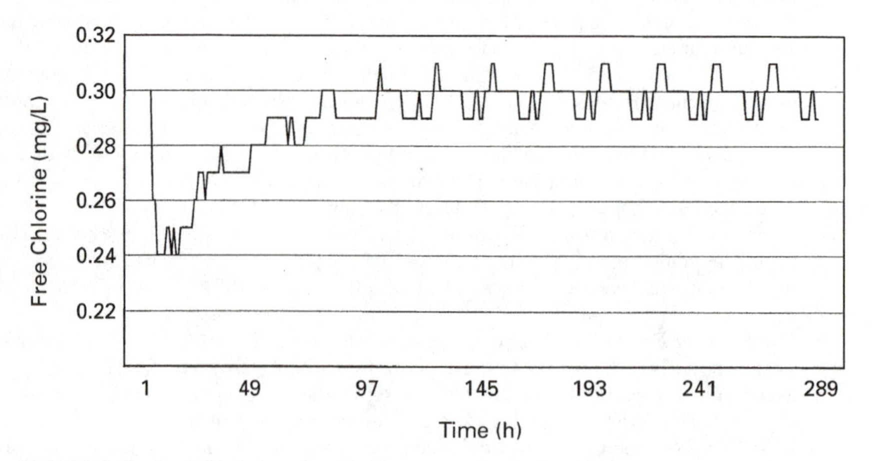

FIGURE 5.15 North Marin example: chlorine concentration, tank 1.

was solved on a UNIX computer cluster consisting of five IBM computers, two RS/6000 Model 390 interactive servers, and three RS/6000 Model 590 mathematical computational processors. The Arizona State University machines run the IBM AIX operating system, which is IBM's version of UNIX. In both instances the time to run a 12-day-long EPANET simulation was about 10 s. The NLP optimization required about one-third of the iterations as the simulated annealing optimization, although there was no attempt to streamline the simulated annealing and it turned out that the best solutions occurred before the final loop started. It is believed that even with some improvement in efficiency the simulated annealing would require at least twice the iterations that NLP requires.

Simulated annealing has been shown to be more flexible and adaptable than NLP optimization. The requirement that many parts of the distribution system such as node quality, pressures, tank levels, and pump operation must have derivatives with respect to pump operation time during each period restricts the flexibility of the method to accept changes to the system or consider a variable pump efficiency. On the other hand, simulated annealing can accept changes to the distribution system since it is concerned with cost calculation and deriving a Markov chain of operation schedules.

The NLP method's efficiency in finding optimum solutions appears to be very sensitive to the lagrangian coefficients used in the outer loop of the optimization. Also, the significant differences in the pump capacities resulted in much more rapid changes in pressures and water quality than changes in operation of the larger pump. This resulted in the preferred changes in the larger pump, which reduced its operating time until any more restrictions would have resulted in an unbalanced solution.

A fundamental *tenant* in using NLP optimization linked to a simulator such as EPANET is the implicit function theorem (Brion, 1991), which states that if D_t is the set of pump durations during periods D_t (the control variable), then $H(D_t)$, the node pressure matrix, and $E(D_t)$, the tank level matrix (the dependant or state values), exist in the neighborhood of D^*, the set of optimum pump operations. This implies that the solutions exist in the neighborhood. However, in running the simulated annealing routines many unbalanced nonfeasible solutions existed in the vicinity of the optimum solution. This becomes important if continuity is assumed in the vicinity of the NLP optimums. This problem does not reveal itself in NLP that includes water quality since the water quality portion of EPANET does not include the solving of equations that could be unbalanced.

Another important issue is global versus local optimum and the concept of frustration. A total of 36 pump operation configurations were found by simulated annealing that have total water quality violations less than 5.89 μg/L and energy costs less than $420 per day, where the water quality violations are the sum of the violations at all demand nodes for the last day of the simulation. There are many nearly equal optimal pump operation schedules that have nearly the same penalty values. This is most likely the reason that simulated annealing found optimum solutions in the next-to-the-last loop. The actual pump operation schedules developed by the two optimization techniques are shown in Fig. 5.16 for the case where both sources had chlorine concentration of 0.5 mg/L.

As previously mentioned, the NLP solution where the sources had identical 0.5-mg/L concentrations resulted in a cost of $399.95 versus $429.53 which appears to be a preferred solution. However, a review of Figs. 5.15 and 5.16 reveals a weakness with the NLP optimization. NLP is optimizing the amount of pumping time for a given period. This can result in several periods where the pumps operate only part of the time. This occurs during periods 2, 3, 8, 10, 12, and 14. In order to operate pumps more continuously, there would need to be some adjustment of the pumping schedule shown in Fig. 5.15, which could impact the ultimate cost of pumping.

5.3.4 Summary and Conclusions—Simulated Annealing Approach

Simulated annealing has been successfully linked with EPANET to optimize the operation of a water distribution system for hydraulic behavior and water quality. The programs developed have successfully considered a variety of optimization objectives. This sharply differs from other optimization techniques which require derivatives and are restricted to one objective function and may need to be modified for each data set. The annealing programming uses the global variables and dynamically allocated memory utilized by EPANET to make the computer memory requirements match the size and scope of the distribution system. The programming utilizes the data set read during the first iteration of EPANET to size the vectors required to store the data. Each implementation imports the data set and carries out the optimization.

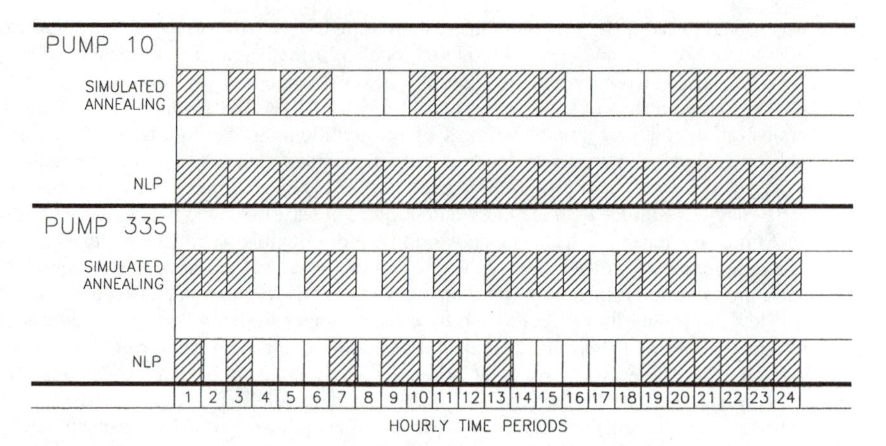

FIGURE 5.16 System 1: simulated annealing and NLP pump operation schedules. 0.5 mg/L free chlorine at both sources and minimum free chlorine at nodes as 0.05 mg/L.

The broadness of the scope of applications and the flexibility to assimilate changes to the distribution system character allow the programming to be adapted to supervisory control and data acquisition (SCADA) systems. For example, let us suppose that a fire event has depleted the storage of water in a system reservoir and it was desired to find the optimal operation of pumps to meet demands and optimally recover storage in the tanks. Simulated annealing could suggest several pump operation schedules that could meet the needs of the system.

The water quality example showed the ability of simulated annealing to suggest pumping schedules to meet water quality objectives. The objective of the example was to obtain desired chlorine residuals in a system served by two water sources using an efficient pump operation. Simulated annealing provides many pump schedules that meet the needs of the system. If conditions change, the basic data file can be modified, perhaps automatically by the SCADA system and a new schedule could be found. If a coliform test sample came back positive, the pump operation schedule can be found to restore chlorine levels. With a minor modification to the program, a variation in chlorine concentration at the source could be considered as part of the effort to find an operating schedule that restored the system to water quality standards as quickly as possible. If flushing is carried out, the only change would be increasing the demand at the node where the flushing occurred in the data set. Simulated annealing could suggest a pumping schedule that would meet the increased demands within the system pressure constraints. Again, with a minor revision, simulated annealing could suggest the optimum flushing flow or could evaluate the efficiency of locating the flushing at alternate locations, again, generating a group of pumping schedules that would meet the problem conditions.

5.4 OPTIMAL OPERATION OF CHLORINE BOOSTER STATIONS

5.4.1 Model Formulation

First we can define the optimization problem for a water distribution system with M pipes, N nodes, K junction nodes, S storage nodes (tanks or reservoirs), and B booster pumps, which are to be operated for T time periods. The statement of the optimization problem is as follows:

Minimize the amount of chlorine used.

$$\text{Minimize } Z = \sum_{b}^{K} \sum_{t} q_{bt} C_b \Delta t_t \qquad \text{g/day} \tag{5.27}$$

where q_{bt} = flow of booster chlorination station b during time period t (gal/h), C_b = concentration of chlorine being injected into the system by station b (mg/L), Δt_i = length of time period t (h), K = (3.785 L/gal)/(1000 mg/g), and $\sum \Delta t_i$ = 24 h, subject to

Conservation of mass for each junction node [Eq. (5.4)]

Conservation of energy for each pipe [Eq. (5.5)]

Water quality constraint [Eq. (5.7)]

Bounds on substance concentrations [Eq. (5.11)]

The above model formulation results in a large-scale nonlinear programming problem with decision variables $(q_{ij})_t$, H_{kt}, y_{st}, q_{pt}, and $(C_{ij})_t$. In the proposed solution methodology, the decision variables are partitioned into two sets: control (independent) and state (dependent) variables. The operation of the chlorine booster station(s) during a time period is the control variable.

The penalty terms for minimum and maximum free chlorine concentration are Eqs. (5.25) and (5.26). The goal is to meet the chlorine residual bonds using the minimum amount of chlorine. This not only decreases the operational costs, but also reduces the amount of trihalomethanes. The solution methodology used is the simulated annealing approach described in Sec. 5.3.2.

5.4.2 Application

The network shown in Fig. 5.17 (Cherry Hill–Brushy Plains, South Central Connecticut Regional Water Authority) is used to illustrate the simulated annealing approach to solve the model formulated in Section 5.4.1 to determine the optimal operation of chlorine booster stations. Boccelli et al. (1998) used this same network to illustrate their methodology. They used linear supposition and linear programming models generating impulse response coefficients using EPANET. This network has six chlorine booster stations designated as 101, 102, … , 106 on the layout in Fig. 5.17. The booster stations were modeled as sources with a fixed concentration of 2000 mg/L and varying flow. This arrangement is similar to a hypochlorinator with a tank holding a chlorine solution at 2000 m/L and a metering pump. The flow rates were in the range of a metering pump, and although they add a small amount of water to the system, they do not impact the system hydraulics.

The hydraulic conditions of the system are constant, with demands and flow rates repeated every 24 h. Simulations were for 7 days (168 h) so that the system could reach steady state, and results of the last day were polled for violations. Because the hydraulics were fixed for this example, the pump and demand characteristics were inputted as pattern multipliers. EPANET designates chlorine booster pumps as sources. Using the [SOURCES] section of the data file, the concentration of the sources is set as 2000 mg/L. The CHG_OP.C subprogram changes the source flow multipliers to match flow rates to the pump operation schedule testing during the trial. A zero flow rate is set as a 0.0000001 multiplier since the EPANET program does not accept a source with zero flow.

Two applications are explained, with the first consisting of one booster station located at the pumping station node 101. The control variable is the flow rate for each of four 6-h periods. Ten flow rates were considered by using a range of 0.0 to 2.0 gallons per minute (gpm) for each of the periods with increments of 0.2 gpm. The chlorine booster station schedule is given in Table 5.2, and the chlorine violations are given in Table 5.3. The one booster station application requires 2813 g/day of chlorine.

The second application considers the six booster station locations, having 4.74×10^{18} possible combinations. This application also set the flow rate for each booster pump for each of the four 6-h time periods. The chlorine booster station schedule is given in Table 5.4, and the chlorine violations are given in Table 5.5. The chlorine usage using six booster stations is 1504 g/day, which is considerably less (a 46 percent reduction) than for one booster station.

5.4.3 Summary and Conclusions—Chlorine Booster Station Operation

The broadness of the scope of applications and the flexibility to assimilate changes to the distribution system character allows the methodology to be applied to SCADA systems. For example, let us suppose that a fire event has depleted the storage of water in a system reservoir and it is desired to

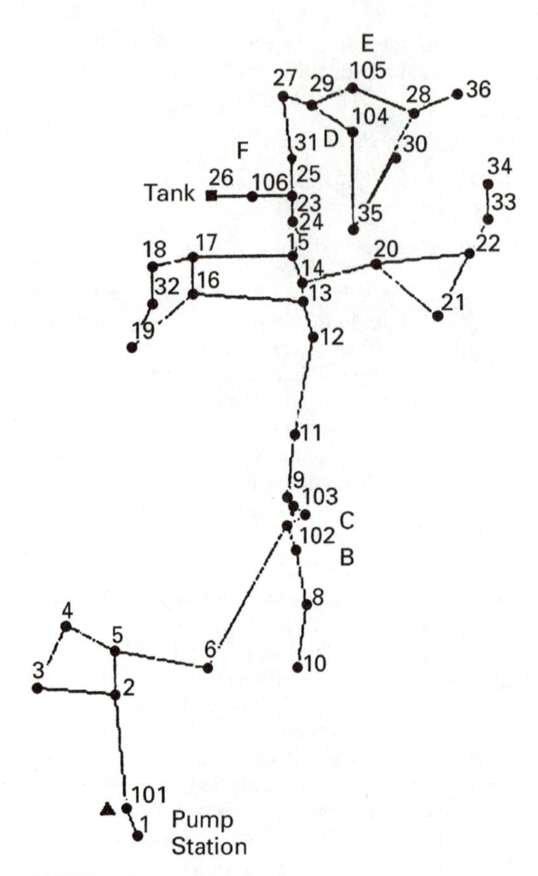

FIGURE 5.17 System layout.

TABLE 5.2 One Chlorine Booster Application: 24-h Pumping Schedule

Zoom loop	6-h 1	6-h 2	6-h 3	6-h 4	Total chlorine, g/day
1	0.6 gram (4542 mg/min)	0 (0)	0.4 gpm (3028 mg/min)	0 (0)	2725
2	0.56 gpm (4239 mg/min)	0.04 gpm (330 mg/min)	0.44 gpm (3331 mg/min)	0.008 gpm (61 mg/min)	2866
3	0.584 gpm (4421 mg/min)	0 (0)	0.44 gpm (3331 mg/min)	0.008 gpm (61mg/min)	2813

TABLE 5.3 One Chlorine Booster Application: Data Points outside of 0.2- to 4.0-mg/L Range

Node	Time period, h	Chlorine concentration, mg/L
23	146	6.1
27	149	5.3

TABLE 5.4 Six Chlorine Boosters Application: 24-h Pumping Schedule

Booster	6-h period 1	2	3	4	Chlorine used, g/day
101	0.0816 gpm (617.7 mg/min)	0.0120 gpm (90.8 mg/min)	0.0960 gpm (726.7 mg/min)	0.072 gpm (545.0 mg/min)	712
102	0.0032 gpm (24.2 mg/min)	0.0120 gpm (90.8 mg/min)	0.0032 gpm (24.2 mg.min)	0.0012 gpm (90.8 mg/min)	83
103	0.0280 gpm (212.0 mg/min)	0.0280 gpm (212.0 mg/min)	0.0432 gpm (327.0 mg/min)	0.0032 gpm (24.2 mg/min)	279
104	0.0032 gpm (24.2 mg/min)	0.0032 gpm (24.2 mg/min)	0.0032 gpm (24.2 mg/min)	0.0032 gpm (24.2 mg/min)	35
105	0.0032 gpm (24.2 mg/min)	0.0032 gpm (24.2 mg/min)	0.0032 gpm (24.2 mg/min)	0.0120 gpm (90.8 mg/min)	59
106	0.0 gpm (0.0 mg/min)	0.060 gpm (454.2 mg/min)	0.0032 gpm (24.2 mg/min)	0.060 gpm (454.2 mg/min)	336
Total					1504

TABLE 5.5 Six Chlorine Boosters: Data Points outside of 0.2- to 4.0-mg/L Range

Node	Time period, h	Chlorine concentration, mg/L
36	144	15.6
36	145	6.6
36	146	5.8
36	147	11.4
36	148	6.88
34	154	0.187
34	155	0.180
8	159	4.07
8	160	4.03
34	164	0.160
34	165	0.188
36	167	5.3

find the best operation of pumps to meet demands and recover the storage tanks. Simulated annealing could easily suggest several pump operation schedules that could meet the needs of the system. If conditions change (such as new demands or closed pipes), the basic data file can be modified, perhaps automatically by the SCADA system, and a new schedule could be computed quite easily. The method described by Boccelli et al. (1998) requires derivations of a new set of impulse functions, which can be quite cumbersome and timely.

5.5 CHALLENGES FOR THE FUTURE

Some of the challenges in model development for the future are listed here.

- In the past few years there have been reported in the literature (1) several models for the optimal design of water distribution systems, (2) several models for reliability analysis, (3) a few models for the optimal operation of water distribution systems, and (4) a few models for optimal reliability and availability based design. See Mays (1989, 2000) for a review of these models. None of these models have adequately interfaced the design, operation, and reliability aspects together.

- There is currently no universally accepted definition or measure of the reliability of water distribution systems. Many models have been reported in the literature for the reliability analysis of water distribution systems. Unfortunately the water distribution industry (engineering community) has not incorporated these methodologies into practice. There are many reasons for this, two of which are that engineers are not familiar with the methodologies and the proper software has not been developed to make the application an easy task.
- An optimization model is needed that would include objective functions such as minimizing energy costs, minimizing water age, minimizing deviations of actual concentrations of a constituent from the desired concentration values, minimization of total pump operation times, and maximizing reliability and availability.
- The model needs to be developed so that it is flexible enough that the user can select the objective function(s), but yet is interfaced with a widely used simulation model such as the EPANET model. A model that selects the optimal pump operation that minimizes water age during operation has never been attempted.
- Utilities have several strategies available to correct problems with excessive loss of disinfectants in distribution networks including: (1) switching to a more stable secondary disinfecting chemical, such as chloramines; (2) pipe replacement, flushing, and relining; (3) making operational changes to reduce the time that water spends in the system; and (4) using booster disinfection. The model suggested above needs to incorporate these strategies in the optimization framework.
- Very few of the previous modeling efforts for optimization have interfaced the design and reliability aspects in a usable form for the practicing community, and none have interfaced the operation and reliability aspects in any kind of form.
- Almost all research efforts to develop optimization models for design and operation have failed to use and rely upon the state-of-the-art hydraulic and water quality simulation models such as EPANET. Many have simplified the hydraulic and water quality processes to fit the optimization methodology selected.
- The biggest challenge of all is to convince the water distribution community that the types of models that have been suggested in the literature and expanded upon above, the operation models and the reliability and availability models, are valid tools that can be used to better design and operate their systems.

5.6 REFERENCES

Bao, Y. X., and L. W. Mays, "New Methodology for Optimization of Freshwater Inflows to Estuaries," *Journal of Water Resources Planning and Management,* ASCE, 120(2):199–217, 1994a.

Bao, Y. X., and L. W. Mays, "Optimization of Freshwater Inflows to the Lavaca-Tres Palacios, Texas, Estuary," *Journal of Water Resources Planning and Management,* ASCE, 120(2):218–236, 1994b.

Boccelli, D. L., M. E. Tryby, J. G. Uber, L. A. Rossman, M. L. Zierolf, and M. M. Polycarpou, "Optimal Scheduling of Booster Disinfection in Water Distribution Systems," *Journal of Water Resources and Planning and Management,* ASCE, 124(2):99–111, 1998.

Boulos, P. F., T. Altman, P. Jarrige, and F. Collevati, "Discrete Simulation Approach for Network-Water-Quality Models," *Journal of Water Resources Planning and Management,* ASCE, 121(1):49–60, 1995.

Brion, L. M. "Methodology for Optimal Operation of Pumping Stations in Water Distribution Systems," Ph.D. thesis, University of Texas, Austin, 1990.

Brion, L. M., and L. W. Mays, "Methodology for Optimal Operation of Pumping Stations in Water Distribution Systems," *Journal Hydraulic Engineering,* ASCE, 117(11):1551–1569, 1991.

Brook, A., D. Kendrick, and A. Meeraus, *GAMS: A User's Guide,* The Scientific Press, Redwood City, Calif., 1992.

Chase, D., and L. Ormsbee, "Optimal Pump Operation of Water Distribution Systems with Multiple Storage Tanks," *Proceedings of the American Water Works Association Comput. Specialty Conference, American Water Works Association,* 1989, pp. 205–214.

Chase, D., and L. Ormsbee, "An Alternate Formulation of Time as a Decision Variable to Facilitate Real-Time Operation of Water Supply Systems," *Proceedings of the 18th Annual Conference of the ASCE Water Resources Planning and Management Division,* ASCE, 1991, pp. 923–927.

Clark, R. M., "Development of Water Quality Models," in M. H. Chandhry and L. W. Mays (eds.), *Computer Modeling of Free-Surface and Pressurized Flows,* Kluwer Academic Publishers, Dordrecht, The Netherlands, 1994a, pp. 553–580.

Clark, R. M., "Applying Water Quality Models," in M. H. Chandhry and L. W. Mays (eds.), *Computer Modeling of Free-Surface and Pressurized Flows,* Kluwer Academic Publishers, Dordecht, The Netherlands, 1994b, pp. 581–612.

Clark, R. M., J. Q. Adams, and R. M. Miltner, "Cost and Performance Modeling for Regulatory Decision Making," *Water,* 28(3):20–27, 1987.

Clark, R. M., W. M. Grayman, J. A. Goodrich, R. A. Deininger, and K. Skov. "Measuring and Modeling Chlorine Propagation in Water Distribution Systems," *Journal of Water Resources Planning and Management,* ASCE 120(6):871–887, 1994.

Clark, R. M., L. A. Rossman, and L. J. Wymer, "Modeling Distribution System Water Quality: Regulatory Implications," *Journal of Water Resources Planning and Management,* 121(6):423–428, 1995.

Cohen, G., "Optimal Control of Water Supply Networks," Chap. 8 in S. G. Tzafestas (ed.), *Optimization and Control of Dynamic Operational Research Models,* Vol. 4, North-Holland Publishing Company, Amsterdam, 1982, pp. 251–276.

Coulbeck, B., M. Brdys, and C. H. Orr, "Exact Incorporation of Maximum Demand Charges in Optimal Scheduling of Water Distribution Systems," *IFAC 10th Triennial World Congress,* Munich, Federal Republic of Germany, 1987, pp. 359–364.

Coulbeck, B., M. Brdys, C. Orr, and J. Rance, "A Hierarchical Approach to Optimized Control of Water Distribution Systems: Part I. Decomposition," *Journal of Optimal Control Applications and Methods,* 9(1):51–61, 1988a.

Coulbeck, B., M. Brdys, C. Orr, and J. Rance, "A Hierarchical Approach to Optimized Control of Water Distribution Systems: Part II. Lower Level Algorithm," *Journal of Optimal Control Applications and Methods,* 9(2):109–126, 1988b.

Coulbeck, B., and C. H. Orr, "Optimized Pump Scheduling for Water Supply Systems," *IFAC 3rd Symposium, Control of Distributed Parameter Systems,* Toulouse, France, 1982.

Dougherty, D. E., and R. A. Marryott, "Optimal Groundwater Management. 1. Simulated Annealing," *Water Resources Research,* 27(10):2493–2508, 1991.

Elton, A., L. F. Brammer, and N. S. Tansley, "Water Quality Modeling in Distribution Networks," *Journal AWWA,* 87(7):44–52, 1995.

Fletcher, R., "An Ideal Penalty Function for Constrained Optimization," *Journal of Institute of Mathematics and Its Applications,"* 15:319–342, 1975.

Goldman, F. E., "The Application of Simulated Annealing for Optimal Operation of Water Distribution Systems," Ph.D. dissertation, Arizona State University, Tempe, 1998.

Goldman, F. E., A. B. Sakarya, L. E. Ormsbee, J. G. Uber, and L. W. Mays, "Optimization Models for Operation," Chap. 16, in L. W. Mays (ed.), *Water Distribution Systems Handbook,* McGraw-Hill, New York, 2000.

Grayman, W., R. A. Dieninger, A. Green, P. Boulos, R. W. Bowcock, and C. Godwin, "Water Quality and Mixing Models for Tanks and Reservoirs," *Journal AWWA,* 88(7):60–73, 1996.

IAPMO, *Uniform Plumbing Code,* International Association of Plumbing and Mechanical Officials, 1994.

Kirkpatrick, S., C. D. Gelatt, and M. P. Vecchi, "Optimization by Simulated Annealing," *Science,* 220(4598):671–680, 1983.

Lansey, K. E., and K. Awumah, "Optimal Pump Operations Considering Pump Switches," *Journal of Water Resources Planning and Management,* ASCE, 120(1):17–35, 1994.

Lansey, K. E., and L. W. Mays, "Optimization Model for Water Distribution System Design," *Journal of Hydraulic Engineering,* 115(10):1401–1418, 1989.

Lansey, K., and Q. Zhong, "A Methodology for Optimal Control of Pump Stations," *Water Resources Infrastructure (Proceedings of 1990 ASCE Water Resources Planning and Management Specialty Conference),* ASCE, 1990, pp. 58–61.

Lasdon, L. S., and A. D. Waren, *GRG2 User's Guide,* Department of General Business, The University of Texas at Austin, 1986.

Lasdon, L. S., A. D. Waren, A. Jain, and M. W. Ratner, "Design and Testing of a Generalized Reduced Gradient Code for Nonlinear Programming," *ACM Transactions on Mathematical Software,* 4(1):34–50, 1978.

Li, Guihua, and L. W. Mays, "Differential Dynamic Programming for Estuariane Management," *Journal of Water Resources Planning and Management,* ASCE, (121)6:455–462, 1995.

Little, K., and B. McCrodden, "Minimization of Raw Water Pumping Costs Using MILP," *Journal of Water Resources Planning and Management,* ASCE, 115(4):511–522, 1989.

Malcolm Pirnie, *Peoria Water Master Plan,* Technical Memorandum No. 3, "Distribution and Flow Model" (draft), 1996.

Mays, L. W. (ed.), *Reliability Analysis of Water Distribution Systems,* American Society of Civil Engineers, New York, 1989.

Mays, L. W., *Optimal Control of Hydrosystems,* Marcel Dekker, New York, 1997.

Mays, L. W. (ed.), *Water Distribution Systems Handbook,* McGraw-Hill, New York, 2000.

Mays, L. W., and Y. K. Tung, *Hydrosystems Engineering and Management.* McGraw-Hill, New York, 1992, pp. 140–157.

Metropolis, N., A. Rosenbluth, M. Rosenbluth, A. Teller, and E. Teller, "Equation of State Calculations by Fast Computing Machines," *Journal of Chemical Physics,* 21:1087–1092, 1953.

Miller, L. H., and A. E. Quilici, *The Joy of C,* John Wiley & Sons, New York, 1997.

Mulligan, A. E., and L. C. Brown, "Genetic Algorithms for Calibrating Water Quality Models," *Journal of Environmental Engineering,* 124(3):202–211, 1998.

Murtagh, B. A., and M. A. Saunders, "A Projected Lagrangian Algorithm and Its Implementation for Sparse Nonlinear Constraints," *Mathematical Programming Study,* 16:84–117, 1982.

Murtagh, B. A., and M. A. Saunders, *MINOS 5.1 User's Guide,* Stanford Optimization Lab., Stanford University, Stanford, Calif., 1987.

Nitivattananon, V., E. C. Sadowski, and R. G. Quimpo, "Optimization of Water Supply System Operation," *Journal of Water Resources Planning and Management,* ASCE, 122(5):374–384, 1996.

Ormsbee, L. E. (ed.), "Energy Efficient Operation of Water Distribution Systems," *A Report by the ASCE Task Committee on the Optimal Operation of Water Distribution Systems,* 1991.

Ormsbee, L. E. and K. E. Lansey, "Optimal Control of Water Supply Pumping Systems," *Journal of Water Resources Planning and Management,* 120(2):237–252, 1994.

Ormsbee, L. E., and S. L. Reddy, "Nonlinear Heuristic for Pump Operations," *Journal of Water Resources Planning and Management,* 121(4):302–309, 1995.

Ormsbee, L. E., T. M. Walski, D. V. Chase, and W. W. Sharp, "Methodology for Improving Pump Operation Efficiency," *Journal of Water Resources Planning and Management,* 115(2):148–164, 1989.

Ostfeld, A., and U. Shamir, "Optimal Operation of Multiquality Networks. I: Steady-State Conditions," *Journal of Water Resources Planning and Management,* 119(6):645–662, 1993a.

Ostfeld, A., and U. Shamir, "Optimal Operation of Multiquality Networks. II: Unsteady Conditions," *Journal of Water Resources Planning and Management,* 119(6):663–684, 1993b.

Pontius, F. W., "Overview of the Safe Drinking Water Act Amendments of 1996," *Journal AWWA,* 88(10):22–33, 1996.

Reklaitis, G. V., A. Ravidran, and K. M. Ragsdell, *Engineering Optimization: Methods and Applications,* Wiley-Interscience, New York, 1983.

Rossman, L. A., *EPANET Users Manual,* Risk Reduction Engineering Laboratory, Office of Research and Development, U.S. Environmental Protection Agency, Cincinnati, Ohio, 1993.

Rossman, L. A., *EPANET Users Guide,* Drinking Water Research Division, Risk Reduction Engineering Laboratory, Office of Research and Development, U.S. Environmental Protection Agency, Cincinnati, 1994.

Rossman, L. A., and P. F. Boulos, "Numerical Methods for Modeling Water Quality in Distribution Systems: A Comparison," *Journal of Water Resources Planning and Management,* 122(2):137–146, 1996.

Rossman, L. A., P. F. Boulos, and T. Altman, "The Discrete Volume Element Method for Network Water Quality Models," *Journal of Water Resources Planning and Management,* 119(5):505–517, 1993.

Rossman, L. A., R. W. Clark, and W. M. Grayman, "Modeling Chlorine Residuals in Drinking-Water Distribution Systems," *Journal of Environmental Engineering,* 120(4):803–820, 1994.

Sadowski, E., V. Nitivattananon, and R. E. Quimpo, "Computer-Generated Optimal Pumping Schedule," *Journal AWWA,* 87(7): 53–63, 1995.

Sakarya, A. B., "Optimal Operation of Water Distribution Systems for Water Quality Purposes," Ph.D. dissertation, Arizona State University, Tempe, 1998.

Sakarya, A. B., and L. W. Mays, "Optimal Operation of Water Distribution Pumps Considering Water Quality," *Journal of Water Resources Planning and Management,* 126(4):210–220, 2000.

Unver, O. L., and L. W. Mays, "Model for Real-Time Optimal Flood Control Operation of a Reservoir System," *Water Resources Management,* Vol. 4, Kluwer Academic Publishers, Dordrecht, The Netherlands, 1990, pp. 21–46.

Vasconcelos, J. J., P. F. Boulos, W. M. Grayman, L. Kiene, O. Wable, P. Biswas, A. Bhari, L. A. Rossman, R. M. Clark, and J. A. Goodrich, *Characterization and Modeling of Chlorine Decay in Distribution Systems,* AWWA Research Foundation and American Water Works Association, Denver, 1996, pp. 258–271.

Wanakule, N., L. W. Mays, and L. S. Lasdon, "Optimal Management of Large-Scale Aquifers: Methodology and Applications," *Water Resources Research,* 22(4):447–465, 1986.

Wood, D. J., *User's Manual—Computer Analysis of Flow in Pipe Networks Including Extended Period Simulations,* University of Kentucky, 1980.

Zessler, U., and U. Shamir, "Optimal Operation of Water Distribution Systems," *Journal of Water Resources Planning and Management,* 115(6):735–752, 1989.

CHAPTER 6

RELIABILITY AND AVAILABILITY ANALYSIS OF WATER DISTRIBUTION SYSTEMS

Kevin Lansey
Department of Civil Engineering and Engineering Mechanics
University of Arizona
Tucson, Arizona

Larry W. Mays
Department of Civil and Environmental Engineering
Arizona State University
Tempe, Arizona

Y. K. Tung
Department of Civil Engineering
Hong Kong University of Science and Technology
Clear Water Bay, Kowloon, Hong Kong

6.1 FAILURE MODES FOR WATER DISTRIBUTION SYSTEMS

To ensure delivery of finished water to the user, the water distribution system must be designed to accommodate a range of expected emergency loading conditions. These emergency conditions may generally be classified into three groups: broken pipes, fire demands, and pump and power outages. Each of these conditions must be examined with an emphasis on describing its impact on the system, developing relevant measures of system performance, and designing into the system the capacity required to handle emergency conditions with an acceptable measure of reliability. *Reliability* is usually defined as the probability that a system performs its mission within specified limits for a given period of time in a specified environment.

Reviews of the literature (see Mays, 1989, 1992, 2000) reveal that there is currently no universally accepted definition or measure of the reliability of water distribution systems. For layer systems with many interactive subsystems, it is extremely difficult to analytically compute the mathematical reliability. Accurate calculation of the mathematical reliability of water distribution systems requires knowledge of the precise reliability of the basic subsystems or components of the water distribution system and the impact on mission accomplishment caused by the set of all possible subsystem (component) failures.

Many researchers, municipal engineers, urban planners, institutes, government agencies, and others have discussed the need to develop explicit measures and methodologies to evaluate water distribution system reliability and performance under emergency loading conditions. Some researchers have proposed candidate approaches using concepts of reliability factors, economic loss functions, forced redundancy in the designs, and so on. All of these approaches have limitations in problem formulation and/or solution technique. Some investigators discuss the need to explicitly incorporate measures of reliability into optimization models to predict system operation under emergency loading conditions. At present, however, no "optimization-reliability" evaluation or design technique with general application has been developed.

The reliable delivery of water to the user requires that the water distribution system be designed to handle a range of expected emergency loading conditions. These emergency conditions can be classified:

1. Fire demands
2. Broken links
3. Pump failure
4. Power outages
5. Control valve failure
6. Insufficient storage capability

A general application methodology that considers the minimum cost and reliability aspects must consider each of these emergency conditions. These conditions should be examined within the methodologies to

1. Describe their importance to the system
2. Develop relevant measures of system performance
3. Design into the system the capacity for the emergency loading conditions with an acceptable measure of reliability

6.1.1 Distribution Repair Definitions

The American Water Works Association Research Foundation (AWWARF) in conjunction with the U.S. Environmental Protection Agency developed a guidance manual entitled *Water Main Evaluation for Rehabilitation/Replacement* (O'Day et al., 1986). Portions of that document will be cited throughout this section. One aspect of the AWWARF manual was the clarification of terminology regarding leak and break reporting.

The lack of consistency between utilities defining leaks and breaks presents difficulties when trying to compare distribution system conditions. The general term *leak* can include both structural failures as well as water loss, which occurs from improperly sealed joints, defective service taps, and corrosion holes. *Breaks,* on the other hand, imply a structural failure of the water main, which results in either a crack or a complete severance of a water main. Breaks cause leakage and water loss; therefore, it can be argued that breaks are a special form of leaks.

Distribution system repairs can be classified by the component which was repaired (main, service, hydrant), the nature of the leak (joint leak, break), and the type of repair. Distribution repair records should identify the specific component repaired. Distribution system components include (O'Day et al., 1986):

- Main: barrel, joint, plug
- Hydrant: branch line, valve, cap barrel
- Line valve: bolts, bonnet, stem
- Service: tap, pipe, curb stop

A set of definitions was developed to help standardize the reporting and logging of distribution repairs (O'Day et al., 1986):

leak repair All actions taken to repair leaks in water mains, line valves, hydrant branches, and service lines are leak repairs.

main leak A water main leak includes all problems that lead to leakage of water from the main. These include joint leaks, holes, circumferential breaks, longitudinal breaks, defective tapes, and split bells. Hydrant service line and valve-related leaks are not considered main leaks.

joint leak Joint leaks represent a loss of water from the joint between adjacent main sections. This is not a structural problem, rather it is a separation of the main sections caused by expansion and contraction, settlement, movement, or movement of joint materials because of pressure or pipe deflection.

main breaks Main breaks represent the structural failure of the barrel or bell of the pipe due to excessive loads, undermining of bedding, contact with other structures, corrosion, or a combination of these conditions. Main break types include (1) circumferential, (2) longitudinal, and (3) split bells, including bell failures from sulphur compound joint materials.

hydrant leak repairs All actions taken to repair hydrant branch lines, hydrant valves, and hydrant barrels are hydrant repairs.

service leak repairs All actions taken to repair leaking taps, corporation stops, service pipes, and curb stops are service leak repairs.

valve leak repairs All actions taken to repair leaks in valve flanges, bonnet, or body are valve leak repairs.

6.1.2 Failure Modes

Although water distribution networks have many features in common with other types of networks (Templeman and Yates, 1984), there are a number of significant differences between them when they are examined from a failure-based point of view. The differences arise primarily as a result of the ways in which water distribution networks can "fail." In its most basic sense, *failure* of water distribution networks can be defined as either the pressure and/or flow falling below specified values at one or more nodes within the network.

Under such a definition, there are two major modes of failure of water distri-bution networks: (1) *performance failure,* i.e., demand on the system being greater than the design value; and (2) *component (mechanical) failure,* which includes the failure of individual network components, e.g., pipes, pumps, and valves. Both these modes of failure have probabilistic bases which must be incorporated into any reliability analysis of networks.

Performance Failure. In the first case, the network fails in a traditional sense of the load exceeding the design capacity. However, it is not necessarily catastrophic, as might be the case of failure of a structural network, e.g., a truss. Rather, either the pressure at one or more nodes may drop below the required minimum and/or the amount of flow available at the node(s) may fall below the required level. The implications of such a failure are difficult to define as they may constitute only a decreased level of services to domestic households over a relatively brief period of time, e.g., 6 h. On the other hand, the failure of the water supply when there is a major fire could result in increased monetary damages and possible loss of life.

The probabilistic aspect of failure in this first case arises from the probability distribution of the loading condition under which the distribution network is expected to operate. In most cases, the probability distributions of the demands, which the network must fulfill, are reasonably well known. The choice of a design value for the demands on the network, for example, a 10-year, 6-h maximum flow, is not a deterministic decision and implies acceptance of a probability of the network not being able to perform to specifications. What has been missing in network analysis to date is an explicit recognition of the probabilistic processes.

The implications of failure as a result of hydraulic loads being greater than design loads are similar to those resulting from component failure; flows and/or supply pressures can fall below the desired levels. The means of improving the reliability of water distribution as measured by the probability of either failure occurring are also similar. Obviously, larger component sizes will provide greater overall flow capacity in the network and therefore reduce the probability that actual demands exceed the design values. In addition, previous work (e.g., Kettler and Goulter, 1985; Mays, 1989) has shown that pipe breakage rates in existing distribution networks are strongly correlated to diameter. Larger-diameter pipes break less frequently than smaller-diameter pipes. Thus, selection of larger-diameter pipes can reduce the probability of component failure while simultaneously improving the flow-exceedance-based measure of reliability.

It is important to note, however, that good performance according to a reliability measure based on one mode of failure does not necessarily mean good performance according to the other. Consider, for example, a large pipe near the source of supply for a particular network. Since this pipe is near the source, a larger percentage of the flow passing to more distant parts of the network will flow through it and, therefore, a large-diameter, large-capacity pipe will be required. This large-diameter pipe will have a low frequency of breakage and would therefore indicate good reliability according to component failure probabilities. The probability that the flow that same pipe is required to carry will be greater than its design capacity may, however, be quite large. If this is the case, its *flow-exceedance* reliability is high. Improvements to both measures of reliability may still be achieved through the selection of larger-diameter pipes.

The improvements to network reliability that are achieved through specification of large pipes can also be achieved through selection of larger capacities for the other important components of a network. By the provision of more pumps, pump station failure is reduced. The additional pumps also provide additional capacity for the pump station thereby reducing the probability that demands on the station will exceed the design capacity. A similar argument can be made for facilities on the network.

Component (Mechanical) Failure. The probability of component failure can be derived from historical failure records and modeled using an appropriate probability distribution (see Sec. 6.2 for time-to-failure analysis). Goulter and Coals (1986) proposed a Poisson probability distribution with parameter λ breaks per kilometer per year (breaks/km/yr) for pipe breakage. The probability of demands being greater than design values can also be derived from historical data and modeled appropriately. The most difficult problem, however, is that even knowing the appropriate probability of the two modes of failure it is difficult to define reliability explicitly. Reduction in either the probability of component failure or the probability of demands being greater than design capacity obviously improves system reliability. The question is, by how much? If a pipe (or any other component) fails, what is the effect on the remainder of the network? Obviously, the more critical the pipe in question to the network, the more widespread and serious is the effect. The actual quantification of the effect is very difficult to assess without extensive simulation of network operation. Similarly, the networkwide implication of a particular demand being greater than the design value for that area is difficult to assess without considerable effort. Furthermore, if network reliability, as defined from measures based on either mode of failure is unsatisfactory, What is the best way to improve it?

There is currently no universally accepted definition or measure of the reliability of water distribution systems. This discussion is a general definition of terms for the reviewer not familiar with the reliability and availability aspects. *Reliability* is usually defined as the probability that a system performs its mission within specified limits for a given period of time in a specified environment. For a large system, with many interactive subsystems (such as a water distribution system), it is extremely difficult to analytically compute a mathematical reliability. Accurate calculation of a mathematical reliability requires knowledge of the precise reliability of the basic subsystems or components and the impact on mission accomplishment caused by the set of all possible subsystem (component) failures. See Cullinane et al. (1992) and Mays (1989, 2000) for complete discussions of the aspects of reliability and availability analysis.

The traditional engineering definition of reliability is the ability of a system to meet demand. Some authors (see Cullinane et al., 1992) prefer to label this concept "availability," while reserving the term reliability for the length of time that a system can be expected to perform without failure. The term reliability in its general sense is defined as any measure of the system's ability to satisfy the requirements placed upon it.

6.1.3 Reliability Indices

A number of different measures of *reliability indices* have been proposed (see Cullinane et al., 1992), but no single one is universally accepted. Each can be useful in reliability analyses, depending on the purpose of the analysis.

Reliability is often defined as the ability of a system to meet demand under a defined set of contingencies. A system may be judged reliable if it can, for example, satisfy demand during a period of drought or despite the failure of a key piece of equipment. Early design criteria were based on such rules of thumb instead of quantitative indices, such as those we will soon discuss. Contingency criteria suffer from arbitrariness in choice of contingency and inconsistency, because different designs able to withstand the same contingency can have different probabilities of failure.

Availability is defined as the probability at a given moment that the system will be found in a state such that demand does not exceed available supply or capacity or that operating conditions are not otherwise unsatisfactory (e.g., low water pressure).

Average availability is the mean probability over a period of time of being found in such a state. In this and Sec. 6.2, the term *availability* will be used in the sense of average availability. The probability of system failure is sometimes defined as 1 minus availability.

Severity indices describe the size of failures. In water supply studies, *reliability* has been defined as the ratio of available annual supply to demand.

Frequency and duration indices indicate how often failures of a given severity occur and how long they last. Such indices have found increasing use in electric utility planning studies.

Economic indices are indices of the economic consequences of shortages, also referred to as *vulnerability.* Letting S_u represent water supply and Q the quantity of water demanded in units of m^3/h, these indices can be phrased as follows:

- *Contingency.* A system is reliable if $P(S_u < Q + \text{contingency}) = 0$.
- *Availability.* The availability of the system is $P(S_u > 0)$.
- *Probability and severity of failure.* The probability of a failure can be defined as $P(S_u < Q - s)$ at a particular moment or over a particular period, where s represents some predefined level of failure severity.
- *Frequency and duration.* The *expected frequency* $E(F)$ (in units of $1/T$) of a failure event of at least severity s in m^3/h is related to the expected duration $E(D)$ by the equation $E(F)E(D) = P(S_u < Q - s)$.

Economic consequences (measured in dollars) are related nonlinearly to the frequency, duration, and severity of events.

Many models have been reported in the literature for the reliability analysis of water distribution systems. Unfortunately the water distribution industry (engineering community) has not incorporated these methodologies into practice. There are many reasons for this, two of which are that engineers are not familiar with the methodologies and the proper software has not been developed to make this an easier task. The list of references includes several of the reliability methodologies that have been reported in the literature: Awumah et al. (1991), Awumah and Goulter (1992), Bao and Mays (1994a,b), Biem and Hobbs (1988), Boccelli et al. (1998), Bouchart and Goulter (1989, 1991, 1992), Cullinane et al. (1989, 1992), Duan and Mays (1990), Duan et al. (1990), Fujiwara and de Silva (1990), Fujiwara and Gannesharajah (1993), Fujiwara and Li (1998), Fujiwara and Tung (1991), Germanopoulis et al. (1986), Goulter (1988, 1992a,b, 1995a,b), Goulter and Bouchart (1990), Goulter and Coals (1986), Goulter and Jacobs (1989), Gupta and Bhave (1994), Hobbs and Biem (1988), Jacobs (1992), Jacobs and Goulter (1988, 1989, 1991), Jowitt and Xu (1993), Kessler et al. (1990), Kettler and Goulter (1983, 1985), Lansey et al. (1989, 1992), Ormsbee and Kessler (1990), Park and Liebman (1993), Quimpo and Shamsi (1991), Shamsi (1990), Shinstine (1999), Su et al. (1987), Tanyimboh et al. (2001), Tanyimboh and Templeman (1993, 1995), Wagner et al. (1988a,b), Wu et al. (1993), Xu and Goulter (1998), Yang et al. (1996a,b).

6.2 COMPONENT (MECHANICAL) RELIABILITY ANALYSIS

A system or its components can be treated as a black box or a lumped-parameter system, and their performances are observed over time. This reduces the reliability analysis to a one-dimensional problem involving time as the only variable. In such cases, the *time to failure* (TTF) of a system or a component of the system is the random variable. It should be pointed out that the term *time* could be used in a more general sense. In some situations other physical scale measures, such as distance or length, may be appropriate for system performance evaluation.

The time-to-failure analysis is particularly suitable for assessing the reliability of systems and/or repairable components. For a system that is repairable after its failure, the time period it would take to have it repaired to the operational state is uncertain; therefore, the *time to repair* (TTR) is also a random variable.

This section will focus on characteristics of failure, repair, and availability of repairable systems by time-to-failure analysis.

6.2.1 Failure Density, Failure Rate, and Mean Time to Failure

Any system will fail eventually; it is just a matter of time. Because of the presence of many uncertainties that affect the operation of a physical system, the time that the system fails to satisfactorily perform as intended is a random variable.

The *failure density function* is the probability distribution that governs the time occurrence of failure. The failure density function serves as the common thread in the reliability assessments in time-to-failure analysis. The reliability of a system or a component within a specified time interval $[0, t]$ can be expressed as

$$p_s(t) = P(\text{TTF} > t) = \int_t^{\infty} f(\tau)\, d\tau \tag{6.1}$$

where TTF is the random time to failure having $f(t)$ as the failure density function. The reliability $p_s(t)$ represents the probability that the system experiences no failure within $[0, t]$. The *failure probability* or *unreliability* can be expressed as $p_f(t) = 1 - p_s(t)$. Note that unreliability $p_f(t)$ is the probability that a component or a system would experience its first failure within the time interval $[0, t]$. Conversely, the failure density function can be obtained from reliability or unreliability as

$$f(t) = -\frac{d[p_s(t)]}{dt} = \frac{d[p_f(t)]}{dt} \tag{6.2}$$

The time to failure is a continuous, nonnegative random variable by nature. Many distribution functions described in Sec. 6.2.1 are appropriate for modeling the stochastic nature of time to failure. Among them, the exponential distribution is perhaps the most widely used. The exponential distribution has been found, both phenomenologically and empirically, to adequately describe the time-to-failure distribution for components, equipment, and systems involving components with mixtures of life distributions.

The *failure rate* is defined as the number of failures occurring per unit time in a time interval $(t + \Delta t)$ per unit of the remaining population at time t. The instantaneous failure rate $m(t)$ can be obtained as

$$m(t) = \lim_{\Delta t \to 0} \left[\frac{N_F(\Delta t)/\Delta t}{N(t)} \right] = \frac{f(t)}{p_s(t)} \tag{6.3}$$

where $N_F(\Delta t)$ is the number of failures in a time interval $(t, t\ \Delta t)$ and $N(t)$ is the number of failures from the beginning up to time t. This *instantaneous failure rate* is also called the *hazard function* or *force of mortality function.*

The failure rate for many systems or components has a bathtub shape, as shown in Fig. 6.1, in that three distinct life periods can be identified. They are *early life* (or *infant mortality*) *period, useful life period*, and *wear—out life period.* In the early life period, quality failures and stress-related failures dominate, with little contribution from wear-out failures. During the useful life period, all three types of failures contribute to the potential failure of the system or component, and the overall failure rate increases with age, because wear-out failures and stress-related failures are the main contributors, and wear out becomes a more dominating factor for the failure of the system with age.

The reliability can be directly computed from the failure rate as

$$p_s(t) = \exp\left(\int_0^t m(\tau)\, d\tau \right) \tag{6.4}$$

Substituting Eq. (6.4) into Eq. (6.3), the *failure density function* $f(t)$ can be expressed, in terms of failure rate, as

$$f(t) = m(t) \left[\exp \left(\int_0^{-t} m(\tau)\, d\tau \right) \right] \tag{6.5}$$

In general, the reliability of a system or a component is strongly dependent on its age. This can be mathematically expressed by the condition probability as

$$p_s(\xi \mid t) = \frac{p_s(t + \xi)}{p_s(t)} \tag{6.6}$$

in which t is the age of the system and up to that point the system has not failed and in which $p_s(\xi \mid t)$ is the reliability over a new mission period ξ, having successfully operated over a period of t.

A commonly used reliability measure of system performance is the *mean time to failure* (MTTF) which is the expected time to failure. The MTTF can be mathematically defined as

$$\text{MTTF} = E(\text{TTF}) = \int_0^{\infty} \tau f(\tau)\, d\tau \tag{6.7}$$

The mean time to failure for some failure density functions is given in Table 6.1.

Repairable Systems. For repairable water resources systems, such as pipe networks, pump stations, and storm runoff drainage structures, failed components within the system can be repaired or replaced so that the system can be put back into service. The time required to have the failed system

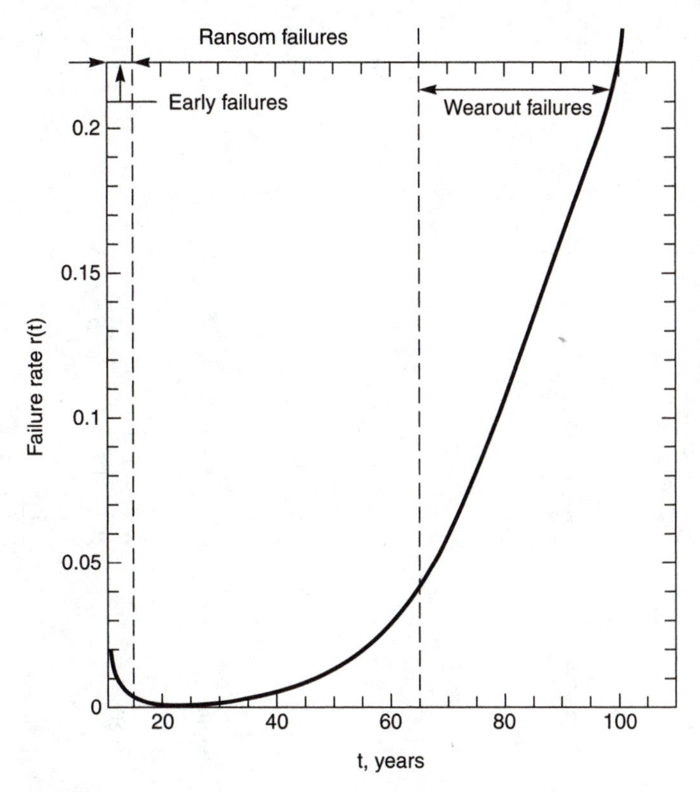

FIGURE 6.1 Failure rate $r(t)$ versus t. (Henley and Kumamoto, 1981)

TABLE 6.1 Mean Time to Failure for Some Failure Density Functions

Distribution	PDF, $f(t)$	Reliability, $p_s(t)$	Failure rate, $m(t)$	MTTF
Normal	$\frac{1}{\sqrt{2\pi}\sigma_T}\exp\left[-\frac{1}{2}\left(\frac{t-\mu_r}{\sigma_T}\right)^2\right]$	$\phi\left(\frac{t-\mu_T}{\sigma_T}\right)$	$\frac{f(t)}{\phi\left(\frac{t-\mu_T}{\sigma_T}\right)}$	μ_T
Lognormal	$\frac{1}{\sqrt{2\pi}\sigma_{\ln T}t}\exp\left[-\frac{1}{2}\left(\frac{\ln(t)-\mu_{\ln T}}{\sigma_{\ln T}}\right)^2\right]$	$\phi\left(\frac{\ln(t)-\mu_{\ln T}}{\sigma_{\ln T}}\right)$	$\frac{f(t)}{\sigma\left(\frac{\ln(t)-\mu_{\ln T}}{\sigma_{\ln T}}\right)}$	$\exp\left(\mu_{\ln T}+\frac{\sigma_{\ln T}^2}{2}\right)$
Exponential	$\beta e^{-\beta t}$	$e^{-\beta t}$	β	$\frac{1}{\beta}$
Rayleigh	$\frac{t}{\beta^2}\exp\left[-\frac{1}{2}\left(\frac{t}{\beta}\right)^2\right], \beta > 0$	$\exp\left[-\frac{1}{2}\left(\frac{t}{\beta}\right)^2\right]$	$\frac{t}{\beta^2}$	$1.253\,\beta$
Gamma	$\frac{\beta}{\Gamma(\alpha)}(\beta t)^{\alpha-1}e^{-\beta t}$	$\int_t^\infty f(\tau)\,d\tau$	$\frac{f(t)}{p_s(t)}$	$\frac{\alpha}{\beta}$
Gumbel	$e^{\pm y} - e^{\pm y}; y = \frac{t-t_0}{\beta}$	$1 - e^{e\pm y}$	$\frac{f(t)}{p_s(t)}$	$x_0 \pm 0.577\,\beta$
Weibull	$\frac{\alpha}{\beta}\left(\frac{t-t_0}{\beta}\right)^{\alpha-1} e - \left(\frac{t-t_0}{\beta}\right)^a$	$\exp\left(\frac{t-t_0}{\beta}\right)^a$	$\frac{\alpha(t-t_0)^{\alpha-1}}{\beta^\alpha}$	$t_0 + \beta\Gamma\left(1+\frac{1}{\alpha}\right)$
Uniform	$\frac{1}{b-a}$	$\frac{b-t}{b-a}$	$\frac{1}{b-t}$	$\frac{a+b}{2}$

repaired is uncertain, and consequently, the total time required to restore the system from its failure to operational states is a random variable.

Like the time to failure, the random time to repair (TTR) has the repair density function $g(t)$ describing the random characteristics of the time required to repair a failed system when failures occur at time 0. The *repair probability* $G(t)$ is the probability that the failed system can be restored within a given time period $[0, t]$,

$$G(t) = P[\mathrm{TTF} \leq t] = \int_0^t g(\tau)\, d\tau \tag{6.8}$$

The repair probability $G(t)$ is one measure of maintainability.

The *repair rate* $r(t)$, similar to the failure rate, is the conditional probability that the system is repaired per unit time, given that the system failed at time 0 and is still not repaired at time t. The quantity $r(t)\, dt$ is the probability that the system is repaired during the time interval $(t, t + dt)$ given that the system fails at time t. The relation between the repair density function, repair rate, and repair probability is

$$r(t) = \frac{g(t)}{1 - G(t)} \tag{6.9}$$

Given a repair rate $r(t)$, the density function and the maintainability can be determined, respectively, as

$$g(t) = r(t) \exp\left(-\int_0^t r(\tau)\, d\tau\right) \tag{6.10}$$

$$G(t) = 1 - \exp\left(-\int_0^t r(\tau)\, d\tau\right) \tag{6.11}$$

The *mean time to repair* (MTTR) is the expected value of the time to repair of a failed system, that is,

$$\mathrm{MTTR} = \int_0^\infty t g(t)\, dt \tag{6.12}$$

The MTTR measures the elapsed time required to perform the maintenance operation and is used to estimate the downtime of a system. It is also a measure for maintainability of a system. Various types of maintainability measures are derivable from the repair density function.

The MTTR is a proper measure of the mean life span of a nonrepairable system. For a repairable system, a more representative indicator for the fail-repair cycle is the *mean time between failures* (MTBF) which is the sum of MTTF and MTTR, that is,

$$\mathrm{MTBF} = \mathrm{MTTF} + \mathrm{MTTR} \tag{6.13}$$

The *mean time between repairs* (MTBR) is the expected value of the time between two consecutive repairs, and it is equal to MTBF.

6.2.2 Availability and Unavailability

A *repairable system* experiences a repetition of repair-to-failure and failure and failure-to-repair processes during its service life. Hence, the probability that a system is in operating condition at any given time t for a repairable system is different than that of a nonrepairable system. The term availability $A(t)$ is generally used for repairable systems, to indicate the probability that the system is in operating condition at any given time t. On the other hand, reliability $p_s(t)$ is appropriate for nonrepairable systems, indicating the probability that the system has been continuously in its operating state starting from time 0 up to time t.

Availability can also be interpreted as the percentage of time that the system is in operating condition within a specified time period. In general, the availability and reliability of a system satisfies the following inequality relationship

$$0 \leq p_s(t) \leq A(t) \leq 1 \tag{6.14}$$

The equality holds for nonrepairable systems. The reliability of a system decreases monotonically to 0 as the age of the system increases, whereas the availability decreases but converges to a positive probability, as shown in Fig. 6.2.

The complement to availability is the *unavailability* $U(t)$, which is the probability that a system is in the failed condition at time t, given that it is in operating condition at time 0. In other words, unavailability is the percentage of time that the system is not available for the intended service in time period $[0, t]$, given it is operational at time 0. Availability, unavailability, and unreliability satisfy the following relationships:

$$A(t) + U(t) = 1.0 \tag{6.15}$$

$$0 \leq U(t) \leq p_f(t) < 1 \tag{6.16}$$

For a nonrepairable system, the unavailability is equal to the unreliability, that is, $U(t) = p_f(t)$.

Determination of availability or unavailability of a system requires full accounting of the failure and repair processes. The basic elements that describe processes are failure and repair processes. The basic elements that describe such processes are the failure density function $f(t)$ and the repair density function $g(t)$. Consider a system with a *constant failure rate* λ and a *constant repair rate* η. The availability and unavailability, respectively, are

$$A(t) = \frac{\eta}{\lambda + \eta} + \frac{\lambda}{\lambda + \eta} e^{-(\lambda+\eta)t} \tag{6.17}$$

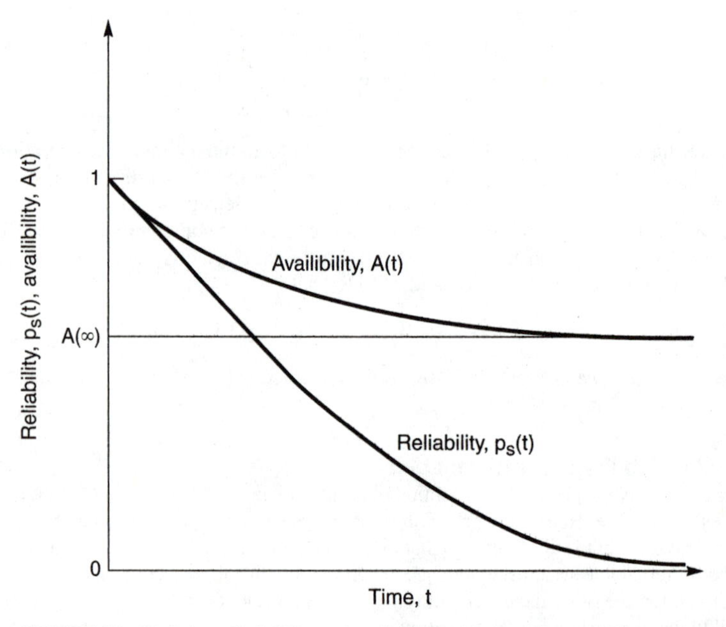

FIGURE 6.2 Variation of reliability and availability with time. (Tung, 1996)

$$U(t) = \frac{\lambda}{\lambda + \eta}\left[1 - e^{-(\lambda+\nu)t}\right] \tag{6.18}$$

As the time approaches infinity ($t \rightarrow \infty$), the system reaches its stationary condition. Then, the *stationary availability* $A(\infty)$ and *stationary unavailability* $U(\infty)$ are

$$A(\infty) = \frac{\eta}{\lambda + \eta} = \frac{1/\lambda}{1/\lambda + 1/\eta} = \frac{\text{MTTF}}{\text{MTTF}} + \text{MTTR} \tag{6.19}$$

$$U(\infty) = \frac{\lambda}{\lambda + \eta} = \frac{1/\eta}{1/\lambda + 1/\eta} = \frac{\text{MTTR}}{\text{MTTF}} + \text{MTTR} \tag{6.20}$$

6.3 METHODOLOGY FOR RELIABILITY AND AVAILABILITY ANALYSIS FOR WATER DISTRIBUTION NETWORKS

6.3.1 Reliability of a System

The methodology presented here summarizes the work of Shinstine et al. (2002). *Reliability* of a water distribution system can be defined using the approach of Goulter (1995):

> The ability of a water distribution system to meet the demands that are placed on it where such demands are specified in terms of:
>
> **1.** The flows to be supplied (total volume and flow rate); and
> **2.** The range of pressures at which those flows must be provided

A useful extension or interpretation of this definition as given by Cullinane et al. (1992) is "the ability of the system to provide service with an acceptable level of interruption in spite of abnormal conditions."

A key feature of this latter definition is that it implies, or introduces, the concept of both the period of time over which the system is unable to meet demands and the particular point in time (i.e, circumstances which caused the system to be unable to supply demands) as being important determinants in defining and calculating reliability. This interpretation of reliability also recognizes that system failure may occur either through the flows associated with the demands not being supplied, and/or the flows associated with the demands being supplied but at pressures lower than the minimum acceptable or minimum specified for the particular circumstances. Of equal importance is that this definition also recognizes that failures should only occur in association with abnormal conditions, e.g., component failure and/or abnormally high demands.

The range of combinations of ways in which failure can occur in a water distribution system constitutes one, and arguably the major, source of the many theoretical and practical difficulties which have been encountered in establishing suitable, i.e., comprehensive and computationally tractable, measures of reliability which can be used in the practical design and operation of water distribution systems. It does have to be recognized, however, that reliability has been an explicit factor in the design and operation of water distribution networks for a considerable time as demonstrated by the existence of looped networks. Looped networks are not the most cost-efficient solution for water supply networks. The presence of loops does, however, add reliability and redundancy to the system by providing excess capacity and alternative paths for the supply of water in the face of failure of one of the system components. The question facing practitioners in the water distribution industry is how to make quantitatively explicit the reliability of supply. This quantitative specification of reliability can then be used by municipal authorities and the like to define compliance levels for the reliability to be provided by the water utility operators both private and public. Such a specification can similarly be used as a point against which the operators can fine-tune and optimize the performance of the system.

Reliability models to compute system reliability for a given system and layout have been developed by Cullinane (1985), Tung (1985), Wagner et al. (1988a,b), Duan and Mays (1990), and Bao and Mays (1990). These approaches allow a modeler to determine the reliability of a system and account for such factors as the probability and duration of pipe and pump failure, the uncertainty in demands, and the variability in the decay of pipes. Notably, Tung (1985) defines system reliability as "probability that flow can reach all the demand points in the network." Among the six reliability methods identified, the cut-set method, defined later in this chapter, appeared to be the most efficient and is easily programmed on a digital computer.

Hydraulic-failure reliability measures have been included in optimization models by Su et al. (1987), Goulter and Bouchart (1990), Lansey et al. (1989), and Duan et al. (1990). Using the minimum cut-set method as the reliability model for the system and the nodes, Su et al. (1987) added a reliability constraint to an optimal network design model.

The second category of reliability measures overcomes the computational problems. Examples include Goulter and Coals (1986) and Shamir and Howard (1981). Goulter and Coals's (1986) model incorporates reliability aspects into a linear programming model. One disadvantage of their model is their assumption that all pipes connecting a node have similar diameters and hence have similar values of failure probability, which is not applicable in real pipe networks. Su et al.'s (1987) model is one example where this problem is solved.

The results from several of the above heuristic measures generally appear consistent with expectations of systems as reliability requirements are increased. However, since the system pressures are not evaluated in these techniques, over- or underdesigned components may result (Cullinane et al., 1992). Therefore, Cullinane et al. (1992) described an approach that is a balance between the two types of measures reported by Mays (1989). Hydraulic failure is verified for several conditions, but not for all as required by other failure-based techniques. Thus, computational burden is reduced, but heuristics and judgment are necessary to define critical conditions.

6.3.2 Methodology

Hydraulic Availability. "*Hydraulic availability*, as it relates to water distribution system design, can be defined as the ability of the system to provide service with an acceptable level of interruption in spite of abnormal conditions" (Cullinane et al., 1992). The evaluation of hydraulic availability relates directly to the basic function of the water distribution system, i.e., delivery of the specified quantity of water to a specific location at the required time under the desired pressure.

Availability is evaluated in terms of developing a required minimum pressure. Generally, for fire conditions the desirable minimum pressure is 137.9 kilonewtons per square meter (kN/m^2) (20 psi) with an absolute minimum pressure of 68.9 kN/m^2 (10 psi). These pressures are required to overcome the head losses between hydrants and fire-engine pumps and are essentially estimates of the net positive suction head (NPSH) required for pump operation. Pressures between 137.9 kN/m^2 (20 psi) and 551.6 kN/m^2 (80 psi) are considered to be desirable pressures under normal daily demands.

Previous works (Su et al., 1987; Goulter and Coals, 1986) proposed the use of a discrete relationship (Fig. 6.3) between availability and pressure. Here, availability during a time period t can be expressed by the following mathematical relationship:

$$\mathrm{HA}_j = \begin{cases} 1 & \text{for } P_j \geq \mathrm{PR} \quad (6.21) \\ 0 & \text{for } P_j < \mathrm{PR} \quad (6.22) \end{cases}$$

where HA_j = hydraulic availability of node j
P_j = pressure at node j
PR = required minimum pressure

The *network hydraulic availability* is the product of the nodal hydraulic availabilities. With this approach, a zero availability index is assigned for all pressure values below the required minimum pressure. For example, if the required minimum pressure is set at 137.9 kN/m^2 (20 psi), a residual of 137.2 kN/m^2 (19.9 psi) results in the same availability index as a residual pressure of 6.9 kN/m^2

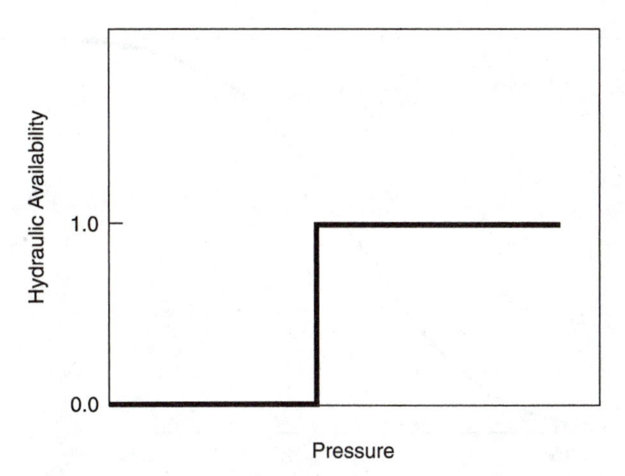

FIGURE 6.3 Hydraulic availability step function. (Shinstine et al., 2002)

(1 psi). Thus, the use of this discontinuous relationship does not adequately represent the engineering reality of the problem.

Cullinane et al. (1992) adopted a more appropriate representation that describes the availability index as a continuous "fuzzy" function. Using a continuous function of this shape, a significant index value may be assigned to pressure values slightly less than the arbitrarily assigned required minimum pressure value PR. The shape of the curve shown in Fig. 6.4 is similar to the cumulative normal distribution, which is mathematically stated as follows:

$$\mathrm{HA}_j = P(\mathrm{PR} \le P_j) = \frac{1}{\sqrt{2\pi}} \int^{(H-\mu_H)/\sigma_H} e^{-t^{5/2}}\, dt = P\left(\frac{H-\mu_H}{\sigma_H}\right) \tag{6.23}$$

where P_j = value of nodal pressure
μ_H = mean nodal pressure
σ_H = standard deviation of pressure

Values of μ_H and σ_H = can be selected to adjust the position and shape of the function, respectively.

Pipe Failure Probabilities. Goulter and Coals (1986) and Su et al. (1987) used similar methods to determine the probability of failure of individual pipes. The probability of failure of pipe *i*, P_i, is determined using the *Poisson probability distribution*:

$$P_i = 1 - e^{-\beta i} \tag{6.24}$$

and

$$\beta_i = r_i L_i \tag{6.25}$$

where β_i = expected number of failures per year for pipe *i*
r_i = expected number of failures per year per unit length of pipe *i*
L_i = the length of pipe *i*

Nodal and System Reliability. The *nodal* and *system reliabilities*, R_{node} and R_S, respectively, are obtained from minimum cut-sets. Su et al. (1987) defines the *minimum cut-set* as "a set of system components (e.g., pipes) which, when failed, causes failure of the system." However, when any one component of the set has not failed, it does not cause system failure (Billinton and Allan, 1983).

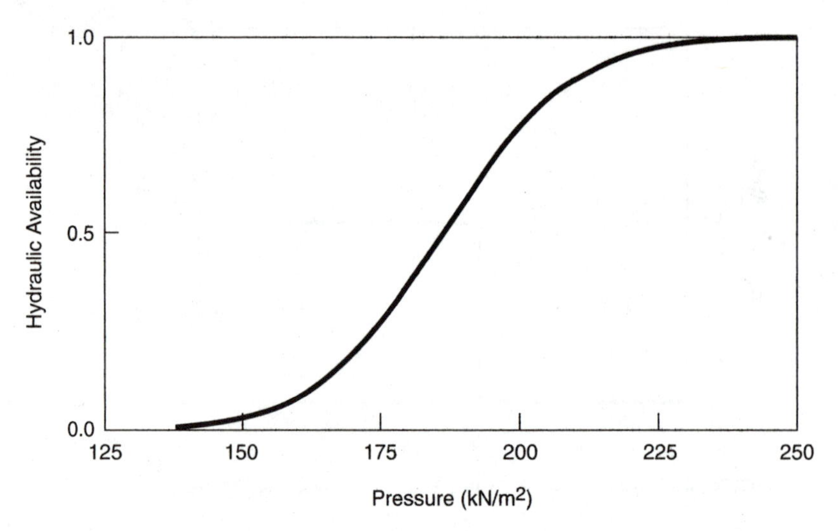

FIGURE 6.4 Continuous hydraulic availability function. (Shinstine et al., 2002)

Assuming that a pipe break can be isolated from the rest of the system, minimum cut-sets are determined by closing a pipe or combination of pipes in the water distribution system and using a hydraulic simulation model (e.g., KYPIPE) to obtain the values of pressure head at each demand node of the system. By comparing these pressure heads with minimum pressure-head requirements, the reliability model can determine whether or not this pipe or combination of pipes is a minimum cut-set of the system or an individual demand node. *A minimum cut-set for a node* is one that causes reduced hydraulic availability at that node, while a *minimum cut-set for the system* is a cut-set that reduces the hydraulic availability for any node in the system. This procedure is repeated until all the combinations of pipes have been considered and hence all minimum cut-sets of the demand nodes and the total system have been determined. The values of system and nodal reliability are then computed. A flowchart of the procedure is shown in Fig. 6.5. Note that prior to hydraulic simulation a test is performed to examine if the failure(s) cause a part of the system to be isolated from all sources and thus have no available supply. This disconnection results in failure of all isolated nodes.

For n components (pipes) in the ith minimum cut-set of a water distribution system, the failure probability of the *ith minimum cut-set* (MC_i) is

$$P(\mathrm{MC}_i) = \prod_{i=1}^{n} P_i = P_1 \cdot P_2 \cdot P_3 \cdot \cdots \cdot P_n \tag{6.26}$$

Using the step function for hydraulic availability and assuming that the occurrence of the failure of the components within a minimum cut-set are statistically independent for a water distribution network with four minimum cut-sets (MC_i), for the system reliability R_s, the failure probability of the system P_s is then defined as (Billinton and Allan, 1983):

$$P_s = P(\mathrm{MC}_1 \cup \mathrm{MC}_2 \cup \mathrm{MC}_3 \cup \mathrm{MC}_4) \tag{6.27}$$

By applying the principle of inclusion and exclusion (Ross, 1985), Eq. (6.27) can be reduced to

$$P_s = P(\mathrm{MC}_1) + P(\mathrm{MC}_2) + P(\mathrm{MC}_3) + P(\mathrm{MC}_4) \tag{6.28}$$

$$= \sum_{i=1}^{4} P(\mathrm{MC}_i) \tag{6.29}$$

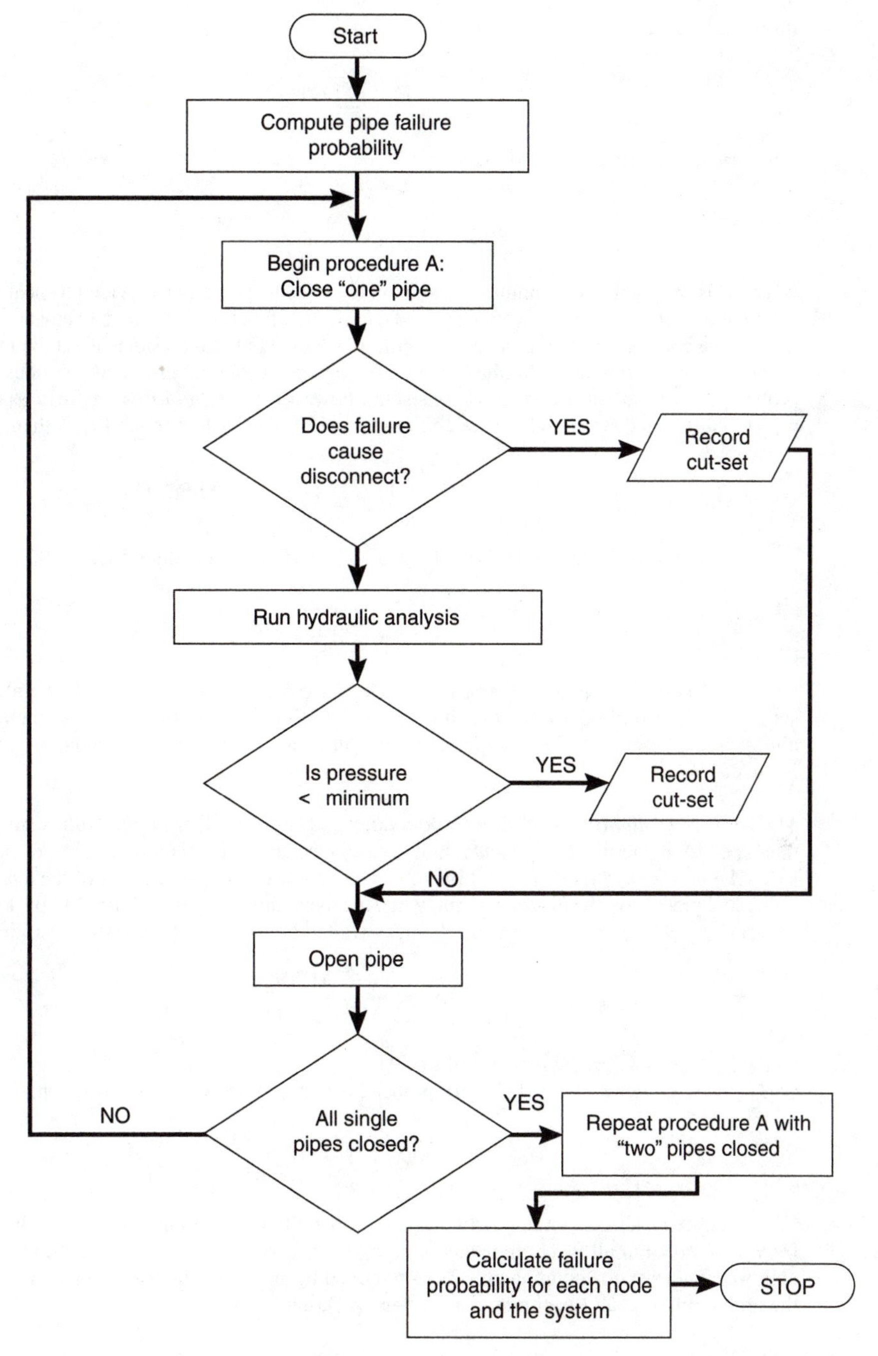

FIGURE 6.5 Minimum cut-set reliability flowchart with step hydraulic availability. (Shinstine et al., 2002)

In general form,

$$P_s = \sum_{i=1}^{M} P(\mathrm{MC}_i) \tag{6.30}$$

The *system reliability* R_s is expressed as

$$R_s = 1 - P_s = 1 - \sum_{i=1}^{M} P(\mathrm{MC}_i) \tag{6.31}$$

where M is the number of minimum cut-sets in the system. To further provide physical significance to the reliability, it is possible to weight the nodal terms as a function of the nodal demand. Nodal reliabilities can be computed with the same relationship including only failures that affect the individual node.

Using the continuous hydraulic availability concept, a true minimum cut-set does not exist. The probability of a cut-set occurring is consistent; however, the *reliability* is defined as the product of pipe reliability and hydraulic unavailability (1 − HA). The system reliability is then,

$$R_s = 1 - P_s = 1 - \sum_{i=1}^{M} (1 - \mathrm{HA}_{\mathrm{net}}{}^{i})\, P(\mathrm{MC}_i) \tag{6.32}$$

where $\mathrm{HA}_{\mathrm{net}}{}^{i}$ is the network hydraulic availability (Fujiwara and de Silva, 1990):

$$\mathrm{HA}_{\mathrm{net}} = \prod_{j=1}^{j} = \mathrm{HA}_j \tag{6.33}$$

where HA_j is the hydraulic availability of node j. If HA equals 1, the failure probability for the cut-set is not included in Eq. (6.32); thus it is identical to Eq. (6.31) for the step-function hydraulic availability case. To compute the system reliability with continuous hydraulic availability, all cut-sets are included.

Mechanical Availability of Piping Components. The *probability of pipe failure* does not consider that pipes are repairable components. In this case, mechanical availability is a more appropriate measure. The *mechanical availability*, MA, of piping can be evaluated in terms of the availability of the individual pipes. For the ith component, when the mean time between failure (MTBF_i) and mean time to repair (MTTR_i) are known, the mechanical availability (MA_i) is (Ross, 1985; Cullinane, 1986)

$$\mathrm{MA}_i = \frac{\mathrm{MTBF}_i}{\mathrm{MTBF}_i + \mathrm{MTTR}_i} \tag{6.34}$$

where $\mathrm{MTBF}_i = 1/[\text{probability of failure, } P_i]$.

Mechanical unavailability (MU_i), the probability that a component is not operational, is then given by

$$\mathrm{MU}_i = 1 - \mathrm{MA}_i = 1 - \frac{\mathrm{MTBF}_i}{\mathrm{MTBF}_i + \mathrm{MTTR}_i} \tag{6.35}$$

For the entire water distribution network, the probability $\mathrm{MA}_{\mathrm{tot}}$ (total availability) that the system is fully operational in all its components (e.g., pipes) is given by the complement to the overall probability that at least one component will be removed from service. In brief, $\mathrm{MA}_{\mathrm{tot}}$ can be evaluated as the probability that all the components are in operating order:

$$\mathrm{MA}_{\mathrm{tot}} = \prod_{i=1}^{\mathrm{mc}} = \mathrm{MA}_i \tag{6.36}$$

where mc is the total number of components.

The probability of a failure of the ith component and all other components remaining operational is given by (Fujiwara and de Silva, 1990; Fujiwara and Tung, 1991)

$$u_i = \mathrm{MA}_{\mathrm{tot}} \frac{\mathrm{MU}_i}{\mathrm{MA}_i} \tag{6.37}$$

Similarly, the *probability of the event of simultaneous failure* of two (the ith and kth components) and all others remaining operational is

$$u_{ik} = \mathrm{MA}_{\mathrm{tot}} \frac{\mathrm{MU}_i}{\mathrm{MA}_i} \frac{\mathrm{MU}_k}{\mathrm{MA}_k} \tag{6.38}$$

Nodal and System Availability. Nodal and system availabilities are computed using a product form similar to equation (6.32). System availability is obtained from (Fujiwara and de Silva, 1990)

$$A_s = \mathrm{HA}_{\mathrm{net}}{}^0 \cdot \mathrm{MA}_{\mathrm{tot}} + \sum_{j=1}^{\mathrm{mc}} \mathrm{HA}_{\mathrm{net}}{}^i \cdot u_i + \sum_{i=1}^{\mathrm{mc}-1} \sum_{k=j-1}^{\mathrm{mc}} \mathrm{HA}_{\mathrm{net}}{}^{ik} \cdot u_{ik} + \cdots \tag{6.39}$$

where mc = total number of mechanical components taken into account
$\mathrm{HA}_{\mathrm{net}}{}^0$ = network hydraulic availability with a fully operational system
$\mathrm{HA}_{\mathrm{net}}{}^i$ = network hydraulic availability with only ith component removed
$\mathrm{HA}_{\mathrm{net}}{}^{ik}$ = network hydraulic availability with only ith and kth components removed

Each term in Eq. (6.39) is the probability of the network being in a certain condition times the hydraulic availability for that condition. Either definition for hydraulic availability can be used in Eq. (6.39). Nodal availabilities are computed with Eq. (6.39) using the HA for the node of concern. Note that compared to Eq. (6.32) the terms are summed so that the total approaches 1 while in Eq. (6.32) the failure conditions are subtracted from 1.

Only first- and second-order cut-sets are assumed to be important given the low failure probabilities, so Eq. (6.39) is truncated at this term. This reduces the number of computations without a significant loss of accuracy as shown in the application. Figure 6.6 illustrates the computational procedure with continuous hydraulic availability (HA) and mechanical availability (MA).

In summary, four measures for analyzing a system have been identified: system reliability [Eq. (6.32)] with step and continuous hydraulic availability and system availability [Eq. (6.39)] using step and continuous hydraulic availability. Section 6.3.3 discusses the values of these measures for two midsize municipal water distribution networks.

6.3.3 Application of Methodology

The Tucson, Arizona, water distribution system is divided into many water service areas (WSAs), with some areas isolated from the main system and some set up to serve specific zones. Water service area WSA-C6 (consisting of 109 pipes and 89 nodes), shown in Fig. 6.7, was chosen to illustrate the methodology for computing system reliability. The pipe failure rates listed in Table 6.2 were determined from the available maintenance and repair data from Tucson Water. The likelihood of failure rate for a given pipe length and type was estimated from the available information on the total length of pipe by size and material. The two sources of data, maintenance database and pipe inventories, were used to provide failure rate per mile per year for each pipe size and material. Based on the maintenance records, the mean time to repair (MTTR) for all pipes was assumed to be 1 day.

The *minimum operational pressure* for Tucson Water is 241.3 kN/m^2 (35 psi) for normal conditions and 137.9 kN/m^2 (20 psi) during fire flows. For an initial reliability analysis, a pressure of 6.9 kN/m^2 (1 psi) was used to identify pipe sets resulting in a geometric disconnection or mechanical failure of the system. Identifying mechanical failure provided a baseline reliability and simplified the process of checking the cut-sets resulting from hydraulic failures.

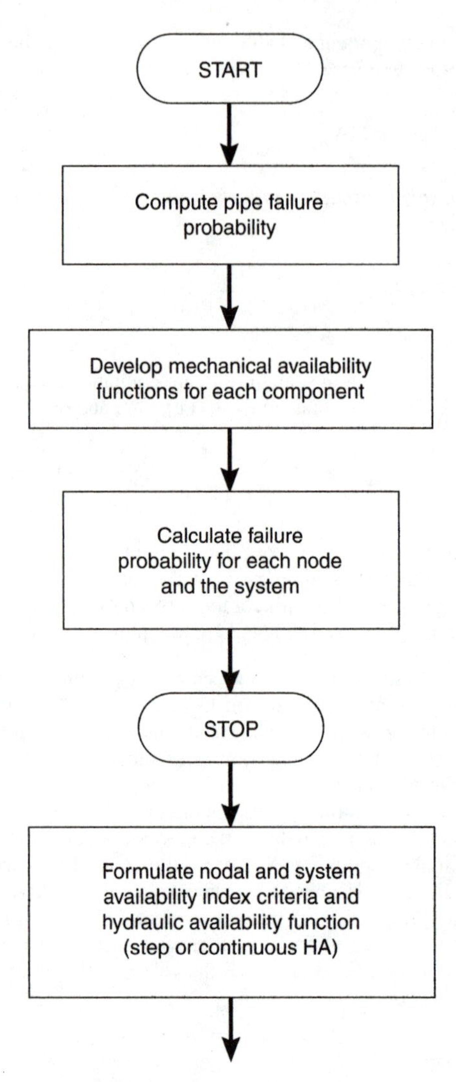

FIGURE 6.6 Steps in computing mechanical availability or hydraulic reliability with pipe failure given step or continuous hydraulic availability. (Shinstine et al., 2002)

When using continuous hydraulic availability, the modeler must define the parameters for the function. Here, two different cases were analyzed. In both cases, the standard deviation was 17.25 kN/m^2 and H_{min} and H_{max} were two standard deviations from the mean pressure requirement. For case 1, mean nodal pressure, minimum nodal pressure (H_{min}), and maximum nodal pressure (H_{max}) were defined as 137.9 kN/m^2, 103.4 kN/m^2, and 172.4 kN/m^2, respectively. The corresponding pressure values for case 2 were 241.3 kN/m^2, 206.8 kN/m^2, and 275.8 kN/m^2, respectively. Figure 6.8 shows the continuous hydraulic availability curves for the two pressure ranges, superimposed on the step function with minimum required pressure (PR) of 241.3 kN/m^2 (35 psi).

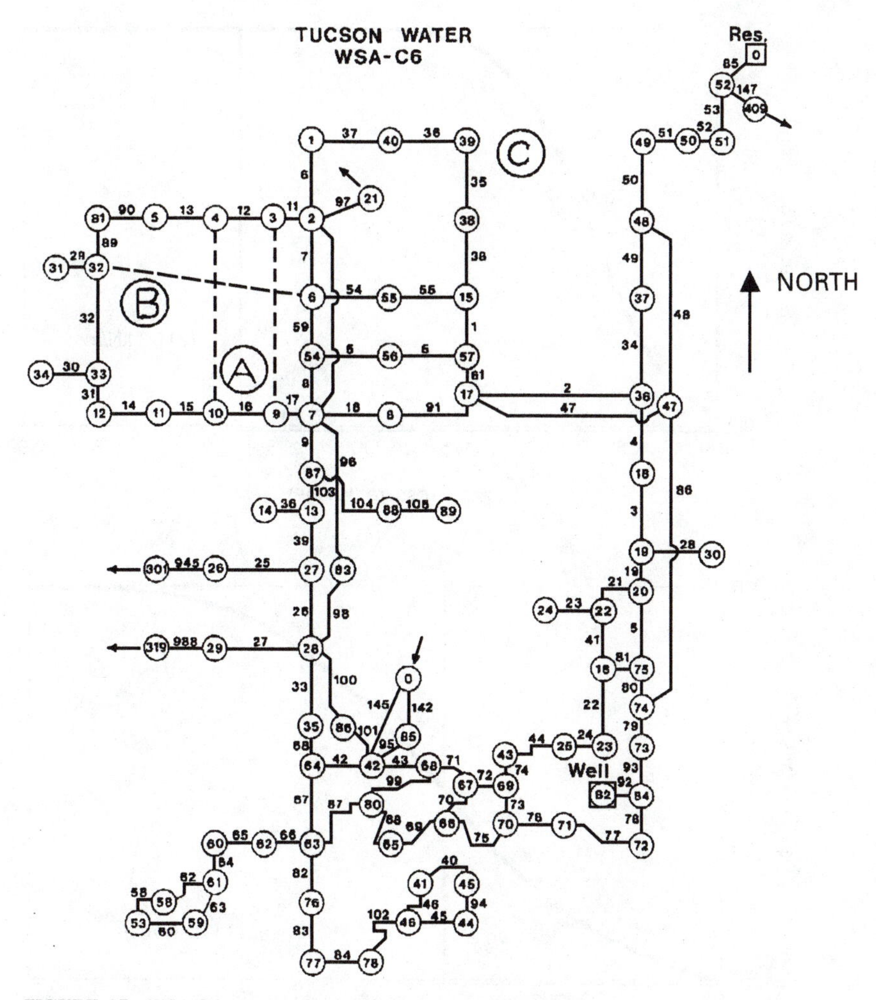

FIGURE 6.7 WSA-C6 water service area in Tucson, Arizona.

TABLE 6.2 Failure Rates and Cumulative Lengths of Pipes (Tucson and St. Louis)

Pipe diameter, m	Failure rate, km/yr		Cum. pipe length, m (WSA-C6)
	Tucson	St. Louis	
0.051	0.022	0.387	402.34
0.102	0.022	0.387	882.70
0.152	0.009	0.225	12,173.10
0.203	0.007	0.171	9,048.30
0.254	0.011	0.130	
0.305	0.005	0.099	4,375.40
0.406	0.007	0.058	3,414.06
0.610	0.002	0.019	2,344.83
0.914	0.005	0.004	226.16

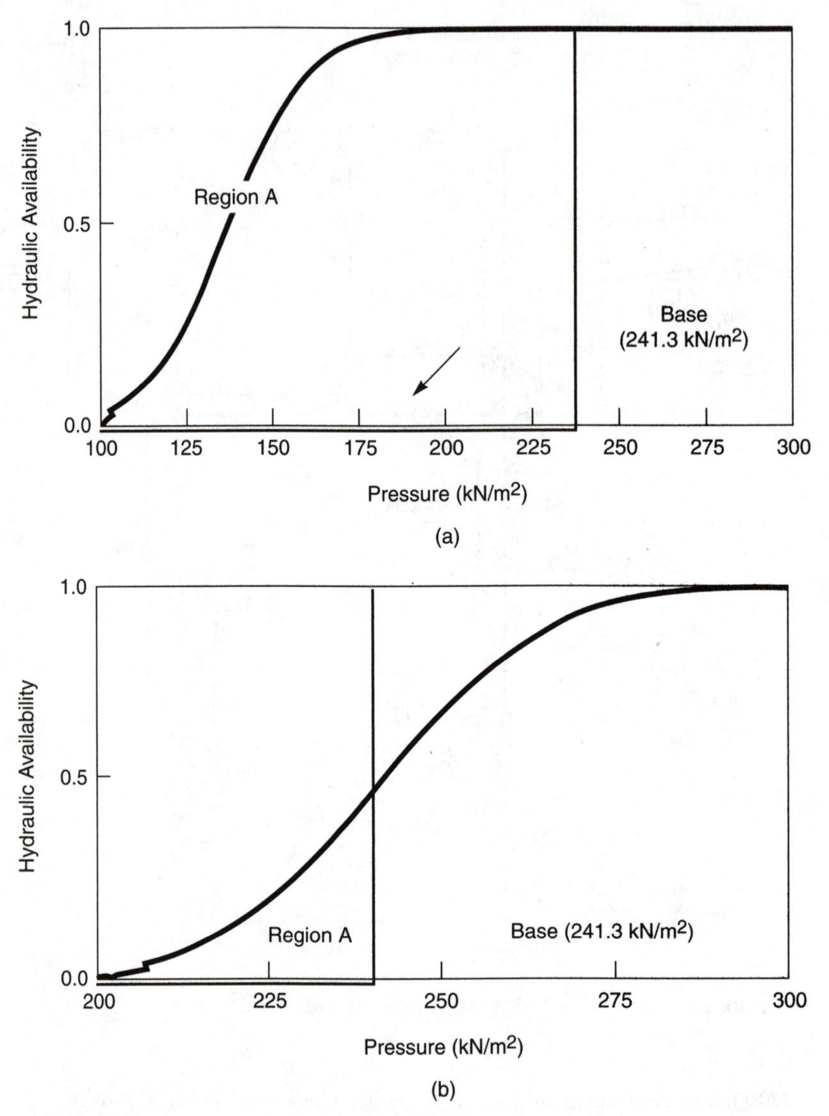

FIGURE 6.8 (*a*) Continuous hydraulic availability function (case 1). (*b*) Continuous hydraulic availability function (case 2). (Shinstine et al., 2002)

As described above, although the pressure heads were less than the minimum required pressure, i.e., the base condition pressure in region A (Fig. 6.8), nodes with pressures in that range were assigned to nonzero hydraulic availability. From the selected parameters and resulting shape of the curves, it is seen that the estimated reliability/availability (R/A) for case 1 will be larger than for case 2, since case 1 has a larger range of nonzero hydraulic availabilities.

Results. Tables 6.2 to 6.4 are used to show the results of applying the methodology to the WSA-C6 water service area in Tucson, Arizona. Four possible measures were analyzed for

TABLE 6.3 WSA-C6 System Reliability Results with Step HA Function

Mechanical reliability (measure 1)				Mechanical availability (measure 3)			
Condition	Cut-sets	System reliability, %	ΔR_s	Condition	Cut-sets	System availability, %	ΔA_s
Pressure, kN/m^2				Pressure, kN/m^2			
6.9	243	96.39231	0.380	6.9	243	99.98991	0.001
137.9	243	96.39231	0.380	137.9	243	99.98991	0.001
241.3	293	96.01107	0	241.3	311	99.98922	0
310.3	944	75.15366	−20.860	310.3	944	99.93055	−0.059
Demand				Demand			
50%	301	96.38246	0.370	50%	391	99.99281	0.004
120%	411	96.00548	−0.010	120%	411	99.93190	−0.057
150%	412	96.00449	−0.010	150%	500	99.93055	−0.059
200%	576	96.00325	−0.010	200%	659	99.93055	−0.059
Roughness				Roughness			
−10%	309	96.00979	−0.001	−10%	309	99.93055	−0.059

Note: ΔR_s and ΔA_s = difference in reliability and availability, respectively, relative to base condition: 241.3 kN/m^2 (35 psi).

TABLE 6.4 System Reliability and Availability Using Step Hydraulic Availability Function

	First-order pipe sets only	First- and second-order pipe sets
Results using Tucson failure rates		
System reliability* (%)	96.15006	96.01107
System availability† (%)	99.97846	99.98922
Results using St. Louis failure rates on Tucson water systems		
System reliability* (%)	18.44431	0
System availability† (%)	99.54907	99.77650

Note: Base condition: 241.3 kN/m^2 (35 psi).
*Using mechanical reliability (measure 1).
†Using mechanical availability (measure 3).

hydraulic reliability and availability in terms of the shape of hydraulic availability (HA) function (step or continuous):

1. Step hydraulic availability with mechanical reliability [Eq. (6.32) with Fig. 6.3]
2. Continuous hydraulic availability with mechanical reliability [Eq. (6.32) with Fig. 6.4]
3. Step hydraulic availability with mechanical availability [Eq. (6.39) with Fig. 6.3]
4. Continuous hydraulic availability with mechanical availability [Eq. (6.39) with Fig. 6.4]

Measures 1 and 2 are described as reliability measures, and measures 3 and 4 are availability terms. Results from the four measures are compared (Tables 6.3 and 6.4), and design proposals (Table 6.5) to improve R/A are discussed for a few cases.

To simulate pipe aging, the roughness coefficient was decreased by 10 percent. The reliability remained the same as the base condition. Since hydraulic failure did not occur until a large demand change was imposed, the lower *C* values did not cause higher head losses that could alter the system R/A.

All results were compared to the base conditions of 241.3 kN/m² (35 psi) and average (normal) demand, with a system reliability for the base conditions of 96.01 percent (Table 6.5). The reliability for the system at the operational pressure of 241.3 kN/m² (35 psi) included conditions resulting in hydraulic and mechanical failure. Noteworthy, at node 37 in WSA-C6, due to its elevation, the pressure runs slightly lower (227.9 kN/m² or 33.1 psi) than the operational minimum of 241.3 kN/m² (35 psi). Therefore, the minimum pressure at this node was set at 206.8 kN/m² (30 psi). At 137.9 kN/m² (20 psi), only mechanical failure was observed. In other words, only cut-sets that geometrically disconnect the system influence the reliability at lower pressures. However, only a 0.4 percent decrease in reliability occurred when considering hydraulic failure at the 241.3-kN/m² (35-psi) requirement.

Pressure requirement, PR, changes had a more significant impact on R/A. The reliability dropped dramatically when the pressure requirement was increased to 36.3 kN/m² (45 psi) (Table 6.3). Many nodes with pressures between 241 and 310 kN/m² (35 and 45 psi) are considered as failing when the PR is increased.

Perhaps the most interesting result from this work is the relatively high values of reliability and, even more so, availability. In all cases, availability was greater than 99.9 percent of the time (364.8 of 365 days). When failure does occur, it may be low pressure or limited periods without service for only a portion of the network. Recall that the mean time to repair after identification of a failure was 1 day. Examination of nodal reliabilities is useful to determine specific system weaknesses.

As expected, the system availability is higher in all cases compared to the system reliability. The probability failure is higher than the probability of being in a failed condition since the pipes are likely

TABLE 6.5 WSA-C6 Design Alternatives

Design alternative	Cut-sets	Reliability, % (measure 1)*	ΔR_{s1}	Reliability, % (measure 2)†	ΔR_{s2}	Availability, % (measure 3)*	ΔA_{s3}
Base	293	96.01107	0	96.39186	0	99.98922	0
A (add pipes between nodes 9-3 & 10-4)	185	96.04030	0.03	96.42105	0.029	99.98922	0
B (add pipe between nodes 6-32)	193	96.03641	0.02	96.41730	0.025	99.9922	0
C (add booster pump adjacent to node 39)	267	96.37476	0.36	96.48752	0.096	99.99027	0.001
A & C (add pipes in the northwest zone & booster pump in north zone)	167	96.39935	0.39	96.49693	0.105	99.99027	0.001

Note: FGN = fixed grade node. ΔR_s and ΔA_s = difference in reliability and availability relative to base condition.
*Base condition: 241.3 kN/m²
†Base condition: mean = 137.9 kN/m², H_{min} = 103.4 kN/m², H_{max} = 172.4 kN/m².

repaired very quickly. In addition, the continuous hydraulic availability relationship results in higher reliability and availabilities compared to the step-function relationship. This result is also expected as described above regarding the hydraulic availability functions.

Equations (6.32) and (6.39) have been truncated at the second-order cut-set term. To see the effect of considering different orders of cut-sets, the system R/A for first- and second-order pipe sets for the step hydraulic availability function are listed in Table 6.4. The reliabilities are essentially the same with and without the second-order pipe sets. This clearly shows that the probability of two pipes failing simultaneously is very small and justifies the assumption that considering at most higher-order failures is unnecessary. However, it is notable and expected that the system reliability decreases with the inclusion of second-order pipe sets. The probability of failure conditions is subtracted from 1 in the reliability relationship [Eq. (6.32)] and more failures occur when second-order cut-sets are included. Availability, on the other hand, sums the conditions that provide acceptable operations, so they increase very slightly in WSA-C6 when second-order sets are included. Because of significant differences in results between first- and second-order pipe sets, consideration of second-order sets is important for reliability measures with high failure rates.

Because of mechanical failure and disconnections on branch portions of the networks with the higher failure rates, the reliability was significantly impacted (to less than 10 percent for both systems). Since it properly considers pipe repair, the availability fell only slightly. Under those conditions, the general network layout would probably not be modified significantly if the network was located in St. Louis rather than Tucson.

System Reliability and Design Changes. The availability of this network is already quite high. Parallel pipes have apparently been added for redundancy and provide additional reliability (e.g., pipes 48 and 86). However, a number of smaller areas are supplied through a single pipe (e.g., southwest downstream of node 63). Examining nodal R/A values, the north zone (nodes 39 and 40) is most affected by hydraulic failure and the northwestern zone is affected by a combination of mechanical and hydraulic failures. Thus, unlike WSA-W1, the weaknesses are dispersed throughout the system. Several alternatives were examined to improve matters on the north side. Referring to Table 6.5, no alternative provided significant improvement with the booster pump ensuring pressure in the north region giving the maximum return. Alternatives A, C, and A & C suggest that the changes are independent of their influence on the system.

6.4 OTHER APPROACHES TO ASSESSMENT OF RELIABILITY

Consider in the instance, a reliability measure based on deficits in volumes of water supplied. In recognition of (1) the need to incorporate both demand failure and component failure (2) the fact that demands are not actually located at nodes as is normally assumed in network analysis and design, and (3) the probability nature of the events that give rise to failure, Bouchart and Goulter (1991, 1992) formulated the following measures for *volume deficits*.

1. Volume deficit arising from "demand variation failure"

$$\mathrm{ED}_d^{\eta} = K \int_{Q_{de}^{\eta}}^{\infty} (Q^{\eta} - Q_{de}^{\eta})\, p(Q^{\eta})\, dQ^{\eta} \tag{6.40}$$

where ED_d^{η} = expected volume of deficit arising from demand at node η being larger than design value Q_{de}^{η} at that node

Q^{η} = demand per unit time at node η

K = factor to convert expected flow rates to expected volumes

Q_{de}^{η} = design demand at node η

$p(Q^{\eta})$ = probability of demand Q_{η} at node η

2. Volume deficit arising from "component (pipe) failure"

$$\mathrm{ED}_f^s = q_d t_f E(\mathrm{NF}) \tag{6.41}$$

where ED_f^s = expected volume of deficit arising from a section of pipe in links failing and having to be isolated for repair

q_d = average demand per unit time in section of pipe isolated

t_f = average duration of repair

$E(\mathrm{NF})$ = average number of failures per unit time of pipe section (equal to length of pipe section between valves multiplied by number of breaks per unit length of pipe)

The expected deficit for the volume deficits due to component (pipe) failure and demand failure now have the same units. Hence, they can be added directly to give the total expected volume of deficit over the network, i.e.,

$$\mathrm{TND} = \sum_{\text{all links}_s} \mathrm{ED}_{f^s} + \sum_{\text{all links}_\eta} \mathrm{ED}_d^\eta \tag{6.42}$$

where TND is the total expected volume of deficit due to both component failure and demand variation failure.

It should be recognized that the statement of reliability in Eq. (6.42) as used by Bouchart and Goulter (1991, 1992) does not address the impacts on the rest of the network caused by the reduction in hydraulic of the network associated with removal of one of the components. For the same reason, it examines the impacts of pump or valve failure (valve failure requires the isolation of the section of the network containing the valve so that the valve can be repaired and has some of the characteristics as pipe failure). The important aspect of the approach is that it recognizes both demand variation failure and component failure in a probabilistic sense. However, the technique is only concerned with purely volumetric deficit and does not take into consideration the ability of a network to meet flow demands at reduced pressures.

An approach for measurement of reliability which does take into account the pressure of delivery formulated and applied by Cullinane (1986) and Cullinane et al. (1992) is based on the concept of availability, which is defined as the percentage of time the pressures in the network are below some preset level. In the case of a node, availability represents the percentage of time the pressure at the node is less then some predetermined level, i.e.,

$$A_j = \sum_{i=1}^{M} \frac{A_i t_i}{T} \tag{6.43}$$

where A_j = nodal availability at node j

A_i = availability during time period i

t_i = length of time period i

M = number of time period i

T = time of simulation = time horizon for analysis of reliability

The network can be defined as

$$A = \sum_{j=1}^{N} \frac{A_j}{N} \tag{6.44}$$

where A is the network availability and N is the total number of nodes. This measure is capable of considering both component (mechanical) and demand variation failure. The technique can also be extended to include the probabilistic aspects of mechanical and demand variation failure though the following expression:

$$\mathrm{AE}_{jl} = \mathrm{R}_{jl} \cdot \mathrm{A}_{jl} + Q_{jl} \cdot A_j \tag{6.45}$$

where AE_{jl} = expected value of availability at node j, considering component l
R_{jl} = probability that component l is operational
Q_{jl} = probability that component l is not operational
A_{jl} = availability of node j with component l operational
A_j = availability of node j with component l nonoperational

An interesting feature of the discussion by Cullinane et al. (1992) of this measure was their observation that "the judgement and experience of the modeler-designer is critical in system evaluation." The importance and relevance of this observation is discussed later in relation to simulation models for assessment of network reliability.

It should be noted that demand variation failure is incorporated in Eq. (6.45) through the R_{jl} term in which failure to meet the predetermined pressure can only occur through a condition arising from demand failure. On the other hand, both component failure and demand failure are recognized in the $Q_j \cdot A_j$ term; component failure specifically through the Q_{jl} term; with the demand failure condition imposed through the A_j term. However, Eq. (6.45) does not explicitly consider the probabilistic aspects of demand failure. Consideration of the probabilistic aspects may be obtained by modifying:

$$\mathrm{AE}_{jl}^{T} = \int_0^{\infty} (R_{jl} A_{jl} + Q_{jl} A_j)\, p(Q^{\eta})\, dQ^{\eta} \tag{6.46}$$

where AE_{jl}^{T} is the expected availability at node j considering demand variation and mechanical failure and all other terms are as described previously.

Although Eq. (6.46) recognizes deficit in pressure as a factor to be addressed in defining network failure, the measure is not able to distinguish between failure conditions involving a large number of failures, each of which results in relatively small decreases in pressure below the minimum acceptable, and failure conditions involving a small number of failures, each of which causes a significant decrease in pressure below the minimum acceptable. For example, a pipe failure that results in a reduction in delivery pressure heads at a node from 30 to 10 m is considerably more serious than a pipe failure of the same duration that results in reduction in delivery pressure head at the same node from 30 to 29 m. Equation (6.46) cannot distinguish between these failures.

One approach to the problem is to use a measure of the following general form:

$$\mathrm{ND} = \sum_{i=1}^{N} q^j H_{\min} - \sum_{i=1}^{N} q_a^j H_{aj} \tag{6.47}$$

$$= Q_T^* H_{\min} - \sum_{j=1}^{N} q_a^j H_{aj} \tag{6.48}$$

where q^i = demand at node j
h = total number of nodes
Q_T^* = total network demand
q_a^j = actual flows supplied at node j
H_{aj} = actual delivery pressure at node j if $H_{aj} > H_{\min}$ set $H_{aj} = H_{\min}$
ND = network deficit

Note that this measure does not consider the stochastic nature of the demands and applies only for known deterministic nodal demands. The primary strength of this measure lies, however, in its ability to recognize explicitly the relationship between flows and pressures. If all flow demands are met, then any deficit is solely due to deficits in supply pressure. If all demand pressures are satisfactory, then any deficit is due to the supply not meeting the flow demand. In both cases, the size of the deficit increases as the amount by which the flow or pressure requirements are not met increases. However, if both flow and pressure requirements are not met simultaneously, i.e., flows less than flow demands are supplied and these flows are only able to be supplied at pressures below minimum acceptable, then the products of the q and H terms in Eq. (6.48) will be even less than if only one aspect of demand was not met. In this sense the expression is very comprehensive in its interpretation of reliability.

However, the units of the measure are volume-pressure, i.e., (m^3/h)(m) and as such do not directly represent reliability as engineers tend to know or assess it. For this reason the expression should be considered as a heuristic measure of reliability wherein it is known that if the deficit determined by the measure decreases, the reliability of the system is improving. Additionally, this expression does not consider the duration or probabilities of the events that give rise to pressure or supplied flow rates being less than the requirements. These failures can be incorporated in the same fashion as they were for the expected availability measure of Eq. (6.46) by replacing the Q_j and A_{jl} terms in that equation by elements from the right side of Eq. (6.48) disaggregated by node, i.e.,

$$\mathrm{ED}_{jl} = R_{jl}[q^j H_{\min}] + Q_{jl}[q_a{}^i H_{aj}] \tag{6.49}$$

with the *total network deficit* given by

$$\mathrm{TND}' = \sum_{j=1}^{n} \mathrm{ED}_{jl} \tag{6.50}$$

where ED_{jl} is the expected deficit at node j measured by the product of flow and pressure deficits and TMD is the total expected network deficit as measured by the product of flow and pressure deficits and without recognition of the stochastic nature of the demands.

Incorporation of demand variation can be achieved for each node by the following expression:

$$\mathrm{ED}_{jl}^{T} = \int_0^\infty \mathrm{ED}_{jl} p(Q^n)\, dQ^n \tag{6.51}$$

Even when these probabilistic aspects have been incorporated in this manner, the measure only gives the average condition; it does not provide any indication of the extent to which the deficit varies around this mean; i.e., again it does not indicate whether the mean is caused by a small number of very large and possibly unacceptable deficits or a large number of acceptable deficits.

This issue aside, the above discussion might indicate that the problem of assessing and measuring reliability in water distribution networks has been essentially solved. However, this is clearly not the case. It is generally accepted in the literature, e.g., Cullinane et al. (1992), Goulter (1992, 1995a) that there is currently no method or measure for assessment of reliability that is both comprehensive in its interpretation of reliability and computationally practical. Goulter et al. (2000) discuss the simulation and analytical models presently available for assessment of reliability in water distribution networks and highlight the strengths and weaknesses of the approaches.

6.5 RELIABILITY-BASED DESIGN OPTIMIZATION MODELS

6.5.1 Framework for Reliability-based Design Model

Su et al. (1987) developed a basic framework for a model that can be used to determine the optimal (least-cost) design of a water distribution system subject to the various design constraints and also subject to reliability constraints. The optimization model for determining pipe diameters can be stated as follows:

Minimize cost subject to

- Conservation of flow and conservation of energy constraints
- Pressure head bounds to satisfy pressures throughout the system
- Reliability constraints [Eq. (6.21)] for both system and demand nodes

The solution procedure used a nonlinear optimization model interfaced with the KYPIPE simulation model so that the hydraulic constraints (conservation of flow and conservation of energy) were actually solved by the simulator, at each iteration of the optimization model, and pressure requirements were guaranteed through a penalty function method for the optimization procedure.

6.5.2 Reliability-based Optimization Model for Operation Considering Uncertainties of Water Quality

Consider a water distribution system with M pipes; N total nodes, of which there are K junction nodes and S storage nodes (tank or reservoir); and P pumps which are operated for time period T. The basic optimization model for water distribution system operation can be stated in general form as

$$\text{Minimize } z = \text{minimize } k \sum_{p,t} \frac{Q_{pt}\text{TDH}_{pt}\text{UC}_t\text{D}_{pt}}{\text{EFF}_{pt}} \tag{6.52}$$

subject to

$$\sum_i (q_{ik})_t - \sum_j (q_{kj})_t - Q_{kt} = 0 \qquad \forall i, j, \in M;\ k = 1, \ldots, K;\ t = 1, \ldots, T \tag{6.53}$$

$$h_{it} - h_{jt} = f(q_{ij})_t \qquad \forall i,j \in M;\ t = 1, \ldots, T \tag{6.54}$$

$$y_{st} = y_{s(t-1)} + \frac{q_{s(t-1)}}{A_s} \Delta t \qquad s = 1, \ldots, S;\ t = 1, \ldots, T \tag{6.55}$$

$$\frac{\delta(C_{ij})_t}{\delta t} = -\frac{(q_{ij})_t}{A_{ij}} \frac{\delta(C_{ij})_t}{\delta x_{ij}} + \Theta(C_{ij})_t \qquad \forall i, j \in M;\ t = 1, \ldots, T \tag{6.56}$$

$$\underline{H}_{kt} \leq H_{kt} \leq \overline{H}_{kt} \qquad k = 1, \ldots, K;\ t = 1, \ldots, T \tag{6.57}$$

$$\underline{y}_{st} \leq y_{st} \leq \overline{y}_{st} \qquad s = 1, \ldots, S;\ t = 1, \ldots, T \tag{6.58}$$

$$\underline{C}_{nt} \leq C_{nt} \leq \overline{C}_{nt} \qquad n = 1, \ldots, N;\ t = 1, \ldots, T \tag{6.59}$$

The objective function [Eq. (6.52)] minimizes the total energy cost as a function of the power rate per kilowatt hour of pumping during time period $t(\text{UC}_t)$, flow from pump p during time period $t(Q_{pt})$, operating point total dynamic head for pump p during time period $t(\text{TDH}_{pt})$, efficiency of pump p in time period t (EFF_{pt}), and the state (on or off) of pump p during time period $t(D_{pt})$, wherein the pump is either on ($D_{pt} = \Delta t$) or off ($D_{pt} = 0$) during the period. Constraint equations (6.53) and (6.54) are the hydraulic constraints: conservation of mass and conservation of energy, respectively. In these equations, $(q_{ij})_t$ is the flow in pipe m connecting nodes i and j at time t, Q_{kt} is the flow consumed (+) or supplied (−) at node k at time t, h_{it} is the hydraulic grade line elevation at node i at time t, $f(q_{ij})_t$ is the functional relation between head loss and flow in a pipe connecting nodes i and j at time t.

Constraint equation (6.55) states that the height of water stored at a storage node for the current time period y_{st} is a function of the height of water stored from the previous time period and the flow entering or leaving the tank during the period Δt. In this equation A_s is the cross-sectional area of a tank and q_s is the flow entering or leaving the tank during period $t - 1$. The water quality constraint equation (6.56) is the conservation of mass of the chlorine in each pipe m connecting nodes i and j in the set of all pipes M. In this equation $(C_{ij})_t$ is the concentration of chlorine in pipe m connecting nodes i and j as a function of distance and time, x_{ij} is the distance along the pipe connecting nodes i and j, A_{ij} is the cross-sectional area of the pipe connecting nodes i and j, $\Theta(C_{ij})_t$ is the rate of reaction of the chemical within the pipe connecting nodes i and j at time t. Equations (6.57) to (6.59) define the lower and upper bounds on the pressure heads at node k at time t, the lower and upper bounds on the depth of water stored at storage node s at time t, and the lower and upper bounds on chlorine concentration at node n at time t, respectively.

Since the minimum and maximum residual chlorine concentrations are uncertain, they are chosen as independent random variables from the viewpoint of operation. Equation (6.59) is replaced by a probabilistic statement known as a *chance constraint.* The chance-constrained formulation of the reduced problem is to optimize the objective function, Eq. (6.52), subject to Eqs. (6.57) and (6.58) and the following chance constraint:

$$P[\underline{C}_{nt} \le C_{nt} \le \bar{C}_{nt}] \ge \alpha_{Ct} \qquad n = 1, \ldots, N; t = 1, \ldots, T \tag{6.60}$$

The storage bound constraint [Eq. (6.58)] has to be considered if the desired minimum and maximum limits of the tank water level storage heights are different than the allowable minimum and maximum tank water elevations which are due to physical limitations of the storage tanks. Otherwise, this constraint is automatically satisfied within the simulation procedure by keeping the storage heights within the given allowable values of the minimum and maximum tank water elevations. Constraint equation (6.60) is expressed as the probability $P(\ )$ that the residual chlorine concentration at each demand node during the time t is between lower and upper bounds with a probability level of α_{Ct}. In general the value of the constraint performance reliability α_{Ct} can be specified and manipulated to consider the effect of uncertainty. The above model [Eqs. (6.52), (6.56), (6.58), and (6.60)] can be transformed into a deterministic form using the concept of the cumulative probability distribution (Mays and Tung, 1992).

The chlorine residual data for water distribution systems are minimal or nonexistent, and it is difficult to assign the appropriate probability distribution for the residual chlorine concentration. The minimum and maximum chlorine residual concentration variables were assumed to follow either the normal or lognormal probability distribution. The deterministic form of the chance constraint expressed by Eq. (6.60) is

$$\mu_{\underline{C}_{nt}} + \sigma_{\underline{C}_{nt}} \cdot z_{(1+\alpha_{Ct})/2} \le C_{nt} \le \mu_{\bar{C}_{nt}} + \sigma_{\bar{C}_{nt}} z_{[1-(1+\alpha_{Ct})/2]} \tag{6.61}$$

for the normal distribution or

$$\exp\{\mu_{\ln \underline{C}_{nt}} + \sigma_{\ln \underline{C}_{nt}} z_{(1+\alpha_{Ct})/2}\} \le C_{nt} \le \exp\{\mu_{\ln \bar{C}_{nt}} + \sigma_{\ln \bar{C}_{nt}} z_{[1-(1+\alpha_{Ct})/2]}\} \tag{6.62}$$

for the lognormal distribution.

The reduced deterministic formulation of the chance-constrained model is expressed by the objective function and constraints, Eqs. (6.52), (6.57), (6.58), and (6.61) or (6.62). The model is a programming problem in which Eq. (6.61) or (6.62) is treated as a simple bound constraint because the right-hand side and the left-hand side are known; in fact, $\mu_{\bar{C}_{nt}}$, $\sigma_{\bar{C}_{nt}}$, $\mu_{\underline{C}_{nt}}$, $\sigma_{\underline{C}_{nt}}$, and α_{Ct} are all specified. The simulated annealing code, by Goldman (1998) and described in Chap. 5, can be used to solve the deterministic form of the chance-constrained model [Eqs. (6.52), (6.57), (6.58), and (6.61) or (6.62)]. This model links the EPANET simulation model with the method of simulated annealing to optimize the operation of water distribution systems for hydraulic and water quality behavior.

6.6 REFERENCES

Awumah, K., and I. Goulter, "Maximizing Entropy-defined Reliability of Water Distribution Networks," *Engineering Optimization,* 20(11):57–80, 1992.

Awumah, K., I. Goulter, and S. Bhatt, "Entropy-based Redundancy Measures in Water Distribution Network Design," *Journal of Hydraulic Engineering,* ASCE, 117(3):595–614, 1991.

Bao, Y. X., and L. W. Mays, "Model for Water Distribution System Reliability," *Journal of Hydraulic Engineering,* ASCE, 116(9):1119–1137, 1990.

Bao, Y. X., and L. W. Mays, "New Methodology for Optimization of Freshwater Inflows to Estuaries," *Journal of Water Resources Planning and Management,* ASCE, 120(2):199–217, 1994a.

Bao, Y. X., and L. W. Mays, "Optimization of Freshwater Inflows to the Lavaca-Tres Palacios, Texas, Estuary," *Journal of Water Resources Planning and Management,* ASCE, 120(2):218–236, 1994b.

Biem, G., and G. Hobbs, "Analytical Simulation of Water System Capacity Reliability. 2: A Markov-Chain Approach and Verification of Models," *Water Resources Research,* 24(9):1445–1458, 1988.

Billinton, R., and R. N. Allan, *Reliability Evaluation of Engineering Systems: Concepts and Techniques.* Pitman Books Limited, London, 1983.

Boccelli, D. L., M. E. Tryby, J. G. Uber, L. A. Rossman, M. L. Zierolf, and M. Polycarpou, "Optimal Scheduling of Booster Disinfection in Water Distribution Systems," *Journal of Water Resources Planning and Management,* ASCE, 124(2):99–11, 1998.

Bouchart, F., and I. C. Goulter, "Implications of Pipe Failure on the Hydraulic Performance of Looped Water Distribution Networks," *Proceedings of the First Caribbean Conference on Fluid Dynamics,* St. Augustine, Trinidad, 1989, pp. 291–297.

Bouchart, F., and I. Goulter, "Reliability Improvements in Design of Water Distribution Networks Recognizing Valve Location," *Water Resources Research,* 27(12):3029–3040, 1991.

Bouchart, F., and I. Goulter, "Selecting Valve Location to Optimize Water Distribution Network Reliability," *Proceedings of the 6th IAHR International Symposium on Stochastic Hydraulics,* J.-T. Kuo and G.-F. Lin (eds.), May 18–20, 1992, Taipei, Taiwan, 1992, pp. 155–162.

Cullinane, M. J., "Reliability Evaluation of Water Distribution System Components," *Proceedings of the Specialty Conference on Hydraulics and Hydrology in the Small Computer Age,* ASCE, New York, 1985, pp. 353–358.

Cullinane, M. J., "Hydraulic Reliability of Urban Water Distribution Systems," in M. Karamoyz et al. (eds.), *Water Forum 1986: World Water Issues in Evolution,* ASCE, 1986, pp. 1264–1271.

Cullinane, M. J., K. E. Lansey, and C. Basnet, "Water Distribution System Design Considering Component Failure During Static Conditions," *Proceedings of the National Conference on Hydraulic Engineering,* ASCE, New York, 1989, pp. 762–767.

Cullinane, M., K. Lansey, and L. Mays, "Optimization-Availability Based Design of Water-Distribution Networks," *Journal of Hydraulic Engineering,* ASCE 118(3):420–441, 1992.

Duan, N., and L. Mays, "Reliability Analysis of Pumping Systems." *Journal of Hydraulic Engineering,* ASCE, 116(2):230–248, 1990.

Duan, N., L. Mays, and K. Lansey, "Optimal-based Design of Pumping and Distribution Systems," *Journal of Hydraulic Engineering,* ASCE, 116(2):249–268, 1990.

Fletcher, R., "An Ideal Penalty Function for Constrained Optimization," *Journal of Institute of Mathematics and Its Applications,* 15:319–342, 1975.

Fujiwara, O., and A. de Silva, "Algorithm for Reliability Based Optimal Design of Water Networks," *Journal of Environmental Engineering,* ASCE, 116(3):575–587, 1990.

Fujiwara, O., and T. Ganesharajah, "Reliability Assessment of Water Supply Systems with Storage and Distribution Networks, *Water Resources Research,* 29(8):2917–2924, 1993.

Fujiwara, O., and J. Li, "Reliability Analysis of Water Distribution Networks in Consideration of Equity, Redistribution, and Pressure Dependent Demand," *Water Resources Research,* 34(7):1843–1850, 1998.

Fujiwara, O., and H. Tung, "Reliability Improvement for Water Distribution Networks through Increasing Pipe Size," *Water Resources Research,* 27(7):1395–1402, 1991.

Germanopoulis, G., P. Jowitt, and J. Lumbers, "Assessing the Reliability of Supply and Level of Service for Water Distribution Systems," *Proceedings of the ICE,* Part 1, 80 (Apr.):413–428, 1986.

Goldman, F. E., "The Application of Simulated Annealing for Optimal Operation of Water Distribution Systems," Ph.D. dissertation, Department of Civil and Environmental Engineering, Arizona State University, Tempe, Ariz., 1998.

Goulter, I. C., "Measures of Internal Redundancy in Water Distribution Network Layouts," *Journal of Information and Optimization Science,* 9(3):363–390, 1988.

Goulter, I. C., "Systems Analysis in Water Distribution Network Design: From Theory to Practice." *Journal of Water Resources Planning and Management,* ASCE, 118(3):238–248, 1992a.

Goulter. I. C., "Modern Concepts of a Water Distribution System: Policies for Improvement of Networks with Shortcomings," in E. Cabrera and F. Martinez (eds.), *Water Supply Systems: State of the Art and Future Trends,* Computational Mechanics Publications, Southampton, U.K., 1992b, pp. 119–138.

Goulter, I. C., "Analytical and Simulation Models for Reliability Analysis in Water Distribution Systems," in E. Cabrera and A. Vela (eds.), *Improving Efficiency and Reliability in Water Distribution Systems,* Kluwer Academic Publishers, Dordrecht, Netherlands, 1995a, pp. 235–266.

Goulter, I. C., "Measures of Internal Redundancy in Water Distribution Network Layouts," *Journal of Information and Optimization Science,* 9(3):363–390, 1995b.

Goulter, I., and F. Bouchart, "Reliability-Constrained Pipe Network Model," *Journal of Hydraulic Engineering,* ASCE, 116(2):211–229, 1990.

Goulter, I., and A. Coals, "Quantitative Approaches to Reliability Assessment in Pipe Networks," *Journal of Transportation Engineering,* ASCE, 112(3):287–301, 1986.

Goulter, I., and P. Jacobs, "Discussion of 'Water Distribution' Reliability: Analytical Method by Wagner, Shamir and Marks," *Journal of Water Resource Planning and Management,* ASCE, 115(5):709–711, 1989.

Goulter, I., T. Walski, L. W. Mays, A. B. Sakarya, F. Bouchart, and Y. K. Tung, "Reliability Analysis for Design," Chap. 18 in L. W. Mays (ed.), *Water Distribution Systems Handbook,* McGraw-Hill, New York, 2000.

Gupta, R., and P. Bhave, "Reliability Analysis of Water Distribution Systems," *Journal of Hydraulic Engineering,* ASCE, 120(2):447–460, 1994.

Henley, E. J., and H. Kumamoto, *Reliability Engineering and Risk Assessment,* Prentice-Hall, Englewood Cliffs, N.J., 1981.

Hobbs, B., and G. Biem, "Analytical Simulation of Water System Capacity Reliability: Modified Frequency-Duration Analysis," *Water Resources Research,* 24(9):1431–1444, 1988.

Jacobs, P., "A Removal Set Based Approach to Urban Water Supply Distribution Network Reliability," Ph.D. thesis, University of Manitoba, Winnipeg, 1992.

Jacobs, P., and I. Goulter, "Evaluation of Methods for Decomposition of Water Distribution Networks for Reliability Analysis," *Civil Engineering Systems,* 5(2):58–64, 1988.

Jacobs, P., and I. Goulter, "Optimisation of Redundancy in Water Distribution Networks Using Graph Theoretic Principles," *Engineering Optimization* 15(1):71–82, 1989.

Jacobs, P., and I. Goulter, "Estimation of Maximum Cut-set Size for Water Network Failure," *Journal of Water Resources Planning and Management,* ASCE, 117(5):588–605, 1991.

Jowitt, P., and C. Xu, "Predicting Pipe Failure Effects in Water Distribution Networks." *Journal of Water Resources Planning and Management,* ASCE, 119(1):18–31, 1993.

Kessler, A., L. Ormsbee, and U. Shamir, "A Methodology for Least-cost Design of Invulnerable Water Distribution Networks," *Civil Engineering Systems,* 1(1):20–28, 1990.

Kettler, A. J., and I. C. Goulter, "Reliability Consideration in the Least-Cost Design of Looped Water Distribution Networks," *Proceedings of the 12th International Conference on Urban Hydrology, Hydraulics, and Sediment Control,* University of Kentucky, Lexington, 1983, pp. 305–312.

Kettler, A. J., and I. C. Goulter, "Analysis of Pipe Breakage in Urban Water Distribution Networks," *Canadian Journal of Civil Engineering,* 12(2):286–293, 1985.

Lansey, K., K. Awumah, Q. Zhong, and I. Goulter, "A Supply Based Reliability Measure for Water Distribution System," *Proceedings of the 6th IAHR International Symposium on Stochastic Hydraulic,* J-T. Kuo and G-F Lin (eds.), May 18–20, 1992, Taipei, Taiwan, 1992, pp. 171–178.

Lansey, K., N. Duan, L. Mays, and Y-K Tung, "Water Distribution Design Under Uncertainties," *Journal of Water Resources Planning and Management,* ASCE, 115(5):630–645, 1989.

Mays, L. W. (ed.), *Reliability Analysis of Water Distribution Systems,* American Society of Civil Engineers, New York, 1989.

Mays, L. W. (ed.), *Water Distributions Systems Handbook,* McGraw-Hill, New York, 2000.

Mays, L. W., and Y. K. Tung, *Hydrosystems Engineering and Management,* McGraw-Hill, New York, 1992.

O'Day, K. D., R. Weiss, S. Chiaveri, and D. Blair, *Water Main Evaluation for Rehabilitation/Replacement,* prepared for American Water Works Association Research Foundation, Denver, and USEPA, Cincinnati, 1986.

Ormsbee, L., and A. Kessler, "Optimal Upgrading of Hydraulic-Network Reliability," *Journal of Water Resources Planning and Management,* ASCE, 116(6):784–802, 1990.

Park, H., and J. Liebman, "Redundancy-Constrained Minimum Design of Water-Distribution Nets," *Journal of Water Resources Planning and Management,* ASCE, 119(1):83–98, 1993.

Quimpo, R. J., and U. Shamsi, "Reliability-based Distribution System Maintenance," *Journal of Water Resources Planning and Management,* ASCE, 117(3):321–339, 1991.

Shamir, U., and C. D. Howard, "Water Supply Reliability Theory," *Journal of the American Water Works Association,* 73(7):379–384, 1981.

Shamsi, U., "Computerised Evaluation of Water Supply Reliability," *IEEE Transactions on Reliability,* 39(1):35–41, 1990.

Shinstine, D. S., "Reliability Analysis Using the Minimum Cut-Set Method for Three Water Distribution Systems," M.S. thesis (Civil Engineering), University of Arizona, Tucson, 1999.

Shinstine, D. S., I. Ahmed, and K. E. Lansey, "Reliability Availability Analysis of Municipal Water Distribution Networks: Case Studies", *Journal of Water Resources Planning and Management*, ASCE, 128(2):140–151, 2002.

Su, Y., L. Mays, N. Duan, and K. Lansey, "Reliability Based Optimization for Water Distribution Systems," *Journal of Hydraulic Engineering,* ASCE, 113(12):589–596, 1987.

Tanyimboh. T. T., M. Tabesh, and R. Burrows, "Appraisal of Source Head Methods for Calculating Reliability of Water Distribution Networks," *Journal of Water Resources Planning and Management,* ASCE, 127(4):206–213, 2001.

Tanyimboh, T. T., and A. B. Templeman, "Optimum Design of Flexible Water Distribution Networks," *Civil Engineering Systems,* 10(3):243–258, 1993.

Tanyimboh, T. T., and A. B. Templeman, "A New Method for Calculating the Reliability of Single-Source Networks," *Developments in Computational Techniques for Civil Engineering,* B. H. V. Topping (ed.), Civil-Comp Press, Edinburgh, U.K., 1995, pp. 1–9.

Templeman, A. B., and D. F. Yates, "Mathematical Similarities in Engineering Network Analysis," *Civil Engineering Systems,* 104(1), 1984.

Tung, Y. K., "Evaluation of Water Distribution Network Reliability," *Proceedings of the ASCE Hydraulics Specialty Conference,* New York, 1985, pp. 359–364.

Tung, Y. K., "Uncertainty and Reliability Analysis," in L. W. Mays (ed.), *Water Resources Handbook,* McGraw-Hill, New York, 1996.

Wagner, J., U. Shamir, and D. Marks, "Water Distribution Reliability: Analytical Methods." *Journal of Water Resources Planning and Management,* ASCE, 114(3):253–275, 1988a.

Wagner, J., U. Shamir, and D. Marks, "Water Distribution Reliability: Simulation Methods," *Journal of Water Resources Planning and Management,* ASCE, 114(3):276–293, 1988b.

Wu, S-J., Y-H. Yoon, and R. Quimpo. "Capacity-Weighted Water Distribution System Reliability." *Reliability Engineering and System Safety,* 42:39–46, 1993.

Xu, C., and I. Goulter, "Probabilistic Model for Water Distribution Reliability," *Journal of Water Resources Planning and Management,* ASCE, 124(4):218–228, 1998.

Yang, S.-L., N. S. Hsu, P. W. F. Louie, and W. W.-G. Yeh, "Water Distribution Network Reliability: Connectivity Analysis," *Journal of Infrastructure Systems,* ASCE, 2(2):54–64, 1996a.

Yang, S.-L., N.-S. Hsu, P. W. F. Louie, and W. W.-G. Yeh, "Water Distribution Network Reliability: Stochastic Simulation," *Journal of Infrastructure Systems,* ASCE, 2(2):65–72, 1996b.

CHAPTER 7
PERFORMANCE INDICATORS AS A MANAGEMENT SUPPORT TOOL

Helena Alegre
National Civil Engineering Laboratory
Lisbon, Portugal

7.1 INTRODUCTION

The urban water supply combines three main characteristics that place it in a rather peculiar situation as an industry: it is an essential service for the health and welfare of the populations, a natural monopoly, and a relevant economic activity. The former characteristic, which is self-explanatory, provides great political and social importance; conversely, it requires that the service provided fits the population needs on a sustainable basis. It is a natural monopoly because it requires the construction and use of expensive transport infrastructures that cannot be used by other service providers, either of the same or of a different nature; subsequently, market competition rules are not applicable. The latter characteristic—*economic relevance*—derives from the fact that it requires high capital investments and produces significant management and operating profits; it has transformed this activity into an attractive business.

These three characteristics are sometimes conflicting. On the one hand, the sustainable development of the urban water supply services requires an efficient management of the undertakings. The systems and technologies have grown in complexity in recent years, and demand and expectations on the service are ever-increasing. On the other hand, the level of development of the urban water supply activity tends to be a step behind other activities with much lower economic relevance, due to the lack of market competition stimulation derived from the monopolistic nature of the service. With the growth of the private participation, the management objectives move toward the shareholders' medium- and long-term recovery of their investments. This may sometimes clash with the consumers' interests.

In summary, there is today a strong call for efficiency and good quality of service. The improvement of performance in terms of organizational and management procedures, aiming to satisfy both public needs and expectations and the environmental policy goals, is therefore a major medium-term target. The development and use of performance indicators is nowadays a hot topic of the water industry agenda worldwide. In fact, the use of standardized performance assessment procedures is a key leverage factor for an enhanced performance, able to identify what activities can be improved, to introduce some artificial competition, and to help in establishing contractual agreements that protect the consumers' interests.

This chapter aims to provide responses for the following questions about performance indicators (PIs):

- What is a PI?
- What are the potential uses and benefits of using a PI system?
- What is the current international state of the art of performance assessment?

- Are there PI systems for water supply services publicly available?
- How is a PI system implemented?

An example is presented based on the PI system recommended by the International Water Association.

7.2 CONCEPT OF PERFORMANCE INDICATOR, CONTEXT INFORMATION, AND UTILITY INFORMATION

Performance indicators are measures of the efficiency and effectiveness of the water utilities with regard to specific aspects of the utility's activity and of the system's behavior (Deb and Cesario, 1997). *Efficiency* is a measure of the extent to which the resources of a water utility are utilized optimally to produce the service, while *effectiveness* is a measure of the extent to which the targeted objectives (specifically and realistically defined) are achieved. Each performance indicator expresses the level of actual performance achieved in a certain area and during a given period of time, allowing for a clear-cut comparison with targeted objectives and simplifying an otherwise complex analysis.

According to the definition recommended by the International Water Association, any performance indicator shall be defined as a ratio between variables of the same nature (e.g., %) or of different natures (e.g., $\$/m^3$ or liters per service connection). In any case, the denominator shall represent a characteristic of the system dimension (e.g., number of service connections or total mains length) to allow for comparisons along time, even though the size of the system evolves, or between systems of different sizes. Variables that may vary substantially from one year to the other, particularly if not under the control of the undertaking, shall be avoided as denominators (e.g., annual consumption, which may be affected by weather or other external reasons), unless the numerator varies in the same proportion.

A global system of performance indicators must comply with the following requirements. It must

- Represent all the relevant aspects of the water utility performance, allowing for a global representation of the system by a reduced number of indicators.
- Be suitable for representing those aspects in a true and unbiased way.
- Reflect the results of the managing activity of the undertaking.
- Be clearly defined, with a concise meaning and a unique interpretation for each indicator.
- Include only nonoverlapping performance indicators.
- Require only measuring equipment that targeted utilities can afford; the requirement of sophisticated and expensive equipment should be avoided.
- Be verifiable, which is especially important when the performance indicators are to be used by regulating entities that may need to check the results reported.
- Be easy to understand, even by nonspecialists—particularly by consumers.
- Refer to a certain period of time (1 year is the basic assessment period of time recommended, although in some cases other periods are appropriate).
- Refer to a well-limited geographic area.
- Be applicable to utilities with different characteristics and stages of development.
- Be as few as possible, avoiding the inclusion of nonessential aspects.

However, the performance indicators cannot be adequately interpreted without taking into account a number of factors that do not depend on managing performance but will affect it. For instance, the cost to produce good-quality drinking water depends on the availability and quality of the raw water, characteristics that are generally beyond the undertaking's control, although something managers must deal with. This context information includes external characteristics (e.g., geographic, demographic, economic, climatic), regarding the region, and internal ones, regarding the water utility and the system on focus (Fig. 7.1). This is particularly important when comparing results with reference numbers from the literature or with other undertakings.

The assessment of PIs requires the availability of reliable utility information. This information can be originated from multiple sources within the organization and be generated in many different ways (e.g., accounting system, metering equipment, information systems). As a general principle, the assessment of the PIs will require only data relevant for the decision making at the various levels within the undertaking, and therefore usually available regardless of its use as PI input. Common characteristics of PI utility data are that they should

- Be absolute values
- Refer to the same period of time and geographic area as the PIs they will be used for
- Be as reliable and accurate as required by the decisions made based on them
- Fit the definition of the PIs they are used for

The utility information used as PI input variables partially overlaps with the context information, particularly with the system profile information.

7.3 USERS, BENEFITS, AND SCOPE OF APPLICATION OF PERFORMANCE INDICATORS

For the managers of water undertakings, demands for higher efficiency and effectiveness can be internal or come from the users of the services (direct users), from the politicians and administrators in the City Hall (indirect users), and from the state regulators and policy-making departments (proactive users). Last, but not least, there is continued and intense awareness from the media and from environmental nongovernmental organizations (NGOs) and other pressure groups (proactive users).

A well-devised system of performance indicators can be useful to any of these entities, having the following potential benefits and uses:

For *water utilities:*

- Facilitates better quality and more timely response from managers
- Allows for an easier monitoring of the effects of management decisions
- Provides key information that supports a pro-active approach to management, with less reliance on apparent system malfunctions (reactive approach)

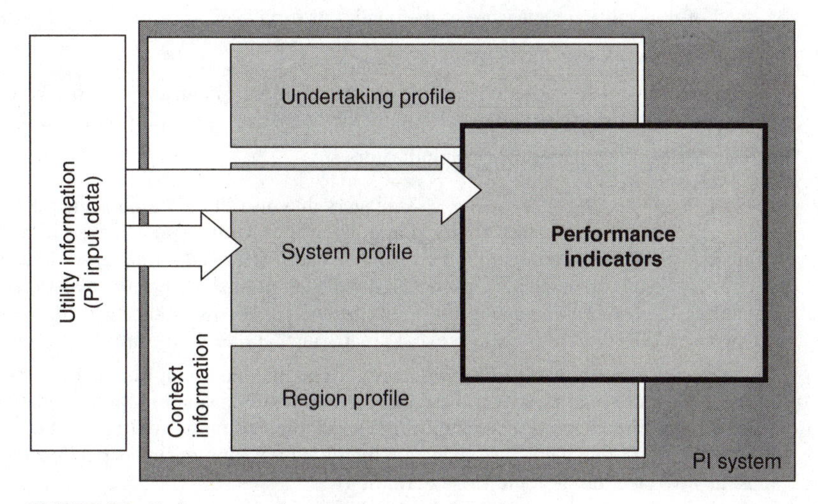

FIGURE 7.1 Performance indicators and context information.

- Highlights strengths and weaknesses of departments, identifying the need for corrective measures to improve productivity, procedures, and routines
- Assists with implementation of a Total Quality Management regime, as a way of emphasizing all-round quality and efficiency throughout the organization
- Facilitates the implementation of benchmarking routines, both internally, for comparing the performance at different locations or systems, and externally, for comparison with other similar entities, thus promoting performance improvements
- Provides a sound technical basis for auditing the organization's workings and predicting the effect of any recommendations made as a result of an audit

For the national or regional *policy-making bodies:*

- Provides an objective means to monitor the undertaking's performance with regard to existing legislation
- Provides a common basis for comparing the performance of water services and identifying possible corrective measures
- Supports the formulation of policies for the water sector, within the integrated management of water resources, including resource allocations, investments, and the development of new regulating tools
- Provides a common reference language to be adopted in the scope of national or international statistic data banks

For *regulatory agencies:*

- Provides key monitoring tools to help safeguard user interests in a monopoly service situation and monitor compliance with contracted goals

For *financing bodies:*

- Provides assistance in assessing investment priorities, project selection, and follow-up

For *quality certifying entities:*

- Provides key monitoring tools

For *auditors*

- Provides a sound technical basis for auditing the organization's workings and can be used to specify some of the recommendations

For *direct users* and *indirect and proactive stakeholders:*

- Provides the means of translating complex processes into simple-to-understand information and of transmitting a measure of the quality of service provided

For *supranational organizations:*

- Provides a very appropriate language for identifying the main asymmetries between regions of the world, their causes and evolution, thereby assisting in defining strategies

The main forms of using performance indicators are

- *Exclusively within the undertaking.* Managers may use PIs to monitor the evolution of performance, comparing the results achieved at a given time with those achieved in previous periods; another form of use is comparing the results actually achieved with preset targets or with reference values from other organizations that have been published. For the two former types of use managers, the undertaking can afford to define and use its own indicators; for the latter, it has to adopt standardized definitions to prevent comparing "apples with pears."
- *In the framework of benchmarking initiatives.* The internal use of PIs can be complemented with metric benchmarking initiatives; these can be held within a small group of undertakings that can agree among them upon a set of indicators, based on a common set of definitions, to be adopted by all the group members. The analysis, interpretation, and eventual publication of the results must always take into account these definitions.

- *As part of a regulatory framework.* The worldwide growth of the private participation in the management of water supply systems demands added responsibility from regulators. The adoption of PI systems to promote artificial competition among water services (comparative regulation or yardstick competition) is gaining new supporters. Also in this case the use of standardized definitions is essential. Normally the PIs used for regulation focus on the quality of service. The increase of efficiency is of the shareholders own interest, and therefore regulators do not have the need to control the internal processes but rather the results.
- *As part of contractual agreements.* Traditionally, the contract agreements between the municipalities and private operators are rather detailed in terms of the financial aspects, but they tend to be less specific in terms of the quality of service to be provided to the consumers. The use of PIs allows scheduled targets to be specified in an objective and auditable way, protecting the consumers' rights. Consequently, PIs are starting to be used in this context. The use of custom PIs is feasible and sometimes necessary, but the use of standardized PIs provides a better assistance for the negotiations, which can be supported on the basis of comparisons with others.
- *As part of quality certification systems.* The monitoring of the organization's internal processes can be undertaken by the use of PIs. As in the previous case, the use of custom PIs is feasible, but the use of standardized PIs is preferable.
- *As part of guaranteed standard schemes.* An increasing number of undertakings have established contracts with their customers that clearly establish their right and duties, based on the so-called guaranteed standards scheme (GSS), which include at least minimum service pressure at the delivery point, maximum time to get a new connection and to repair an existing one, maximum time of written responses, and appointment times to attend customers' premises. In this case the undertakings generally report back to the customers the application of the GSS by monitoring corresponding PIs.
- *In the scope of statistic reports publicly available.* Another form of using PIs is in statistics published by waterworks associations or by regional or international organizations (e.g., World Health Organization, Asian Development Bank, World Bank). Traditionally, this type of performance monitoring was limited to statistical data and economical indicators, but recently reports have started to incorporate other types of indicators. The practice shows that incoherent information is frequently found in these reports, due to the lack of precise definitions. The standardization of definitions is therefore very important in these cases as well.

Table 7.1 defines the main forms of use of a PI system for each type of user.

7.4 STATE OF THE ART OF PERFORMANCE ASSESSMENT

7.4.1 Overview

In the early nineties, the International Water Supply Association (IWSA) selected the topic Performance Indicators for one of its world congresses. No abstracts were submitted under this topic, and it had to be cancelled. The subject did not seem to raise much interest. However, just 3 or 4 years later, the response to an inquiry held in the scope of the IWSA to about 150 senior members of water utilities from all over the world clearly showed that PIs and unaccounted-for water are by far the two topics of greatest interest in the scope of the water transmission and distribution systems. Such rapid evolution deserves some thought.

Independently from their nature (private, public, or combined) and geographic extent, all water utilities comply with a managing logic whose main philosophy may be stated as follows: *greater satisfaction of a greater number of consumers and concerned entities, with the best use of the available resources* (Faria and Alegre, 1996). In terms of the utilities, this is equivalent to stating *greater efficiency and effectiveness of the management.*

However, the need to improve the efficiency and effectiveness is not new and does not explain by itself the current interest in the assessment of PIs. A very brief analysis of some key events in the world can help to understand this trend.

TABLE 7.1 Scope of Application of PI Systems

	Exclusively within the undertaking	In the framework of benchmarking initiatives	As part of a regulatory framework	As part of contractual agreements	In the scope of quality certification systems	Guaranteed standard schemes	In the scope of statistic reports publicly available
Water utilities	✓	✓	✓	✓	✓	✓	✓
Policy-making bodies		✓					✓
Regulatory agencies		✓	✓				
Financing bodies		✓		✓			
Quality certifying entities					✓		
Auditors	✓	✓	✓		✓		
Direct users and indirect and proactive stakeholders		✓				✓	✓
Supranational organizations		✓					✓

7.4.2 Influence of the Growing Private Participation in Undertakings Management

It is not the aim of this section to discuss the comparative advantages and disadvantages of the public and private water services. However, it does aim to present the reasons why the use of PIs is becoming more and more important nowadays.

Most water services in the world are public. The main reason for this is the nature of the service: human life requires the consumption of water, and the health of the population is greatly affected by the quality and availability of this water. It is therefore a public service. However, many countries came to the conclusion that the management of public companies is not efficient, as these companies are often underfunded, and there is a need to change the situation. Enhancing the private sector participation was the solution pointed out in many of these cases. An extreme situation of full privatization occurred in England and Wales. Although there are no parallels to this extreme case, examples of this trend can be found in many countries all over the world. The increasing number of systems run by French companies in Europe, America, the Far East, and Africa is a clear demonstration of this trend.

The requirement to report systematically the performance achieved by means of a framework of PIs is not a general practice in most existing cases. The most relevant exception is what happens in England and Wales, where the utilities must assess and publish their levels of service. The contracts focus mainly on financial aspects and on the physical conditions of the assets, and the private companies are subject to the same legislation as the public companies in terms of service requirements. This option goes in line with the tradition in France, where private companies have for many years been responsible for the management of important water supply systems, in coexistence with public companies. Only very recently have these companies started to demonstrate some interest in the use of PIs. However, it has to be taken into account that the French privatization process was very slow, and there was time enough to stabilize the procedures. Nowadays, these processes tend to be much more rapid, there are normally several companies competing when there is a call for tenders for the management of water and sewerage systems, and there is a consensus of the benefits arising from clear rules and objectives. This is another very important reason that explains the increasing interest for PIs.

Last, but not least, many managers of public undertakings want to demonstrate that they are not less efficient and effective than the private companies. The systematic use of PIs is a good means of reaching this target.

7.4.3 Objective-Oriented Management and Benchmarking

Another factor that has greatly influenced this trend is closely related with the evolution of management procedures in most industrial sectors. Management techniques have changed all over the world, and nowadays there seems to be a general agreement that the implementation of objective-oriented management procedures is an indispensable step for the success of most companies. This approach requires the establishment of clear objectives to be achieved within given deadlines, the comparison between targets and results, and the correction of the causes for the deviations, so that the company's efficiency may improve. Performance indicators are a rather powerful tool in this context, as they allow for clear and quantified comparative measures. Some companies have realized that if they compare themselves to the best ones, and correctly identify the reasons for any discrepancies, they can improve their performance significantly. This is how benchmarking appeared and was successfully used in many industrial sectors (for example, photocopying companies, and car manufacturers). Benchmarking is starting to become popular in the water industry as well, and it is obvious that the comparison between different companies requires the use of standardized PIs.

7.4.4 Lessons Arising from the U.K. Privatization Process

It was mentioned earlier that the privatization process held in England and Wales is a rather special one. It has advantages and disadvantages; it has supporters, but also detractors. One of the criticisms often pointed out about the system is the fact that the British companies spend enormous amounts of money in publicity campaigns to improve their external image and keep the customer happy, instead of reinforcing the investment for improving their physical assets. However, it has to be recognized that the history of water services in England and Wales since the 1970s has had a major impact on the water industry in the world. For this reason, its analysis is fundamental for any country interested in enhancing the private sector partnership.

In 1970 the British government decided to merge several hundreds of medium- and small-size water and wastewater companies into 10 water authorities. The scale effect thus achieved allowed the British water industry to undergo a tremendous technical evolution. The new bodies could afford the human expertise and equipment that the previous companies could not reach due to their small size. During this phase many modern management procedures were implemented, and the efficiency and effectiveness achieved were assessed by the use of PIs.

However, many of the existing systems were rather old, heavy investments were required, and the water authorities could not solve the difficulties without external financial support. It was within this context that the government decided to privatize all water and wastewater services in England and Wales in 1990. Regardless of its advantages and disadvantages, this was an unparalleled initiative worldwide, at least to the author's knowledge. It was a major challenge, particularly due to the monopolistic nature of the industry. There had been some unsuccessful experiences with private companies in the English water sector, and the British government, under criticism from many sides, tried to protect itself by creating a legal framework that would ensure an efficient control of the situation (the 1990 Water Act is the key legal instrument of that time). Before the private players started to perform, all the "rules of the game" were defined. New regulatory bodies were created, and the new water service companies had to report regularly the levels of service achieved to the Office of the Water Services (OFWAT). Therefore, a clear and coherent way to assess those levels of service had to be defined. The previous work already developed by some water authorities inspired the new assessment system. Three main alternatives were available:

- Allow the private companies to act just as the previous public authorities, subject only to the existing legislation
- Create mechanisms to control the activity developed inside the company
- Create mechanisms to control the output of the company

The first option, adopted by other countries in privatization processes, was not acceptable due to the reasons previously presented. The second alternative had some major disadvantages:

- The private companies did not accept easily continuous interference in their activities and therefore would not support this option.
- It is a mistake to support any control mechanisms on data that cannot be easily verifiable; in order to control all the internal procedures in a reliable way, the regulators would need to have huge teams, and the system would hardly be effective.

The third alternative, selected by the British government, has some important advantages:

- The companies can freely keep their "business secrets," provided that the service actually delivered is good enough.
- The consumers, together with the media, help to control the whole process, as actual managing partners.
- The number of variables to monitor and control is much smaller and these variables are verifiable by any audit, whenever appropriate.

The decision to limit the reporting requirements to the *output indicators,* in terms of the actual service delivered, is one of the key reasons why the British experience is so relevant to other countries. This means that the companies were free to adopt the management procedures they considered most appropriate, provided that the tariffs were fair and the service provided to the consumers was good enough. This apparently simple decision represents a major milestone in the history of privatization of water supply systems. Nothing similar had been implemented before. This decision was based on the principle that the regulator should focus its attention on the effectiveness of the utility. It is a target of any private agency to improve its efficiency, in order to increase its profits. Therefore, the regulator does not need to care too much about this side of performance, particularly because the physical assets belong to the private managing body. When this is not the case, complementary measures are necessary, such as the need for periodic independent audits of the physical conditions of the assets.

The presentation of the British experience within this context does not mean it is necessarily an example to follow. On the contrary, it is a singular experience, implemented due to a number of circumstances that are unlikely to happen in any other country. Therefore, it cannot be directly exported elsewhere. However, there are some lessons that any country should learn from it, particularly if the country is going through the process of enhancing the private sector participation in the water industry. In the author's view, the major input of the British experience to the outside world is

- The implementation of a clear regulatory framework applicable to all the water service companies, that ensures that the key needs of the population are safeguarded
- The systematic publication of the results achieved with regard to the quality of service provided to the population by each company, contributing to the creation of a new culture of "open doors" and transparency
- The implementation of competition mechanisms in a sector of natural monopoly
- The implementation of procedures for data quality control and the publication of the degree of confidence associated with each result reported
- The decision of controlling only the output performance, leaving companies with a reasonable degree of freedom to implement their own management strategies
- The capacity for focusing the control mechanisms in a relatively small number of PIs
- The recognition of the increasing role and rights of the consumer

The information published yearly by OFWAT is divided into two main reports: Report on Levels of Service for the Water Industry in England and Wales and Report on the Cost of Water Delivered and Sewage Collected. The former report also receives inputs from two other regulatory bodies, the Drinking Water Inspectorate, which controls drinking water quality, and the Environment Agency, which enforces environmental quality standards and reports on compliance, particularly with regard to the sewerage companies. These reports focus on the following PIs, designated by levels of service:

- Properties at risk of low pressure
- Properties subject to unplanned supply interruptions of 12 h or more
- Population subject to flooding incidents
- Properties at risk of flooding (sewerage services)
- Billing contacts not responded to within 5 working days
- Written complaints not responded to within 10 working days
- Bills not based on meter reading[1]

The process held in England and Wales had a great leverage effect worldwide. Most of the other existing initiatives related to performance assessment and regulation got its influence in one way or another.

7.4.5 The IWA PI System

The International Water Association (IWA)—the largest international association in the water and wastewater field, representing about 130 countries—recently developed a system of PIs for water services, which is currently becoming a reference in the industry. This system is a powerful and timely management tool for water services undertakings, independent from their level of development and climatic, demographic, and cultural characteristics of the regions. It aims to cover the full range of management, water resources, personnel, physical, operational, quality of service, and financial PIs. It aims to be a reference PI language covering the full range of management.

The IWA PI system includes water resources, personnel, physical, operational, quality of service, and financial indicators. It also includes the definition of the information required to establish the utility profile, system profile, and region profile. While this system is being implemented and field-tested by almost 70 undertakings worldwide, a similar system is under development for the wastewater services.

The IWA project on PIs began as long ago as 1997 with establishment of an IWSA[2] task force under the operations and maintenance committee (Alegre et al., 1997). The work has since progressed through more than 20 scientific and technical meetings in Europe, South America, and Africa and with the benefit of helpful written comments and suggestions from over 100 experienced managers, practitioners, and researchers in more than 50 countries from the five continents. In July 2000 this task force produced *Performance Indicators for Water Supply Services* in the IWA Manual of Best Practices Series (Alegre et al., 2000). The primary aim is that the manual should be a useful management tool in water undertakings at any stage of development and regardless of local demographic, climatic, or cultural conditions.

Water undertakings are the main users of the manual, but the position of the industry as a monopoly essential service supplier capable of making significant impacts on the natural environment means that the manual is equally capable of use by national and regional policy makers, environmental and financial regulators, watchdog organizations, and the general public as a means of gauging individual aspects of water supplier performance. The manual has been constructed with these needs in mind and consists of a 160-page textbook and a CD-ROM containing the SIGMA Lite software.

The IWA-PI system is currently becoming a reference in the industry (Merkel, 2001). Many undertakings are adopting it directly as an internal management tool. In other cases it was the starting point for performance assessment approaches, as reported by Denmark, the Czech Republic, Australia, Germany, South Africa, Sweden, and Portugal (Merkel, 2001), often in modified and adapted versions. This system is publicly available and is the basic reference for the following sections of this chapter.

7.4.6 The World Bank Benchmarking Toolkit

It was noted previously that PIs are a powerful tool for financial agencies, as a means for assessing investment priorities, project selection, and subsequent investment follow-up. In fact, agencies such as the World Bank and the Asian Bank seem to be fully aware of this fact and started the development and use of PIs long ago. *Water and Wastewater Utilities Indicators* (Yepes and Dianderas, 1996) is a World Bank publication focusing on water and wastewater services indicators and is one of the former relevant publications on this topic.

More recently the World Bank published a new system of indicators aiming to support benchmarking initiatives. The *Benchmarking Water and Sanitation Utilities* (BWSU), designed primarily for developing regions, aims at facilitating the sharing of cost and performance information between utilities and between countries by creating a network of linked web sites, through a global partnership effort. The BWSU includes 27 indicators on water consumption and production, unaccounted-for water, metering practices, pipe network performance, cost and staffing, quality of service, billings and collection, financial performance, and capital investment. In order to assist users in comparing PI values, three explanatory factors—utility size, range of service, and extent of private sector involvement—are also provided (*www.worldbank.org/html/fpd/water/topics/uom_bench. html*). Currently, the project counts around 130 undertakings from Africa, North and South America, Australia, and Europe (Merkel, 2001).

7.4.7 The Asian Development Bank Data Book

In 1997 the Asian Development Bank (McIntosh and Iñiguez, 1997) compared in a well-structured way the performance of 50 utilities of the Asian and Pacific regions. The report is divided into three parts: part I: sector profile; part II: regional profiles; and part III: water utility and city profiles.

A multidimensional analysis procedure was developed: each indicator was analyzed individually for the set of utilities, in terms of trend analysis, and each company and city is presented and analyzed through the whole range of indicators. Information includes both explanatory information and indicators. The indicators used to summarize the results achieved are given in Table 7.2.

7.4.8 The Water Utility Partnership for Capacity Building in Africa

The *Water Utility Partnership for Capacity Building in Africa* (WUP) is a joint program initiated by the *Union of African Water Suppliers* (*www.uade-wup.org/*)—Abidjan, Côte d'Ivoire, the Regional Center for Low Cost Water and Sanitation (CREPA)—Ouagadougou, Burkina Faso, and the *Center for Training, Research and Networking for Development* (TREND)—Kumasi, Ghana. Though established

TABLE 7.2 Indicators Used by the Asian Development Bank

Indicator	Unit	Indicator	Unit
Private sector participation	Short description	Metering	%
Production/population	$m^3/d/c$	Operating ratio	—
Coverage	%	Staff/1000 connections ratio	—
Water availability	Hours	Management salary	US$
Consumption	Liters/connection/day	New connection	US$
Unaccounted-for water	%	Accounts receivable	Months
Nonrevenue water	%	Grant financing	%
Average tariff	US$/$m^3$	Commercial financing	%
Water bill	US$/month	Local bond financing	%
Power/water bill ratio	—	Capital expenditure/connection	US$
Public taps	Yes/no	Annual report	None, type script, or glossy covered report

in 1995, the program was launched in 1996 with the support of the World Bank. The European Union, France, Sweden, and the United Kingdom currently fund WUP as well. The basic idea leading to the creation of the WUP was to build a partnership among *African Water Supply and Sanitation Utilities* (WSSUs) and other key sector institutions to create opportunities for sharing of experiences and capacity building. The WUP intends to achieve its objectives through five specific but very closely linked projects, one of them titled *Service Providers' Performance Indicators and Benchmarking Network* (SPBNET).

The purpose of this project is to provide utilities with sustainable arrangements for compiling and sharing performance data and to develop an understanding of how the data can be used for benchmarking. Twenty-one utilities from 15 countries provided their performance information, and the results have been published by the WUP. The project is now being extended to cover the rest of the utilities in Africa (about 100 committed to participate). Specific activities include

- Development of a software package for use in performance data analysis
- Collection of performance data from utilities
- Production of a data bank on the performance of utilities in Africa
- Training of utilities personnel on benchmarking and its application to the sector

The IWA PI system and the World Bank benchmarking toolkit were the two basic references for the system developed. A spreadsheet application has been developed for data collection and PI assessment.

7.4.9 The Six Scandinavian Cities Group

A similar initiative is under way in the Nordic countries (Adamsson, 1997; Adamsson et al., 2000), involving in a first stage a group of six cities: Stockholm, Gothenburg, Malmö, Copenhagen, Oslo, and Helsinki. Water supply and wastewater services in Scandinavia are normally provided by the local authorities that also operate and own the assets. In small communities in Denmark, Finland, and Norway the water supply is often managed by consumer cooperatives, while the municipality provides sewerage. There is a growing interest in increasing cooperation between neighboring municipalities.

The cooperation among the six cities started in the late seventies between the planning departments in the utilities. It turned out to be very efficient by exchange of experiences and resulted in effective master plans. Later on, overall management issues were discussed and the managing directors were involved and started yearly 2-day meetings with a well-prepared agenda.

In 1995 the six-cities group decided to start a joint cooperation project with the aim to develop PIs that would facilitate comparisons between the cities and give a better base for discussions with the politicians and give impulses for development of the utilities. Their PI system covers

- *Customer satisfaction*—PIs and measuring methods reflecting the customers' expectations and appraisal of the water services

- *Quality*—Quality-related PIs complementing the economical PIs and the customer satisfaction PIs
- *Availability*—PIs describing the reliability in operation of the entire system
- *Environment*—PIs illustrating the utilities' environmental achievements
- *Organization/personnel*—PIs describing efficiency and the relation between in-house work and external services
- *Economy*—PIs comparing costs on an overall level

The selected PI structure was based on the objectives, targets, and PIs used in the six cities. The group discussed which PIs could be of interest on the management level and added some new PIs that were considered to be of interest for the overall assessment of the long-term development of the water and wastewater services.

The selected PI set has been tested yearly on the six cities' activities since 1996. The test showed that despite country boundaries, different languages, and different currencies, it was possible to compare the results. The tests have shown that

- Definitions of data and PIs are most important.
- Many of the PIs should include a description of the trend during a 5- to 10-year period.
- Local conditions and accounting principles have a great influence on the PI values.
- During the process of testing, the existing PIs were improved and new PIs developed.
- The interest for working with PIs increased in the utilities.
- The PIs can be used as a supplement to the annual accounts.

This project encouraged the Swedish Water and Wastewater Association to initiate a study with the aim to survey the performance assessment of its members. This system is based on the IWA PI system.

7.4.10 The Dutch Contact Club for Water Companies

A rather interesting example can be found in the Netherlands (van der Willigan, 1997). In 1992, the 13 larger Dutch water companies created a "contact club for water companies" and a common standard to be used for benchmarking has been introduced. These companies agreed to define indicators to assess their own performance and compare the results achieved once a year. There are no goals set or yearly targets. According to van der Willigen (1997), the following main categories of performance indicators are being considered:

- *General*—Includes connections, personnel, and length of main pipes
- *Production*—Includes internal production, external production, external delivery, delivery to distribution, unaccounted-for water, net sales, sales in volume, and sales per connection
- *Costs*—Includes production, distribution, sales, general, total, income, and result and is broken down into costs total, per m^3 sales, and per connection
- *Personnel*—Includes connection per person, production, distribution, sales, general, salary per person, absenteeism, short-term absenteeism, and long-term absenteeism

7.4.11 An Engineering Approach for Performance Assessment

Coelho (1997) and Cardoso et al. (2000) address the engineering side of the problem by using a flexible framework based on an array of performance indices. The performance evaluation system they developed is a technical analysis tool, designed to shift the focus of management and operations of water systems to a wider, more rigorous, performance-oriented view. It is based on a system of penalty curves, flexible enough to accommodate individual sensitivities and interpretations. The methodology allows for a rapid gain in sensitivity to the network's behavior, providing a standardized means of diagnosis, and is a valuable aid for planning, design, operation, and rehabilitation of water distribution systems.

Similarly to the other type of performance indicators, these also contemplate quantitative indicators based on the analysis, from specific viewpoints, of the network's characteristics and behavior. However, the former type of performance analysis is mainly based on sound records, such as damage rate or maintenance costs, or on data obtained from systematic or ad hoc field surveys. This latter type requires the use of hydraulic and water quality dynamic simulation tools, in order to infer the system's behavior if a given remedial solution was implemented.

The performance indicators developed by the author are classified under three generic terms, although the methodology is applicable to other viewpoints:

- *Hydraulic performance*—Focuses on pressure head and pressure head variation
- *Water quality performance*—Focuses on concentrations and travel times
- *Reliability performance*—Focuses on network topological reliability

The limit values for every one of these performance indicators must be defined case by case by the undertaker, according the local characteristics, legal constraints, and its management strategy.

The performance evaluation can be accessed on an empirical and nonsystematic basis or in a more systematic way. In this latter situation, three types of entities, established for each aspect to be analyzed, can define the framework:

- A relevant *state variable,* that is, the quantity that translates the network behavior or properties at the network element level, from the point of view taken into consideration. The network element is the node or the pipe, as conventionally used in network analysis, from which all modeling assumptions are inherited. Examples of such state variables are nodal head, to check compliance with the pressure requirements; damage rate, to be compared to reference values defined for each type of material and class of diameter; or nodal concentration of a particular constituent as compared to the water quality guidelines.
- A *penalty function,* which scores the values of the state variable against a scale of index values (6—optimum through 0—no service). The penalty curves are designed by the user and translate as much as possible to a commonsense grading of performance in the addressed domain.
- A *generalizing function,* used for extending the element-level calculation across the network, producing zonal or networkwide diagnoses. The indices are intended to have both local and networkwide meaning. In the case, for example, of a pressure index, the proposed generalizing function is the weighted average across the network. Other types of operators may be used, such as those that focus on maximum or minimum values.

The technique is applied both to extended period operational scenarios and to a range of load factors, displaying appropriate *dispersion bands* (25 percent stepped percentiles).

The above considerations about the nature of the indices imply the need for a careful selection of the variables that may be eligible for such a treatment. In hydraulic terms, for instance, the variables used to assess a system's energy performance are total pumping energy, total energy dissipated, and the difference between the actual potential energy and the minimum potential energy required for supplying every node with the minimum allowable pressure.

7.4.12 Status of Performance Assessment in the United States

From the applied research point of view, there are two projects launched by the American Water Works Research Foundation that became an international reference. The former, a report entitled Distribution System Performance Evaluation, defines distribution system performance measures, develops measurement procedures and techniques, and establishes national target levels for these measures. It also develops guidelines for utility managers to routinely assess the overall condition of their distribution system in order to identify system and investment needs (AWWARF, 1995). The latter, a report entitled Performance Benchmarking for Water Utilities, determines areas of water utility operations and management that would be appropriate for developing benchmarks to evaluate utility performance. It includes suggestions that water utility managers would be able to use to evaluate and improve the performance of their individual operations (AWWARF, 1996).

From the practical point of view, and according to Paralez (2001), 53 percent of the 900 utilities who replied to a questionnaire held by the Government Accounting Standards Board (GASB survey) in 1998–1999 use PI measures. However, there is a clear gap among those who assess performance measures and those who use them as part of the decision-making process. In fact, only about half of those who have PIs report them to elected officials, and only 25 percent report outcome measures to internal management. A more recent custom survey (5 percent of respondents) confirms the same trend. A good number of undertakings state that their measurement efforts do not enjoy practical use or are not accepted by the staff. Parelez (2001) considers that the following issues need to be actively addressed in order to overcome the shortcomings identified:

- Those that gather data should be those using the data.
- Field staff (labor) should be involved in both gathering and using the data; i.e., maintenance, repair, refurbishing, decision making.
- The distribution and understanding of performance data should be enhanced.
- An added effort is needed for the integration of information systems.
- Maximizing data availability and use should be emphasized.
- Appropriate quality control should be applied to all data.
- The acceptability and practical use of information should be encouraged.

7.4.13 Other Initiatives

Many water utilities are already using their own PIs. However, the most interesting experiences to be analyzed from the conceptual viewpoint are the ones regarding groups of utilities. Beyond the benchmarking processes previously referred to, such as in the Nordic countries and in North America (involving the United States and Canada), others are currently starting to be developed and implemented in other areas of the world. In Africa, for instance, a benchmarking process conducted by Umgeni Water, from South Africa, has already produced its results.

A study that cannot go without reference is a report published in 1996 by the *Malaysian Water Association* (MWA, 1996). It is an excellent document that suggests that the PIs are organized under three generic terms, as follows:

- *Physical*—Gives an indication of the size and coverage of the physical aspect of the water supply and the extent of the underground assets and is useful for future planning.
- *Service*—Focuses on consumer services and productivity in an environment where consumers' expectations are increasing; they measure the utility's output against staff resources, assessment of level of service such as staff ratios, interruptions to mains supply, quality of water, water losses, and consumers' complaints on an annual basis.
- *Financial*—Assesses the financial, human, and physical aspect of efficiency, covering costs and the economic utilization of assets, work force, plant, and equipment.

The MWA continued its activity within this topic, publishing a report in 2001 (MWA, 2001). The MWA is also participating in an IWA PI field-testing project with the case study of Penang.

Also from the scientific point of view the topic of PIs has been attracting the attention of researchers in different places of the world. Apart from the studies previously mentioned, several Ph.D. theses have already been published on the topic. There are, for instance, the cases of Ancarani (1999), who focused on the economic aspects of the performance assessment, of Guérin-Schneider (2001), a very good work on the use of PIs from the regulatory point of view, and of Cabrera Jr. (2001), who designed a system of performance assessment embracing a methodology for metric and process benchmarking. The two latter studies have a very close link to the IWA PI system.

One aspect of performance assessment and benchmarking is communication. The Water Research Centre, in the United Kingdom, publishes periodically the on-line newsletter *Watermarque* (*Watermarque Magazine: Your Guide to Water and Wastewater Industry Benchmarking*). It is issued bimonthly via e-mail, provides updates on global regional, national, and company initiatives; reports on case studies and on benchmarking awards; provides conference details; and includes an on-line searchable archive of back issues. It is published six times per year for subscribers. Subscription is free at *www.wrcplc.com/pbngroup/pbngroup.nsf/htmlmedia/index.html.*

7.4.14 Summary of Performance Assessment Projects for Water Supply (WS) and Wastewater (WW) Services

Table 7.3 summarizes the main international and national initiatives related to performance assessment. It includes not only the initiatives reported in the previous section but also other ongoing initiatives. It is partially based on the information provided by Merkel (2001).

7.5 THE IWA PI SYSTEM FOR WATER SUPPLY SERVICES

7.5.1 Highlights of the IWA PI System

In order to meet the criterion that the PI system be independent of the state of development or internal organizational structure of any particular undertaking, the indicators have been framed in terms of the principal management objectives common to all undertakings. These are structured in six separate categories of performance: 2 water resources indicators, 22 personnel indicators, 12 physical indicators, 36 operational indicators, 25 quality-of-service indicators, and 36 financial indicators.

In recognition of the differing states of development and management sophistication that may apply in different undertakings, the manual includes a tentative grading of level of priority of implementation, to be improved during the ongoing field test. However, this is just a general guidance, and it is up to each undertaking to define the importance and applicability of every indicator for the organization. Any subsets can easily be selected, according to the utility's needs and objectives. Conversely, if the degree of detail in considered insufficient, users are naturally free to split the existing indicators in subcategories or add their own indicators, provided that they keep in mind that these new PIs are not standard and therefore are not suitable for future comparisons among undertakings. However, as noted in Sec. 7.2, the interpretation of an undertaking's performance cannot be assessed without taking its own context into account, as well as the most relevant characteristics of the system and the region. This is why the IWA PI system includes the definition of context information, structured within the profiles of the undertaking, system, and region.

These profiles contain the figures that many good managers know by heart and frequently use to present their companies. The undertaking profile outlines the framework of the organization. The system profile focuses mainly on the water volumes managed, on the physical assets, on the technological means used, and on the customers. The region profile will be relevant for comparisons between undertakings, allowing for a better understanding of the demographic, economic, geographic, and environmental context.

7.5.2 Listing of the IWA PIs and Guidance on Their Relative Importance

The current stage of development aims to include the PIs relevant at the top-management level of a water supply undertaking. Those aim to incorporate all the relevant aspects required to express management objectives and results in terms of an undertaking's performance.

The tentative grading of level of priority of implementation referred to in Sec. 7.4 includes three categories, plus one category of complementary indicators. Complementary indicators may be needed only at the departmental level and are generally much more organization dependent and therefore are not included in the IWA PI system.

- *Level 1 (tagged "L1").* A first layer of indicators that provide a general management overview of the efficiency and effectiveness of the water undertaking.
- *Level 2 (tagged "L2").* Additional indicators, which provide a better insight than the level 1 indicators for users who need to go further in depth.
- *Level 3 (tagged "L3").* Indicators that provide the greatest amount of specific detail but are still relevant at the top-management level.
- *Complementary (not included in this report).* Further indicators, which provide a greater amount of specific detail than level 3 indicators or which are for specific use at the departmental level (which tend to be undertaking dependent).

TABLE 7.3 Summary of Performance Assessment Projects for Water Supply (WS) and Wastewater (WW) Services

Organization/country	Scope	Projects/products
		International initiatives
IWA	WS	*Performance Indicators for Water Supply Services,* Manual of Best Practices Series (published in July 2000).
		http://www.iwapublishing.com/template.cfm?name = isbn1900222272
		http://www.sigmalite.com
		International field test of the IWA PI system (2000–2003).
IWA	WW	*Performance Indicators for Water Supply Services,* Manual of Best Practices Series (scheduled for Dec. 2002).
Asian Develop. Bank	WS	Report: *Second Water Utilities Data Book–Asian and Pacific Region (1997)*
The World Bank	WS+WW	Project: "Benchmarking Water and Sanitation Utilities."
		http://www.worldbank.org/html/fpd/water/topics/uom_bench.html
WUP	WS+WW	Service Providers' Performance Indicators and Benchmarking Network (SPBNET): Questionnaire in MS Excel format with 184 data entries for the assessment of PI indicators of about 100 African utilities (*http://www.wupafrica.org/*).
World Health Organization	WS+WW	2000 Water and Sanitation Assessment: the WHO and UNICEF Joint Monitoring Program for Water Supply and Sanitation (JMP) provides a snapshot of water supply and sanitation worldwide at the turn of the millennium using information available from different sources. From 2001 the JMP database—for both historic data and future projections—will be periodically updated.
		http://www.who.int/water_sanitation_health/Globassessment/GlobalTOC.htm
ISO	WS+WW	The International Standardization Organization established in late 2001 is a work group for the preparation of standards for performance assessment of water supply and wastewater services.
Six Scandinavian cities group	WS+WW	Six-cities PI group: Copenhagen, Gothenburg, Helsinki, Malmö, Oslo, and Stockholm established a systematic routine for PI comparison: in January the group meets and discusses the work plan for the year; in February the process starts with sending out forms for collection of data for the previous year; during March/April, data is collected and the new PI values are assessed and analyzed; in May a preliminary report is produced; in June the reports are discussed within the PI group; at the yearly six-city meeting in September/October, the report and proposals for continued work are presented for the managing directors of each utility.
		National initiatives
Argentina	WS+WW	Ongoing project for the establishment of a common platform of PI for WS and WW to be adopted by the Argentinean regulators, as a basis for yardstick competition; other countries of the region are starting to cooperate as well.
		Participation in the IWA-PI field testing (ETOSS, La Plata, Trelew, and Tucumán).
		Participation in the core team of the IWA-PI-WW project.
Australia	WS+WW	Annual performance reports from WSAA, NMU (AWA), NSW, Vicwater, Queensland Performance Monitoring System, ANCID (e.g., AWA, 2000).

Brazil	WS+WW	ABES Quality National Award: this award, created in August 1995, aims to stimulate the practice of managerial models compatible with the world trends, acknowledgment of successful experiments which utilize the methodology and promote the exchange of the best practices, making possible the quality improvement in the life of Brazilian populations through the development of the sanitation sector (*http://www.pnqs.com.br/english/index.html*).
		Participation in the IWA-PI project (SABESP—São Paulo, COPASA—Belo Horizonte, and CAESB—Brasília).
Czech Republic	WS	Case study in 1997–1999 based on a preliminary version of the IWA PI system (WS).
Germany	WS+WW	Three annual performance assessment projects (WS): 30 large-scale, 270 medium or scale, 20 bulk water suppliers; IWA field test (WS) with 13 utilities, Bavarian WS performance assessment project with around 100 utilities, several WW benchmarking projects (e.g., AV+EG/LV, VkY, and others).
Denmark	WS+WW	Participation in the six Scandinavian cities group, in the seven Danish cities group (WS), 18 Danish wastewater plants (WW), PA project of DWSA and DWWA.
France	WS+WW	Quality management projects (WS); participation in the IWA-PI project (SAGEP-Paris); workshops on performance assessment; Ph.D. thesis on the use of PI for regulation in France (Guérin-Schneider, 2001); development of national standards (AFNOR) on performance and quality of service.
Finland	WS+WW	Participation in the six Scandinavian cities group and in the 10 Finnish Water and Wastewater Utilities.
Italy	WS	Benchmarking initiative of Italian waterworks under FEDERGASAQUA; participation in the IWA PI project (Turin, SIDRA, AGAC, and SOGEAS).
Malaysia	WS	*Performance Indicators for Water Supply* (MWA, 1996) and *Malaysia Water Industry Guide 2001* (MWA, 2001).
Mexico	WS	Project of the IMTA, Mexican Institute of Water Technology: the objective is the definition of a short number of indicators and of applicability subsets, depending on the size and level of development of the utility.
Netherlands	WS+WW	Reports produced by all Dutch WS companies (annual reports include only economic indicators; triannual reports include water quality, service, and environmental and economic indicators); first pilot projects on performance assessment for WW services held in 2001.
Portugal	WS+WW	PI-Waters project (National Civil Engineering Laboratory, LNEC): the objectives are: (i) to reinforce the implementation of the IWA PI system at a national level; (ii) to test the PI application within different national contexts; (iii) to share experiences at the national level; (iv) to promote a broader application of the PI system by the other Portuguese water utilities. Products: Portuguese versions of the IWA PI manual and of SIGMA software.
		Participation in the core team of the IWA PI project.
		Participation in the IWA PI project with 17 undertakings from Algarve, Cávado river catchment, Douro region, Águas de Portugal, S.A. (public holding), Barreiro, Beja, Castelo Branco, Esposende, Figueira da Foz, greater Lisboa region, Loures, Luságua (private holding), Oeiras, Amadora, Sintra, Vila Nova de Gaia,and IRAR, the national regulator.
Poland	WS+WW	Quality management projects, private “leadership” projects (e.g., SNG).
Romania	WS+WW	A set of 19 PI is being assessed systematically for some Romanian cities (Cluj, Constanta, Arad, Oradea, Bucharest).

TABLE 7.3 Summary of Performance Assessment Projects for Water Supply (WS) and Wastewater (WW) Services (*Continued*)

Organization/country	Scope	Projects/products
		National initiatives
Spain	WS+WW	Development of the software SIGMA for the assessment of the IWA PIs.
		Design of System for the Management Assessment of Urban Water Supply Systems (Cabrera, 2001); participation in the core team of the IWA PI project.
		Participation in the IWA PI project with 12 undertakings (AEAS VI, Alicante, Barcelona, Castellón, Córdoba, Huelva, Madrid, Murcia, Sevilla, Valencia, Vigo, and Zaragoza).
South Africa	WS + WW	Annual reports from the Department of Water Affairs (WS + WW), ZA Assn. of Water Boards (WS), Water Research Commission (Guidelines for Benchmarking Water and Sanitation Services for Local Authorities).
		Participation in the IWA PI project with Rand Water (Johannesburg)
Sweden	WS + WW	Six Scandinavian cities group, DRIVA, VASAM, VA-Plan 2050.
		Participation in the IWA PI project with Stockholm and Gothenburg.
United Kingdom	WS + WW	OFWAT performance assessment projects and periodic reports (e.g., OFWAT, 2000).
		Participation in the IWA PI project with Bristol, Hatfield, North East and East of Scotland.
United States	WS + WW	Several projects sponsored by the AWWARF (AWWARF, 1995 and 1996; Deb and Cesario 1997) and WERF, initiatives of utilities (e.g., WRWBG). Initiatives to improve effectiveness and efficiency (Westerhoff et al., 1998).

Indicators that conceptually have a given level of importance but that are not easily assessed in a reliable way are classified in a lower-level grade. This full listing of indicators and of the respective preliminary guidance of level of implementation priority is shown in Tables 7.4 to 7.9.

7.5.3 Data Reliability and Accuracy

The methodology proposed by the IWA was adopted first in England and Wales by the OFWAT, where economic regulation of the water supply companies is highly structured and detailed. As

TABLE 7.4 Water Resources Indicators

Indicator	Unit	Suggested level
Inefficiency of use of water resources	%	L1
Resources availability ratio	%	L2

TABLE 7.5 Personnel Indicators

Indicator	Unit	Suggested level
Total personnel		
Employees per connection	No./1000 connections	L1
Personnel per main function		
Management and support personnel	No./1000 connections	L2
Financial, commercial personnel	No./1000 connections	L2
Customer service personnel	No./1000 connections	L2
Technical activities personnel	No./1000 connections	L2
Planning & construction personnel	No./1000 connections	L3
Operations & maintenance personnel	No./1000 connections	L3
Water resources, catchment, and treatment personnel	No./10^6 m^3/year	L3
Transmission, storage, and distribution personnel	No./10^2 km	L3
Laboratory personnel	No./1000 tests	L3
Meter maintenance personnel	No./1000 meters	L3
Other personnel	No./10^6 m^3/year	L3
Personnel qualification		
University degree personnel	%	L3
Personnel with basic education	%	L3
Other personnel	%	L3
Personnel training		
Total training	Days/employee/year	L3
Internal training	Days/employee/year	L3
External training	Days/employee/year	L3
Personnel health and safety		
Working accidents	No./employee/year	L3
Absenteeism	Days/employee/year	L3
Absenteeism due to working accidents or disease	Days/employee/year	L3
Absenteeism due to other reasons	Days/employee/year	L3

TABLE 7.6 Physical Indicators

Indicator	Unit	Suggested level
	Treatment	
Treatment availability	%	L1
	Storage	
Impounding reservoir capacity	Days	L2
Transmission and distribution storage capacity	Days	L2
	Pumping	
Standardized energy consumption	Wh/m^3 at 100 m	L2
Reactive energy consumption	%	L3
Energy recovery	%	L3
	Transmission and distribution network	
Valve density	No./km	L3
Hydrant density	No./km	L3
Meters		
District meter density	No./1000 service connections	L3
Customer meter density	No./service connection	L2
Metered customers	No./customer	L3
Metered residential customers	No./customer	L3

the authors of the manual note, within this structure is a carefully developed scheme for grading the confidence that can be placed on the reliability and accuracy of inputs. The testing results currently available seem to demonstrate that this confidence-grading scheme is adequate for any other country.

The confidence-grading scheme includes a four-grade measure of the reliability of the data (sound records, analysis result, forecast, etc.) and a seven-grade classification for their respective accuracy. The interpretation to be adopted for the questions on data reliability and data accuracy is shown in Table 7.10.

The confidence grades will be an alphanumeric code, which couples the reliability band and the accuracy band, for instance:

A2. Data based on sound records, etc. (highly reliable, band A), which is estimated to be within ±5 percent (accuracy band 2).

C4. Data based on extrapolation from a limited sample (unreliable, band C), which is estimated to be within ±25 percent (accuracy band 4).

The reliability and accuracy bands would form the matrix of confidence grades shown in Table 7.11.

7.5.4 Organization of the IWA PI Manual

The structure of the IWA PI manual is shown in Fig. 7.2. Sections 1 to 3 of the document cover general matters such as concepts and uses of PIs, structure of the manual, and, separately, a detailed list and full explanation of the definitions used in the subsequent text. Special attention is paid to the definitions related to water balance, main functions of the undertaking, and financial terminology and conventions.

TABLE 7.7 Operational Indicators

Indicator	Unit	Suggested level
	Inspection and maintenance	
Pumping inspection	%/year	L2
Storage tank cleaning	%/year	L2
Network inspection	%/year	L2
Leakage control	%/year	L2
Active leakage control repairs	%/year	L2
Hydrant inspection	%/year	L3
Instrumentation calibration		
System flow meters	%/year	L3
Meter replacement	%/year	L2
Pressure meters	%/year	L3
Water-level meters	%/year	L3
On-line water quality monitoring equipment	%/year	L3
Electrical equipment inspection		
Electrical equipment inspection by number	%/year	L3
Electrical equipment inspection by power	%/year	L3
Vehicle availability	Vehicles/km	L3
	Mains, service connection, and pumps rehabilitation	
Mains rehabilitation	%/year	L1
Mains relining	%/year	L2
Replaced or renewed mains	%/year	L2
Replaced valves	%/year	L2
Service connection rehabilitation	%/year	L1
Pumps rehabilitation		
Pump refurbishment	%/year	L2
Pump replacement	%/year	L2
	Water losses	
Water losses	m^3/connection/year	L1
Apparent losses	m^3/connection/year	L3
Real losses	Liters/connection/day when system is pressurized	L1
Infrastructure leakage index	—	L3
	Failures	
Mains failures	No./100 km/year	L1
Service connection failures	No./1000 connections/year	L1
Hydrant failures	No./1000 hydrants/year	L2
Power failures	Hours/pumping station/year	L2
	Metering	
Customer reading efficiency*	%	L1
Residential customer reading efficiency*	%	L1
	Water quality monitoring	
Tests performed (quantity compliance)	%	L1
Aesthetic	%	L2
Microbiological	%	L2
Physical-chemical	%	L2
Radioactivity	%	L3

*These indicators shall be used in alternative.

TABLE 7.8 Quality of Service Indicators

Indicator	Unit	Suggested level
Service		
Households and businesses supply coverage*	%	L1
Buildings supply coverage*	%	L1
Population coverage*	%	L1
With service connections	%	L2
Public taps and standpipes	%	L2
Public taps and standpipes		
Distance to households	m	L1
Quantity of water consumed	Liter/person/day	L1
Population per public tap or standpipe	Persons/tap	L2
Pressure of supply adequacy	%	L2
Continuity of supply	%	L1
Water interruptions*	%	L2
Interruptions per connection*	No./1000 connections	L2
Population experiencing restrictions to water service*	%	L2
Days with restrictions to water service*	%	L2
Quality of supplied water (quality compliance)	%	L1
Aesthetic	%	L2
Microbiological	%	L2
Physical-chemical	%	L2
Radioactivity	%	L3
New connection efficiency	%	L2
Connection repair efficiency	%	L2
Customer complaints		
Service complaints	No. complaints/ connection/year	L1
Pressure complaints	%	L2
Continuity complaints	%	L2
Water quality complaints	%	L2
Restrictions or interruptions	%	L2
Billing complaints	No. complaints/ customer/year	L1
Other complaints and queries	No. complaints & queries/customer/year	L2
Response to written complaints	%	L2

*These indicators shall be used in alternative.

Section 4 is titled "Data Reporting" and considers the means of assessing the reliability of the contextual and PI information, by use of the confidence-grading scheme presented above.

Section 5 describes the operating context under three headings, profiling the undertaking itself, its operating systems, and the operating region.

Section 6 of the IWA manual is devoted entirely to the tabulation of all 133 suggested indicators together with their respective priority level, base concept, and processing rule in the form of a simple equation. The background data definitions and processing rules are very fully elaborated in an appendix, which accounts for over half of the entire manual. Figures 7.3 and 7.4 show the organization of this chapter, containing the identification, classification, definition, and processing rules

TABLE 7.9 Financial Indicators

Indicator	Unit	Suggested level
Annual costs		
Unit total costs	US$/m^3	L2
Unit running costs	US$/m^3	L1
Unit capital costs	US$/m^3	L3
Composition of annual running costs per type of costs		
Internal work force costs ratio	%	L3
External services costs ratio	%	L3
Imported (raw and treated) water costs ratio	%	L3
Energy costs ratio	%	L3
Other costs ratio	%	L3
Composition of annual running costs per main function of the water undertaking		
Management and support costs ratio	%	L3
Financial and commercial costs ratio	%	L3
Customer service costs ratio	%	L3
Technical services costs ratio	%	L3
Composition of annual capital costs		
Depreciation costs ratio	%	L3
Interest expenses costs ratio	%	L3
Annual revenue		
Unit annual revenue	US$/m^3	L2
Sales revenues	%	L2
Other revenues	%	L2
Annual investment		
Average unit investment	US$/m^3	L2
Average annual investments for new assets	%	L3
Average annual investments for assets replacement	%	L3
Average water charges (without public taxes)		
Average water charges for direct consumption	US$/m^3	L1
Average water charges for exported water	US$/m^3	L1
Efficiency indicators		
Total cost coverage ratio	—	L1
Operating cost coverage ratio	—	L1
Delay in accounts receivable	Months equivalent	L2
Investment ratio	%/year	L2
Contribution of internal sources to investment = CTI	%	L1
Average age of tangible assets	%	L2
Average depreciation ratio	%	L3
Late payments ratio	%	L2
Leverage indicators		
Debt service coverage ratio = DSC	%	L2
Debt equity ratio	—	L2
Liquidity indicator		
Current ratio	—	L1

TABLE 7.9 Financial Indicators (*Continued*)

Indicator	Unit	Suggested level
Profitability indicators		
Return on net fixed assets	%	L2
Return on equity	%	L2
Water losses indicators		
Nonrevenue water by volume	%	L1
Nonrevenue water by cost	%	L3

TABLE 7.10 Definition of the Confidence-Grading Scheme

Data reliability	Definition
A—Highly reliable	Data based on sound records, procedures, investigations, or analyses that are properly documented and recognized as the best available assessment methods.
B—Reliable	Generally as in band A, but with minor shortcomings; e.g., some of the documentation is missing, the assessment is old, or some reliance on unconfirmed reports or some extrapolations are made.
C—Unreliable	Data based on extrapolation from a limited sample for which band A or B is available.
D—Highly unreliable	Data based on unconfirmed verbal reports and/or cursory inspections or analysis.
Data accuracy	**Definition**
1—Error (%): [0; 1]	Better than or equal to ±1%
2—Error (%): [1; 5]	Not band 1, but better than or equal to ±5%
3—Error (%):[5; 10]	Not bands 1 or 2, but better than or equal to ±10%
4—Error (%): [10; 25]	Not bands 1, 2, or 3, but better than or equal to ±25%
5—Error (%): [25; 50]	Not bands 1, 2, 3, or 4, but better than or equal to ±50%
6—Error (%): [50; 100]	Not bands 1, 2, 3, 4, or 5 but better than or equal to ±100%
Error (%): >100	Values which fall outside the valid range, such as >100%

TABLE 7.11 Matrix of Confidence Grades

	Reliability bands			
Accuracy bands, %	A	B	C	D
[0; 1]	A1	++	++	++
[1; 5]	A2	B2	C2	++
[5; 10]	A3	B3	C3	D3
[10; 25]	A4	B4	C4	D4
[25; 50]	++	++	C5	D5
[50; 100]	++	++	++	D6

Note: 11 indicates confidence grades that are considered to be incompatible.

of the PIs. Figures 7.5 and 7.6 show the organization of the appendix of the IWA manual, containing the definition of every variable required.

The main text concludes with a section on the relative importance of the PIs to water undertakings, regulators, and users. All 133 indicators are retabulated into the appropriate priority levels for each of the separate user groups.

IWA Manual of Best Practice—Table of Contents

1. Introduction
2. About this document
3. Definition
4. Data reporting
5. Context information
6. Performance indicators
7. Relative importance of the performance indicators

Appendix. Data definition and processing rules

FIGURE 7.2 Structure of the PI manual.

6.4 Operational indicators

INDICATOR *Level of importance* (*)	(unit)	CONCEPT Processing rule
INSPECTION AND MAINTENANCE		
Op1 - Pump inspections *L2*	(%/year)	Σ (nominal power of pumps and related ancillaries subjected to *inspection* during the year) / Σ (nominal power of pumps) x 100 *Op1 = D5/C5 x 100*
Op2 - Storage tank cleaning *L2*	(%/year)	Volume of storage tank cells cleaned during the last five years /total volume of storage tank cells x 100 / 5 *Op2= D6/C2 x 100*
Op3 - Network inspection *L2*	(%/year)	Length of transmission and distribution mains where at least valves and other fittings were *inspected* during the year / total mains length x 100 *Op3 = D7/C8 x 100*
Op4 - Leakage control *L2*	(%/year)	Length of mains subject to *active leakage control* / total mains length x 100 *Op4 = D8/C8 x 100*
Op5 - Active leakage control repairs *L2*	(%/year)	Number of leaks detected and repaired due to *active leakage control* / total mains length x 100 *Op5 = D9/C8 x 100*
Op6 - Hydrant inspection *L3*	(%/year)	Number of hydrants *inspected* during the year / total number of hydrants x 100 *Op6 = D10/C31 x 100*

...

FIGURE 7.3 General outline of the PI definitions within the manual.

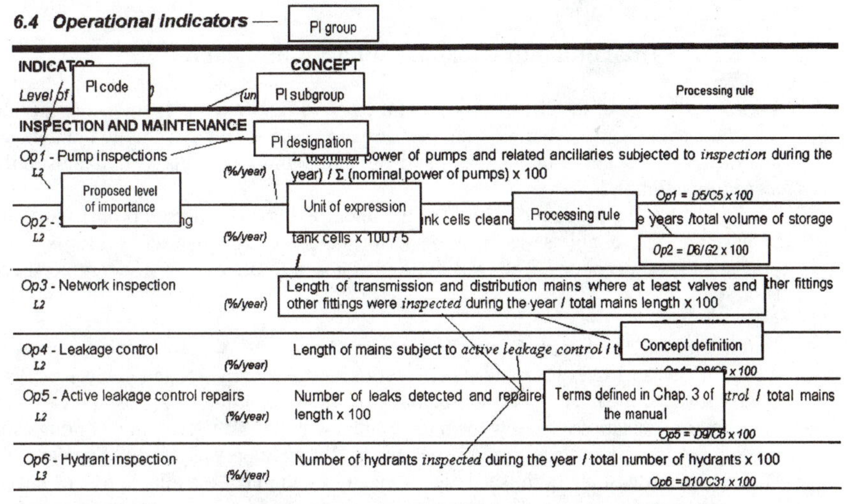

FIGURE 7.4 PI identification and classification, description, terms, and processing rules.

A5 - IMPORTED RAW WATER		
UNIT OF EXPRESSION: m³/day	PERIOD: *[dd.mm].yy-1 – [dd.mm].yy*	VALID VALUES: ≥ 0
DEFINITION: Total annual volume of raw water transferred from other water supply systems.		
PROCESSING RULE: Input data (Reliability: *[targeted reliability]* ; Accuracy: *[targeted accuracy]*)		
COMMENT:		
RESPONSIBILITY: *[service responsible for this data (e.g. Operational indicators team)]*		
USED FOR VARIABLES: *[A2], A20, A24 (A20), A26*		
USED FOR INDICATORS: *[WR1 (A2)], WR2, Ph2 (A20), Ph3 (A20), Op22 (A20), Op24 (A24), Fi36 (A26), F137 (A24)*		

FIGURE 7.5 General outline of variable definition.

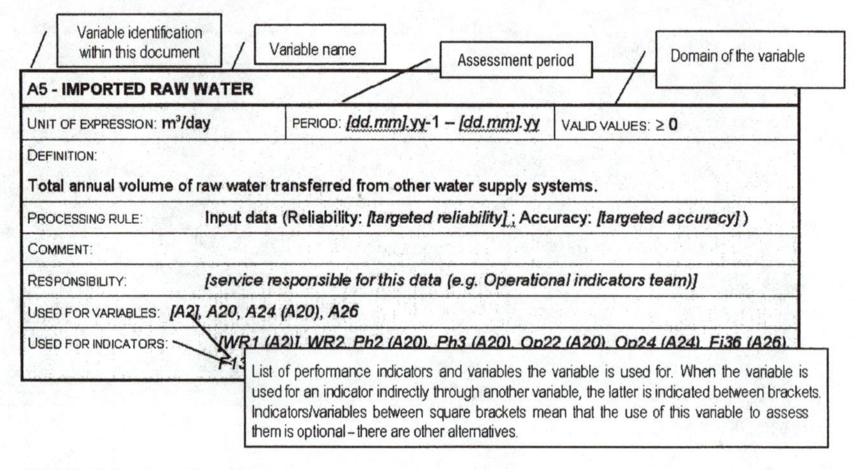

FIGURE 7.6 Variable definition, classification, and description.

7.5.5 A Partial In-depth Look into the IWA PI System

Viewpoint. One of the characteristics of the IWA PI system that is frequently pointed out as a comparative advantage to other existing performance assessment systems is the data structure adopted and the level of detail of the indicator and variable definitions. This section aims to illustrate this with a specific example.

As water is the product delivered to the undertaking customers, a proper monitoring of the water balance and the reliable assessment of performance in terms of water losses is fundamental to support managers' decisions. This was the topic chosen to provide an in-depth perspective of the IWA PI system.

Definition of Water Supply System Inputs and Outputs.[3] A well-established water balance is fundamental for water losses assessment. The definitions, terminology, and choices of PIs for water losses used in this document are based substantially on the work of the IWA Task Force on water losses, with some minor adaptations to allow them to be fitted within the overall context of this wider PI report. The reading of the Blue Pages on Losses from Water Supply Systems: Standard Terminology and Performance Measures (published electronically by the IWA; available free of charge for members) is recommended.

Figure 7.7 illustrates the principal inputs and outputs to a typical water supply system sequentially, from raw water intake to consumption by customers. Some systems will of course be simpler and will not have all the features shown.

The water balance requires estimates of volumes of water to be made at each measurement point applicable to the system under consideration. Where there are actual meters, data from these would normally be used, but in the absence of meters, a "best estimate" based on other related available data and application of sound engineering judgment may be required. The water balance is normally computed over a 12-month period and thus represents the annual average of all components.

Definitions relating to Figs. 7.7 and 7.8, are given below. Because of widely varying interpretations of the term *unaccounted-for water* (UFW) worldwide, the IWA does not recommend use of this term. If the term is used at all, it should be defined and calculated in the same way as *nonrevenue water* in Fig. 7.8.

Water abstracted The annual volume of water obtained for input to water treatment plants (or directly to the transmission and distribution systems) that was abstracted from raw water sources.

Raw water, imported or exported The annual volumes of bulk transfers of raw water across operational boundaries. The transfers can occur anywhere between the abstraction point and the treatment plants.

Treatment input The annual volume of raw water input to treatment works.

Water produced The annual volume of water treated for input to water transmission lines or directly to the distribution system. (The annual volume of water that is distributed to consumers without previous treatment shall also be accounted for in *water produced.*)

Treated water, imported or exported The annual volumes of bulk transfers of treated water across operational boundaries. The transfers can occur anywhere downstream treatment. [The annual volume of water (if any) that is abstracted and delivered to consumers without any treatment shall also be accounted for as *treated water* in the scope of the water balance.]

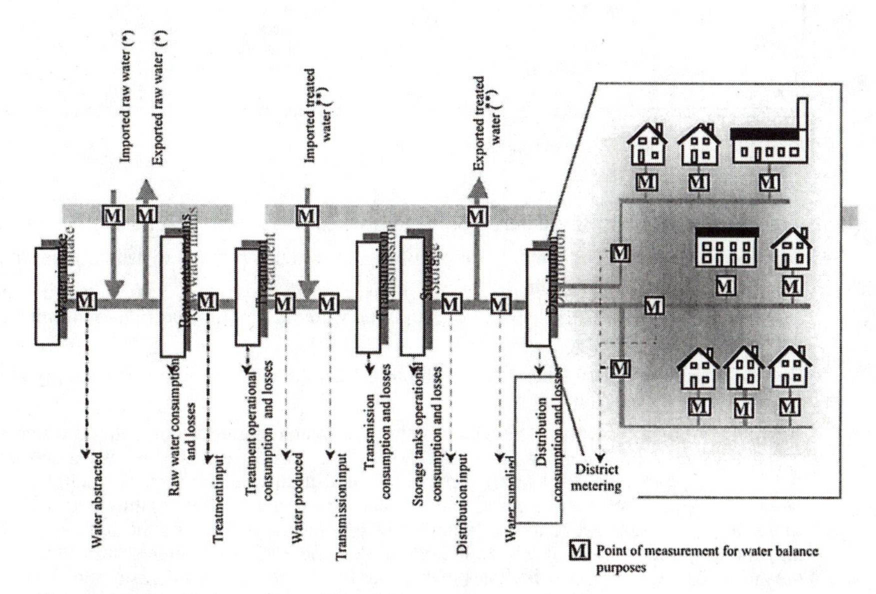

FIGURE 7.7 Definition of water supply system inputs and outputs.

<table>
<tr><th>A</th><th>B</th><th>C</th><th>D</th><th>E</th></tr>
<tr><td rowspan="10">System input volume

[m³/year]</td><td rowspan="4">Authorized consumption

[m³/year]</td><td rowspan="2">Billed authorized consumption

[m³/year]</td><td>Billed metered consumption (including water exported)
[m³/year]</td><td rowspan="2">Revenue water

[m³/year]</td></tr>
<tr><td>Billed unmetered consumption
[m³/year]</td></tr>
<tr><td rowspan="2">Unbilled authorized consumption

[m³/year]</td><td>Unbilled metered consumption
[m³/year]</td><td rowspan="8">Nonrevenue water

[m³/year]</td></tr>
<tr><td>Unbilled unmetered consumption
[m³/year]</td></tr>
<tr><td rowspan="6">Water losses

[m³/year]</td><td rowspan="2">Apparent losses

[m³/year]</td><td>Unauthorized consumption
[m³/year]</td></tr>
<tr><td>Metering inaccuracies
[m³/year]</td></tr>
<tr><td rowspan="4">Real losses

[m³/year]</td><td>Real losses on raw water mains and at the treatment works (if applicable)
[m³/year]</td></tr>
<tr><td>Leakage on transmission and/or distribution mains
[m³/year]</td></tr>
<tr><td>Leakage and overflows at transmission and/or distribution storage tanks
[m³/year]</td></tr>
<tr><td>Leakage on service connections up to the measurement point
[m³/year]</td></tr>
</table>

FIGURE 7.8 Components of water balance.

Transmission input The annual volume of treated water input to a transmission system.

Distribution input The annual volume of treated water input to a distribution system.

Supplied water The *distribution input* minus *treated water exported.* (When it is not possible to separate transmission from distribution, supplied water is the *transmission input* minus *treated water exported*).

System input volume The annual volume input to that part of the water supply system to which the water balance calculation relates.

Authorized consumption The annual volume of metered and/or nonmetered water taken by registered customers, the water supplier, and others who are implicitly or explicitly authorized to do so by the water supplier, for residential, commercial, and industrial purposes. It includes *water exported. Notes:* (1) Authorized consumption may include items such as fire fighting and training, flushing of mains and sewers, street cleaning, watering of municipal gardens, public fountains, frost protection, and building water. These may be billed or unbilled, metered or unmetered, according to local practice. (2) Authorized consumption includes leakage and waste by registered customers that are unmetered.

Water losses The difference between system input volume and authorized consumption. Water losses can be considered as a total volume for the whole system, or for partial systems such as raw water mains, transmission, or distribution. In each case the components of the calculation would be adjusted accordingly. Water losses consist of real losses and apparent losses.

Real losses Physical water losses from the pressurized system, up to the point of measurement of customer use. The annual volume lost through all types of leaks, bursts, and overflows depends on frequencies, flow rates, and average duration of individual leaks. *Note:* Although physical losses after the point of customer flow measurement are excluded from the assessment of *real losses,* they are often significant (particularly where customers are unmetered) and worthy of attention for demand management purposes.

Apparent losses Accounts for all types of inaccuracies associated with production metering and customer metering, plus unauthorized consumption (theft or illegal use). *Note:* Underregistration of production meters, and overregistration of customer meters, leads to underestimation of *real losses.* Overregistration of production meters, and underregistration of customer meters, leads to overestimation of *real losses.*

Nonrevenue water The difference between the annual volumes of *system input* and *billed authorized consumption. Nonrevenue water* includes not only the *real losses* and *apparent losses,* but also the *unbilled authorized consumption.*

Water Balance Components[3] Figure 7.8 shows the recommended standard format and terminology for water balance calculations for one or more sections of a water supply system (e.g., raw water mains, transmission, distribution). If original water balance data is available in any alternative format or terminology, they must be reassembled into the components in Fig. 7.8 in volume per year, before attempting to calculate any PIs. *Note*: Consumption of water by registered customers who pay indirectly through local or national taxation is deemed to be billed authorized consumption for the purposes of the water balance.

The steps for calculating nonrevenue water and water losses are as follows:

Step 1. Define *system input volume* and enter in column A.

Step 2. Define *billed metered consumption* and *billed unmetered consumption* in column D; enter total in *billed authorized consumption* (column C) and *revenue water* (column E).

Step 3. Calculate the volume of *nonrevenue water* (column E) as *system input volume* (column A) minus *revenue water* (column E).

Step 4. Define *unbilled metered consumption* and *unbilled unmetered consumption* in column D; transfer total to *unbilled authorized consumption* (column C).

Step 5. Add volumes of *billed authorized consumption* and *unbilled authorized consumption* in column C; enter sum as *authorized consumption* (column B).

Step 6. Calculate *water losses* (column B) as the difference *between system input volume* (column A) and *authorized consumption* (column B).

Step 7. Assess components of *unauthorized consumption* and *metering inaccuracies* (column D) by best means available; add these and enter sum in *apparent losses* (column C).

Step 8. Calculate *real losses* (column C) as *water losses* (column B) minus *apparent losses* (column C).

Step 9. Assess components of *real losses* (column D) by best means available (night flow analysis, burst frequency/flow rate/duration calculations, modeling, etc.); add these and cross-check with volume of *real losses* in column C.

Particular difficulty is experienced in completing the water balance with reasonable accuracy where a significant proportion of customers are not metered. Authorized consumption in such cases should be derived from sample metering of sufficient numbers of statistically representative individual connections of various categories and subcategories and/or by measurement of total flows into discrete areas of uniform customer profile, also of various categories and subcategories. In the latter method, subtraction of leakage demand from total input is necessary, leakage being determined by analysis of the subcomponents of night demand, adjusting for diurnal pressure variation as appropriate. The confidence grade allocated to the authorized consumption (see Sec. 7.5.3) should reflect the rigor of the investigations.

In this document it is the intention that performance related to management of water losses should be measured from three different viewpoints: financial, technical, and water resources. If the water balance calculations proceed no further than step 3 in Fig. 7.8, as is the case in the simplest traditional water balances, the only performance indicator that can be calculated is the financial one of nonrevenue water by volume (%). The task force on water losses emphasized the importance of completing the calculation up to step 8 (and preferably step 9) in Fig. 7.8—in particular, the importance of attempting, by the best means available, the separation of water losses into apparent and real losses. This allows for the calculation of a required range of water resource, operational, and financial PIs as shown in Fig. 7.9 and Table 7.12.

Definition of the Water Losses Indicators. Table 7.13 reproduces the definition of all the water loss indicators listed above. Each variable identified in the processing rule is defined in detail in forms such as illustrated in Sec. 7.5.4. Table 7.14 summarizes their meanings.

7.5.6 The SIGMA Lite Software

The SIGMA Lite software (see Fig. 7.10) was prepared by the Instituto Tecnológico del Agua (ITA), Spain, as a simplified version of its professional PI evaluation system, SIGMA Pro. Both versions of the software are based on the IWA PI framework. SIGMA Lite is available free of charge through the Internet and can be downloaded at the SIGMA web site: *http://www.sigmalite.com.*

Although dating from 1997 as an independent initiative, the SIGMA project joined the IWA task force on PI in mid-1999, as a follow-up of an IWA invitation to ITA to prepare a freeware version of the program that would be the appropriate complement to the paper document. The Lite version features:

- A stand-alone PI evaluation system, independent of the database systems.
- The complete set of performance indicators from the IWA proposal.
- A user-friendly graphic interface, with intuitive management of the indicators and variables involved.

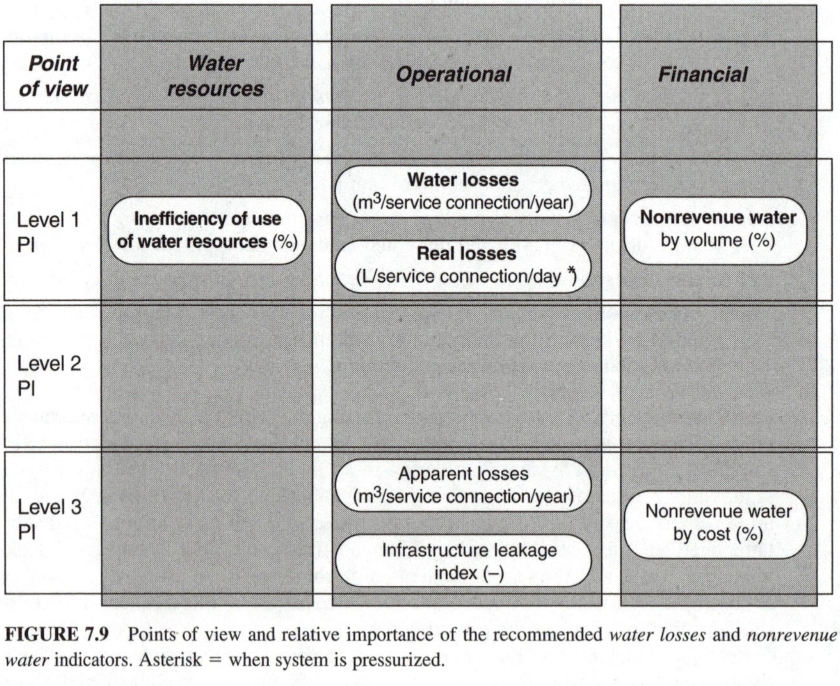

Point of view	*Water resources*	*Operational*	*Financial*
Level 1 PI	**Inefficiency of use of water resources** (%)	**Water losses** (m³/service connection/year) **Real losses** (L/service connection/day *)	**Nonrevenue water** by volume (%)
Level 2 PI			
Level 3 PI		Apparent losses (m³/service connection/year) Infrastructure leakage index (–)	Nonrevenue water by cost (%)

FIGURE 7.9 Points of view and relative importance of the recommended *water losses* and *nonrevenue water* indicators. Asterisk = when system is pressurized.

TABLE 7.12 Performance Indicators for Water Losses and Nonrevenue Water

Indicator	Recommended units	Comments
Inefficiency of use of water resources	Real losses as % of system input volume	Unsuitable for assessing efficiency of management of distribution systems
Water losses	m^3/service connection/year*	Same units as authorized consumption
Apparent losses	m^3/service connection/year*	Same units as authorized consumption
Real losses	Liters/service connection/day* when system is pressurized	Allows for intermittent supply situations
Infrastructure leakage index	Ratio of real losses to technical achievable low-level annual real losses	Technical achievable low-level annual real losses are equal to the best estimate of so-called unavoidable average real losses (UARL). They include system-specific allowance for connection density, customer meter location on service, and current average pressure
Nonrevenue water by volume	Volume of nonrevenue water as % of system input volume	Can be calculated from simple water balance
Nonrevenue water by cost	Value of nonrevenue water as % of annual cost of running system	Allows separate values per m^3 for components of nonrevenue water

*Where service connection density is less than 20 per km of mains, use "/km mains/" instead of "/service connection/" for these performance indicators.

- Easy operation—calculating PI values is intuitive and fast; chances of calculation errors are greatly reduced.
- Compatibility with MS-Excel for further interpretation and processing of the results.

(See Fig. 7.11.) The Lite version aims to be a good *tryout* tool to discover the potential of the IWA PI system. Additionally it is a powerful educational tool.

The professional version of SIGMA, developed as an independent project of ITA, has not only the basic features of SIGMA Lite, but also a good number of complementary ones. For instance, it allows for the customization of the IWA PI system or even the construction of a completely new set of PIs. It also allows the user to store indicators allowing for time-based comparisons, and report and chart customization features will be available.

7.5.7 International Field Test of the IWA PI System (2000–2003)

The IWA PI system resulted from a careful analysis of the potential uses of each indicator and hopefully corresponds to the most current management needs. However, there is no doubt room for improvement. With the publication of the *Performance Indicators for Water Supply Services* in the Manual of Best Practice Series, the IWA launched a new initiative aiming to promote the use of this manual and field-test it in a broad number of real situations. That work had been ongoing since July 2000 and was planned to continue until July 2003 when the results were to be used as the basis for a revised edition of the manual and of the software.

The IWA provided some seed money to make this initiative possible, and a volunteer core team assures the coordination.[4] The participating undertakings are committed to use the system for about 2 years during the course of the project, to reply to three inquiries, one per year, and to share their hands-on experience of using the IWA PI system. Conversely, this is a unique opportunity for them to start using the system counting on technical support from the IWA, while contributing and influencing an IWA-led development process, "by the industry, for the industry," not driven by external financial or commercial interests.

TABLE 7.13 Definition of the Water Losses Indicators

Indicator	Level of importance*	Unit	Concept	Processing rule
WR1: Inefficiency of use of water resources	L1	%	Real losses/water abstracted and imported water × 100 *This indicator must not be used as a measure of efficiency of management of the transmission and/or the distribution system.*	*WR1* = (*A*24)/(*A*4 + *A*5 + *A*8) × 100
Op22: Water losses	L1	m^3/connection/year	Water losses/number of *service connections* *If service connections density $<$20/km of mains (e.g., transmission networks), then this indicator should be expressed in m^3/km of water mains/year.*	Op22 = *A*20/C32
Op23: Apparent losses	L3	m^3/connection/year	Apparent losses/number of *service connections* *If service connections density $<$20/km of mains (e.g., transmission networks), then this indicator should be expressed in m^3/km of water mains/year.*	Op23 = *A*23/C32
Op24: Real losses	L1	Liters/connection/day when system is pressurized	Real losses × 1000/(number of service connections × 365 × *T*/100) (*T* = % of year system is pressurized†) *If service connections density $<$20/km of mains (e.g., transmission networks), then this indicator should be expressed in liters/km of water mains/day*	Op24 = *A*24 × 1000/(C32 × 365 × *D*29/365)
Op25: Infrastructure leakage index	L3	—	Real losses (Op)/ technical achievable low-level annual real losses (when system is pressurized) *The technical achievable low-level annual real losses are equal to the "best estimate" of so-called unavoidable annual real losses (UARL), which can be calculated with the equation derived by the Water Losses Task Force (see AQUA Dec. 1999 and IWA Blue Pages "Losses from water supply systems"):* UARL (liters/service connection/day) = (18 × *Lm*/*Nc* + 0.7 + 0.025 *Lp*) × *P* *This equation, based on empirical results of international investigations, recognizes separate influences on real losses from:* *Length of mains Lm in km (C6)* *Number of service connections Nc (C32)* *Average length of service connections Lp in m (C33)* *Average operating pressure in m (D31)* *Well-managed systems are expected to have low values of this infrastructure leakage index—close to 1.0—while systems with infrastructure management deficiencies will present higher values.*	*Op*25 = *Op*24/(18 × *C*6/*C*32 + 0.7 + 0.025 × *C*33) × *D*31

Fi36: Nonrevenue water by volume	L1	%	Nonrevenue water/system input volume × 100	Fi36 = A26/(A7 + A8) × 100
Fi37: Nonrevenue water by cost	L3	%	Valuation of nonrevenue water components/annual running costs × 100 *This is the sum of separate valuations for unbilled authorized consumption, apparent losses and real losses. With respect to real losses, the attributed unit cost (G50) will be the highest of the variable components of bulk supply charge or long-run marginal cost (LRMC) for own sources. It is usually worthwhile to calculate and review the three components of Fi37 separately.*	Fi37 = ((A18 + A23) × G49 + A24 × G50)/G2 × 100

*L1 = level 1 indicator; L2 = level 2 indicator; L3 = level 3 indicator.

†The term "% of year system is pressurized" is only relevant for intermittent supply networks. For permanent supply networks it assumes the value 1.

TABLE 7.14 Synthesis of Variables Used

A4 ***Water abstracted*** (m^3/year): The annual volume of water obtained for input to water treatment plants (or directly to the transmission and distribution systems) that was abstracted from raw water sources.

A5 ***Imported raw water*** (m^3/year): Total annual volume of raw water transferred from other water supply systems.

A7 ***Water produced*** (m^3/year): Total annual volume of water treated for input to water transmission lines or directly to the distribution system.

A8 ***Imported treated water*** (m^3/year): Total annual volume of treated water imported from other water undertaking or system.

A18 ***Unbilled authorized consumption*** (m^3/year): Total annual amount of unbilled water consumed.

A20 ***Water losses*** (m^3/year): The difference between *system input volume* and *authorized consumption.* Water losses can be considered as a total volume for the whole system, or for partial systems such as raw water mains, transmission, or distribution. In each case the components of the calculation would be adjusted accordingly.

A23 ***Apparent losses*** (m^3/year): Total annual amount of water unaccounted for due to metering inaccuracies and unauthorized consumption.

A24 ***Real losses (m^3/year): Total annual amount of physical water losses from the pressurized system, up to the point of customer metering.***

A26 ***Nonrevenue water*** (m^3/year): Difference between the annual volumes of system input and billed authorized consumption (including exported water).

C6 ***Mains length*** (km): Total transmission and distribution mains length (service connections not included).

C32 ***Service connections*** (number): Total number of service connections.

C33 ***Average service connection length*** (m): Frequently water undertakings do not have detailed accurate information to assess the total service connections length. In these cases, a qualitative assessment will be adopted.

D29 ***Time system is pressurized*** (hours): Amount of time of the year the system is pressurized.

D31 ***Average operating pressure*** (kPa): Average operating pressure at the delivery points when system is pressurized.

G2 ***Annual running costs*** (US\$/year): Total annual operations and maintenance costs + internal work force costs − capitalized costs of self-constructed assets.

G49 ***Average water charges for direct consumption*** (US\$/m^3): Water sales revenue for direct consumption/billed water.

G50 ***Attributed unit cost for real losses*** (US\$/m^3): Highest of the variable components of bulk supply charge or long-run marginal cost (LRMC).

Although the initial target was to have 10 to 15 case studies from different regions and with different characteristics, the number of volunteers exceeded the more optimistic perspectives, and the project currently incorporates 70 undertakings from 19 countries (from Europe, Africa, South America, Middle East, and the Asia-Pacific region). Figure 7.12 presents the geographic locations of the participants. These include organizations of different type (utilities, regulators, holding companies), size, activity (water only or water and other activities; bulk supply or distribution), asset ownership, and type of operations (public or private).

FIGURE 7.10 SIGMA Lite CD-ROM.

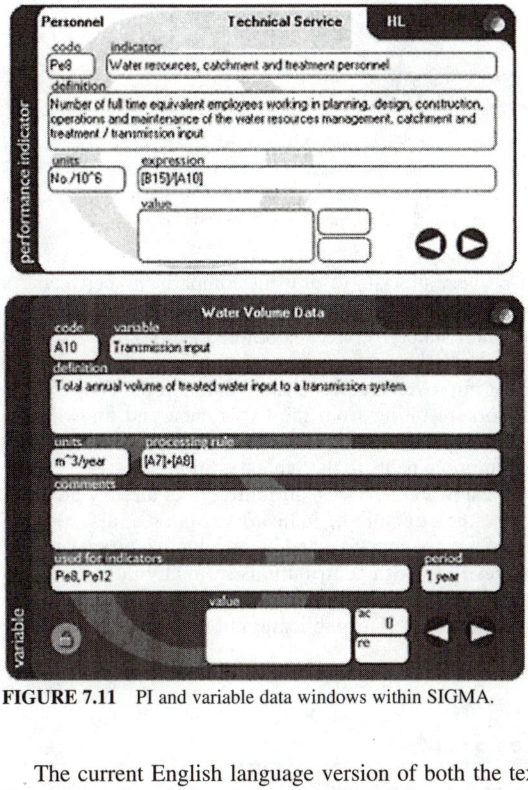

FIGURE 7.11 PI and variable data windows within SIGMA.

The current English language version of both the text and the software is being translated into Portuguese, Spanish, and German. Translation into French is under consideration, and a simplified Czech version is available.

Given the complexity and variety of situations, any general PI system needs to be sufficiently flexible to accommodate each case, allowing for customization. However, on the other hand, definitions

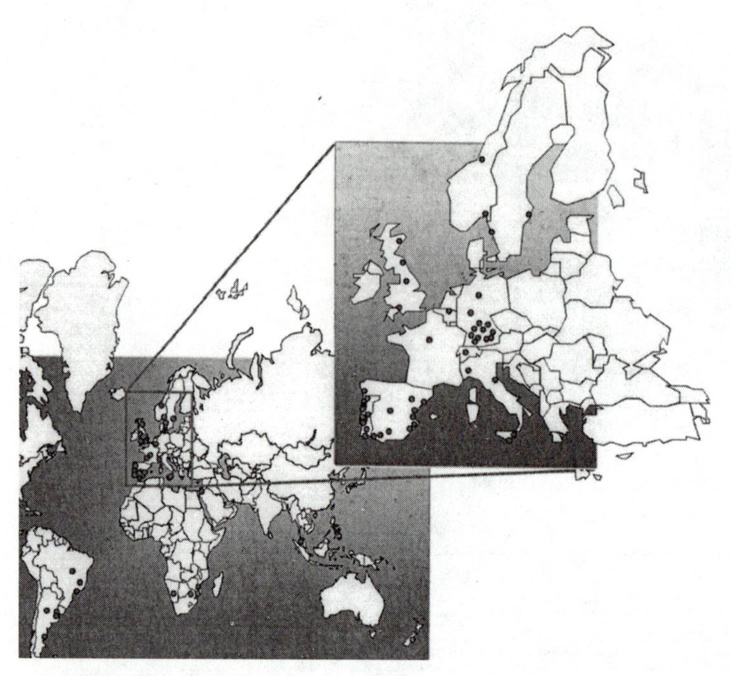

FIGURE 7.12 Participating countries. Number of entities per country: Argentina (4), Belgium (1), Brazil (3), France (1), Germany (13), Israel (1), Italy (4), Malaysia (1), Mozambique (2), Namibia (1), New Zealand (1), Norway (2), Portugal (15), South Africa (1), Spain (12), Sweden (2), Switzerland (1), UK/England (3), UK/Scotland (1).

and processing rules must be very precise, to prevent comparisons between "pears and apples." These are underlying principles adopted for the IWA PI system.

The IWA PI system for water supply services is tailored as a tool to support strategic management. It aims to be an evolving tool. Given the rapid evolution in this area of knowledge, such a system will always have room for improvement. If the first edition of the IWA PI manual is already the output of a rather mature work resulting from the experience and know-how from many IWA members, the contributions arising from 2 years of use by such a broad group of undertakings will certainly provide significant improvements to the second edition.

The IWA PI system for wastewater services, currently in an already advanced stage of development, is supported on the same structure of indicators (Matos et al., 2001). The methodology adopted combines the work of a core team, assisted by an ad hoc international advisory group that provides comments and suggestions, with contributions from any interested person. The working documents are available at any time to any interested person who requests them, the process of evolution being completely transparent. The publication of the corresponding manual is scheduled for 2003.

7.6 IMPLEMENTATION OF A PI SYSTEM

7.6.1 Phases of Implementation

The implementation of a PI system within an organization must be carefully planned. The recommended implementation phases are synthesized in the schematics of Fig. 7.13. The following subsections refer to each one of these phases.

7.6.2 Definition of the Strategic Performance Assessment Policy

Objective Definition. The implementation of any PI system has to be objective-oriented. The number and definition of the PI system to be adopted shall reflect directly the specific objectives of its use. The indicators adequate for a global and synthetic performance assessment for the top managers of a big organization are different from the indicators adequate for a more in-depth analysis, or for a thematic use, such as water losses or rehabilitation.

Definition of the Future Scope of Application. The PI definitions to be adopted will depend on the domain within which the indicators will be compared and interpreted; the greater the group of undertakings involved is, the smaller is the degree of flexibility for creating ad hoc indicators; however, it is advisable to use definitions nationally or internationally accepted whenever

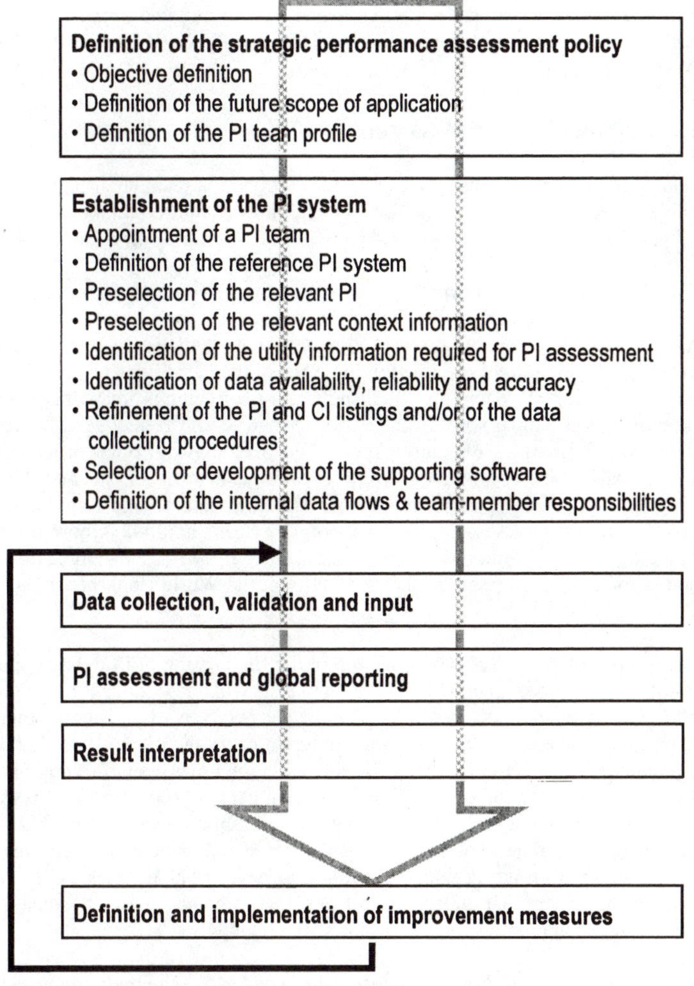

FIGURE 7.13 Phases of implementation of a PI system.

possible, even though the immediate use is confined to the undertaking or to a small group of undertakings.

Definition of the PI Team Profile. According to the specific objectives and taking into account the organization size resources, the decision makers should define the appropriate profile of the future PI team. A successful implementation of a PI system within an organization requires that the project manager be a senior staff member, if necessary assisted by less experienced collaborators; the involvement and active participation of the decision makers who are going to use the PI values is also indispensable; it is strongly recommended that representatives of the departments who need to provide data are members of the PI team, so that they can participate in all phases of the process; in fact, experience has shown that one of the important benefits of the implementation of PI systems is the sharing of experiences and viewpoints between staff members with different responsibilities and backgrounds; when the PIs are selected, it is recommended that the name of a person or sector is assigned to each data item as being responsible for its provision, validation, and input. Experience shows that the assistance of an experienced external consultant during the implementation period can be beneficial, particularly if it is a coordinated initiative of a group of undertakings (e.g., six Scandinavian cities group project; Portuguese PI project; Waters, German PI projects; WUP PI project).

7.6.3 Establishment of the PI System

Appointment of a PI Team. Based on the profile defined, a team shall be appointed and briefed about the objectives and intended scope of application. The future collaborators within the undertaking shall also be informed and motivated for the process from a very early stage.

Definition of the Reference PI System. Before selecting the specific PI to be adopted, the organization managers need to decide whether they prefer to use an ad hoc system, developed expressly for the objective defined, or an existing external PI system (e.g., the IWA PI system). The former option has the advantage of providing more freedom in terms of the definitions to be adopted; however, the development of a dedicated system has the disadvantages of being a time-consuming and expensive task, preventing future comparisons with the outside world and reflecting the internal experience, leaving aside the contributions of many experienced persons from other organizations. The latter option has the disadvantage of imposing several constraints in terms of data and PI definitions; however, existing systems have the major advantage of being matured and tested and allowing future comparisons with reference levels or with other organizations in a much more reliable way. Added advantages are the cost and time of implementation, since these systems already exist, and software is available for some of them (e.g., the IWA PI system, the World Bank benchmarking toolkit, and the WUP PI system).

Preselection of the Relevant PI. The selection of the final listing of PI to be adopted is sometimes an iterative process. The experience of the decision makers will allow for selecting a preliminary listing that seems to fit the defined objectives. In some cases the choice is trivial, but in other cases it is more complex than it may appear. It is important to analyze a priori the information that a given indicator can provide and check if it actually fits the objectives. Sometimes even different units of expression lead to rather different information. Another factor that cannot be undervalued is the confidence of the data required to assess the indicators. It may be preferable to use a less ambitious indicator that is based on reliable data than another theoretically better indicator which requires data that is not available or reliable within the organization: the cost/benefit balance has to always be taken into account. It is recommended that undertakings use parsimony and start with a small number of indicators. Later on they can embrace new indicators step by step.

Identification of Data Availability, Reliability. and Accuracy. The preliminary listing of PI allows the team to identify all the required input data and work with the data providers in order to assess the data's availability, reliability, and accuracy. Whenever a variable is not available or is inaccurate, decision makers and data providers have to agree on what to do: improve the situation or accept that the indicators using this data cannot be assessed or are inaccurate.

Refinement of the PI Listing and Data Collecting Procedures. The assessment of data availability, reliability, and accuracy allows for a consolidation of the PI listing, as well as for some improvements of the data measurement and collection procedures, when appropriate. Another task at this stage is the definition of the frequency of assessment of the indicators. In general, the indicators for overall assessment of the undertaking are assessed once per year, but there are indicators and objectives for which different frequencies are advisable.

Selection or Development of the Software. Although the processing rules of each indicator tend to be as simple as a short number of algebraic operations, the total number of variables manipulated justifies the use of an adequate software application. It is advisable that the software incorporates analysis of consistency procedures, in order to minimize the errors (e.g., range of valid values and cross-coherence between some data values). The software application can be developed for the specific case or the undertaking or be a software of general use. The spreadsheet application made available by the World Bank (WB benchmarking toolkit) and the software SIGMA Lite (IWA PI system) are available free of charge and can be used if the PIs selected are part of the WB indicators or the IWA indicators, respectively.

Definition of the Internal Data Flows and Team Member Responsibilities. A responsible person has to be assigned to each data item required by the PI system. Each variable measurement (or estimate) must be compatible with the preset PI frequency, be assessed for reliability and accuracy, and be input into the PI system. If the same person or team does not hold all of these tasks, it is recommended that the procedure and the responsible person for each step of the process be specified.

The synergy with the existing undertaking information systems should be maximized. Data providers need to feel that the PI system is part of the organization's normal activity, and that there are major advantages in using the same data (i.e., defined and expressed in the same way) for the various existing applications requiring similar information. However, the effective integration of the existing data banks is part of a more general strategic objective of most utilities, and not a PI-specific problem.

7.6.4 Data Collection, Validation, and Input

Taking into account the periods of reference for the PI assessment, the PI team requests the data values to the respective responsible. Depending on the internal procedure agreed upon, data input could be performed directly by each responsible person or collectively by the PI team.

7.6.5 PI Assessment and Global Reporting

PI assessment is just a fingertip task, provided that the data are correctly input and an adequate software application is available. Reports should be customized taking into account the intended use of the information. It is important that the reports include not only the PIs but also the relevant context information. General reports can be limited to tables containing the results, but can also include graphical representations showing the PI evolution with time, presenting sets of related PIs in the same graphic, or comparing PIs with preset targets or reference thresholds, for instance.

7.6.6 Results Interpretation

The success of implementing a PI system depends on the effective use of the results as a decision-making tool. The objective is to interpret the results (i.e., to decide whether a given result is satisfactory or not). If the results are not satisfactory, a diagnosis needs to be established and improvement measures identified.

The process of results interpretation requires comparison of the results to a reference value or a reference threshold, taking into account the context information, and analyzing at each time a group of related indicators and not single indicators individually. The sources of reference values are of five main types:

- Equivalent results of previous years
- Targets established in the master plans
- Targets established within legal, contractual, or regulatory frameworks
- Results from other operational units of the same undertaking (e.g., valid for holdings managing more than one undertaking or for undertakings with more than one system or operational unit)
- Results from other undertakings within metric benchmarking initiatives
- Reference values publicly available

The internal comparison is in general the most important and effective. However, it starts to be feasible only after some years of systematic use of the PI system. The comparison with other undertakings or with publicly available reference values is therefore a need in the first years. On the other hand, sharing experiences with others has a significant leverage effect that should not be ignored.

As managers may be interested in keeping some of the indicators confidential, comparisons of these can be held within small carefully selected groups or provided to an independent external entity that collects the data of the various undertakings, processes them, and reports the results of the group in a form in which individual identification is not possible.

7.6.7 Definition and Implementation of Improvement Measures

The definition and implementation of improvement measures, although not being a part of the performance assessment itself, is so important that it must be discussed. Based on the results interpretation, weaknesses are identified and a diagnosis should be established. From these, decision makers can establish short-, medium-, or long-term targets for improvement and implement the remedial measures. The systematic use of the PI system will later on allow for monitoring of the results and comparing them with the initial targets. Deviations shall be carefully interpreted in order to introduce corrective measures.

7.7 EXAMPLE OF THE USE OF SIGMA LITE FOR DATA INPUT AND INDICATORS ASSESSMENT

7.7.1 Scope of the Example

This section exemplifies the use of the software SIGMA Lite. It contains a step-by-step procedure:

- Input the context information.
- Select the PI to be assessed.
- Input the respective data values.
- Assess the PI.
- Produce PI reports.

A small group of indicators related to personnel is used for this scope:

- Pe1: Employees per connection
- Pe2: Management and support personnel
- Pe3: Financial and commercial personnel
- Pe4: Customer service personnel
- Pe5: Technical services personnel
- Pe6: Planning and construction personnel
- Pe7: Operations and maintenance personnel

7.7.2 Starting to Use SIGMA Lite

The installation of SIGMA is self-explanatory, and the user just needs to reply to the questions asked. A wizard is available the first time the program is run. If the user wishes to reset SIGMA to see the wizard again and input new values, then he or she has to select Tools, Options, Data, or Reset SIGMA from the SIGMA Lite menu. The first step is user identification (Fig. 7.14).

7.7.3 Context Information

The next step is the input of the context information (Fig. 7.15). This information is not used for the assessment of the indicators, but exclusively to help interpret the results. Therefore, the user does not have to fulfill the entire set of tables, but rather select the information considered to be relevant.

In the case of personnel, the most critical context information relates to the type of undertaking and of the service provided, and the percentage of outsourcing for each main function within the organization. In fact, personnel indicators are assessed based on the permanent and temporary personnel of the undertaking, but do not incorporate the work force included as part of contracted services (e.g., construction). The analysis of the personnel indicators cannot be held without the background knowledge of how much of the service is outsourced.

Online help and a system of wizards guide the user, as exemplified in Figs. 7.16 and 7.17. The context information can be edited at any time, at a subsequent stage.

7.7.4 PI Selection and Data Input

The next step is the specification of the indicators that the user selected. The SIGMA window shows which indicators have been selected. If only a subset is to be used, it is necessary to choose Cancel Selection (see Fig. 7.18). Then the user can tick all the relevant indicators, either from all the indicators or group by group.

After the indicators have been selected, the user introduces the values of the corresponding variables, automatically selected by the program, as shown in Figs. 7.19 and 7.20. To fill in the reliability and

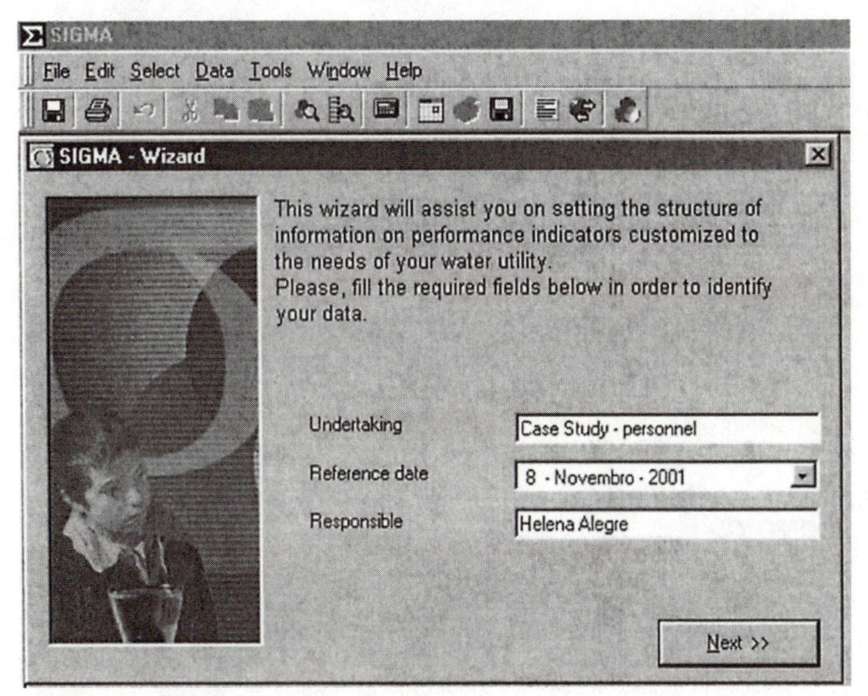

FIGURE 7.14 First SIGMA window: user identification.

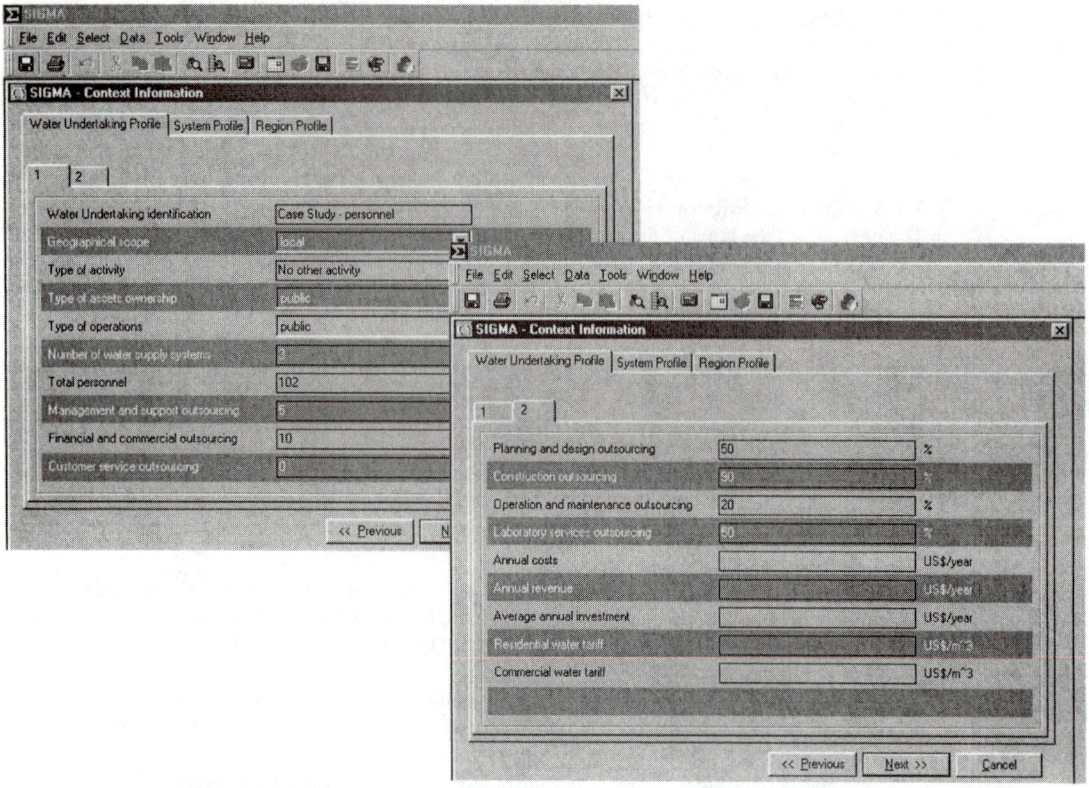

SIGMA
File Edit Select Data Tools Window Help
SIGMA - Context Information
Water Undertaking Profile | System Profile | Region Profile
1 | 2

Water Undertaking identification	Case Study - personnel
Geographical scope	local
Type of activity	No other activity
Type of assets ownership	public
Type of operations	public
Number of water supply systems	3
Total personnel	102
Management and support outsourcing	5
Financial and commercial outsourcing	10
Customer service outsourcing	0

<< Previous | N

SIGMA
File Edit Select Data Tools Window Help
SIGMA - Context Information
Water Undertaking Profile | System Profile | Region Profile
1 | 2

Planning and design outsourcing	50	%
Construction outsourcing	90	%
Operation and maintenance outsourcing	20	%
Laboratory services outsourcing	50	%
Annual costs		US$/year
Annual revenue		US$/year
Average annual investment		US$/year
Residential water tariff		US$/m^3
Commercial water tariff		US$/m^3

<< Previous | Next >> | Cancel

FIGURE 7.15 Context information.

accuracy values, pick a cell, press the right button of the mouse, and select from the table shown at the bottom of Fig. 7.20.

7.7.5 PI Assessment and Reporting

The next step is the assessment of the indicators. The user can select Data, Evaluate from the menu (Fig. 7.21), or press the Evaluate icon from the tool bar. A window with a log file appears with the statistics (Fig. 7.22). If the program detects any errors, such as missing variables to assess the selected indicators, these are listed, and the user is guided to the location where the error occurred in order to correct it.

To see the report, press the Spreadsheet icon. It is separated into three parts: context information, variables, and performance indicators (Figs. 7.23 to 7.25). The results are presented in a spreadsheet-like format that can be exported to other applications.

7.8 EXAMPLE OF THE USE OF THE IWA PI SYSTEM FOR WATER LOSSES CONTROL

7.8.1 Scope of the Example

This section presents an example of assessment of the PI selected by a given undertaking to support the process of water losses control. It aims to highlight that PIs can be used not only for the overall assessment of the company but also for specific uses.

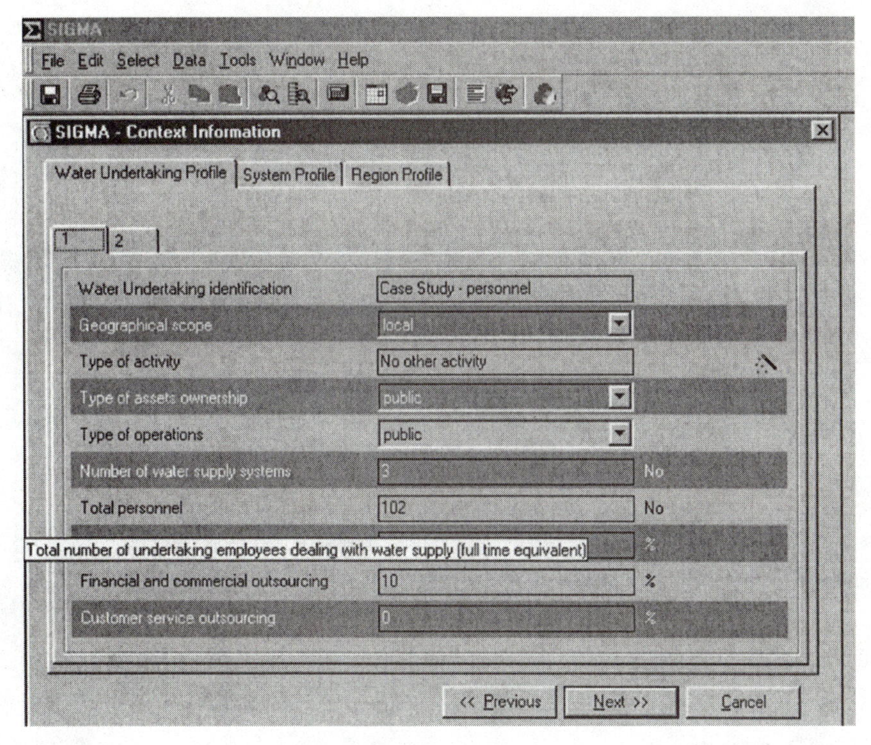

FIGURE 7.16 Example of online help.

7.8.2 Objective Definition and Form of Use of the Indicators

In this example the objective is to identify whether or not a given network is problematic in terms of water losses. Managers are interested in comparing their situation with other networks of the same region, in order to be able to prioritize interventions. Not only will the operational point of view be taken into account, but also the water resources and financial ones. Improvement of customer metering is part of the undertaking policy, in order to minimize the unmetered consumption. The indicators will be used exclusively internally to support managers' decision making. However, comparison with external reference values will be held whenever feasible.

7.8.3 Appointment of the PI Team and Definition of the Reference PI System

The team has been established in order to create an effective group and to involve decision makers and data providers. A part-time project manager with 10 years of professional experience and a background on operations of water supply networks was appointed. A younger full-time assistant was contracted. One top manager assured his participation in periodic meetings. The contribution of field staff, customer service, and financial and commercial personnel was assured by their participation in periodic meetings. In a later stage of the process, specific responsibilities will be committed to these staff members.

7.8.4 Selection of the Indicators

As a starting point, the undertaking managers analyzed the water losses indicators recommended by the IWA PI system, with emphasis to the level 1 indicators (Sec. 7.5.5), because the managers had agreed to start with a small number of indicators. None of the IWA PI level 2 or level 3 indicators were considered necessary to select at this stage.

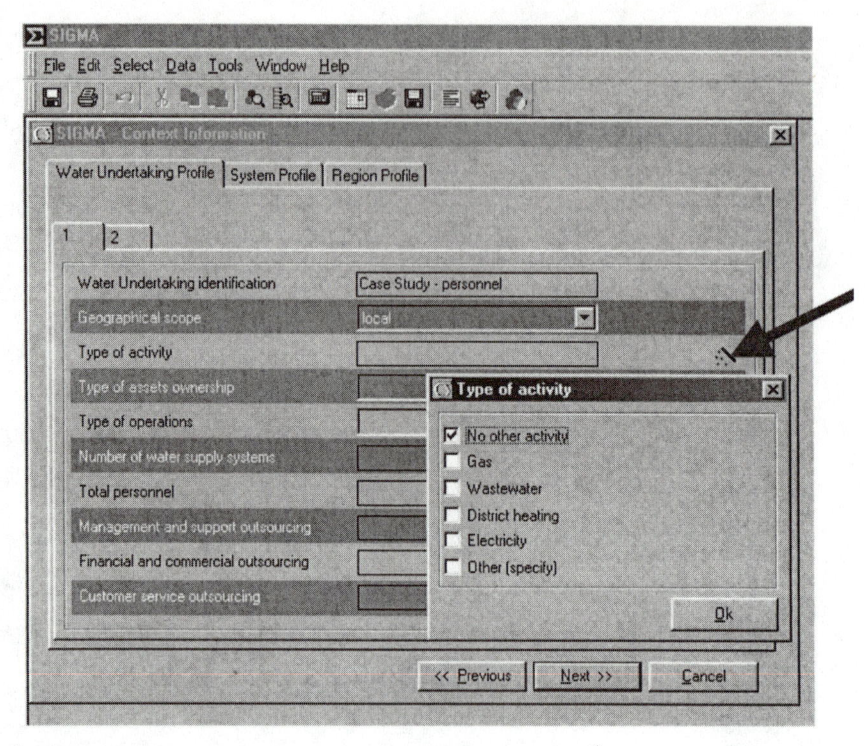

FIGURE 7.17 Example of wizard.

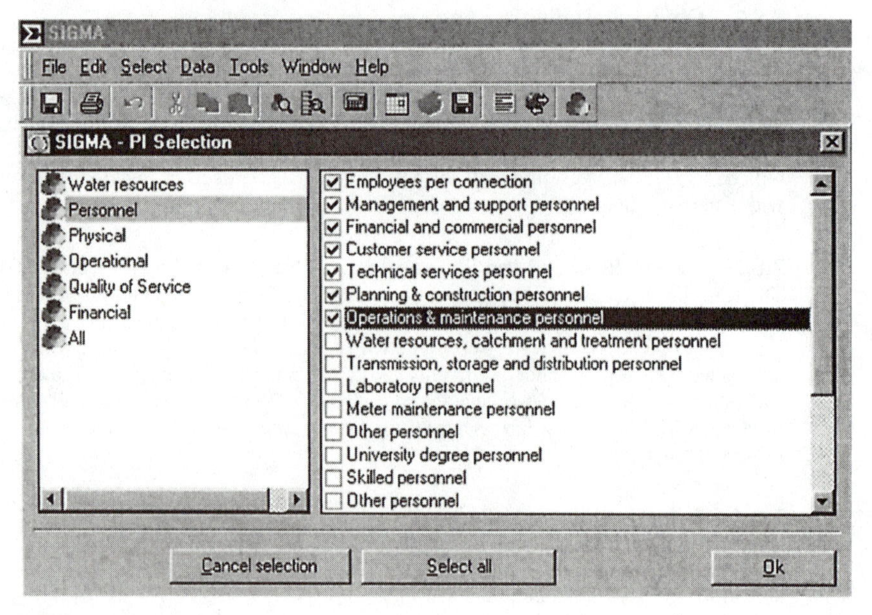

FIGURE 7.18 Selection of the indicators.

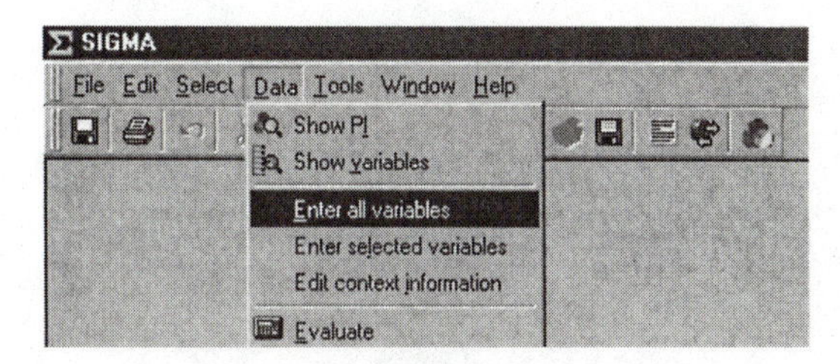

FIGURE 7.19 Selection of data entry.

SIGMA - [Insert values]

File Edit Select Data Tools Window Help

Code	Variable	Units	Value	Reliability	Accuracy
B1	Total personnel	No	102	A	1
B2	Management and support personnel	No	16	A	1
B3	Administration personnel	No	5.5	A	1
B4	Strategic planning personnel	No	3	A	1
B5	Legal affairs personnel	No	1.5	A	1
B6	Human resources personnel	No	5	A	1
B7	Public relations personnel	No	1	A	1
B8	Quality management personnel	No	0	A	1
B9	Other supporting activities personnel	No	1	A	1
B10	Financial and commercial personnel	No	12	A	1
B11	Customer service personnel	No	16	A	1
B12	Technical services personnel	No	58	A	1
B13	Planning and construction personnel	No	10	A	2
B14	Operations and maintenance personnel	No	48	A	2
C32	Service connections	No	32000	B	2

Confidence grades matrix

Accuracy bands (%)	Reliability bands			
	A Highly reliable	B Reliable	C Unreliable	D Highly unreliable
[0; 1]	A1	++	++	++
]1; 5]	A2	B2	C2	++
]5;10]	A3	B3	C3	D3
]10; 25]	A4	B4	C4	D4
]25; 50]	++	++	C5	D5
]50; 100]	++	++	++	D6

FIGURE 7.20 Data input.

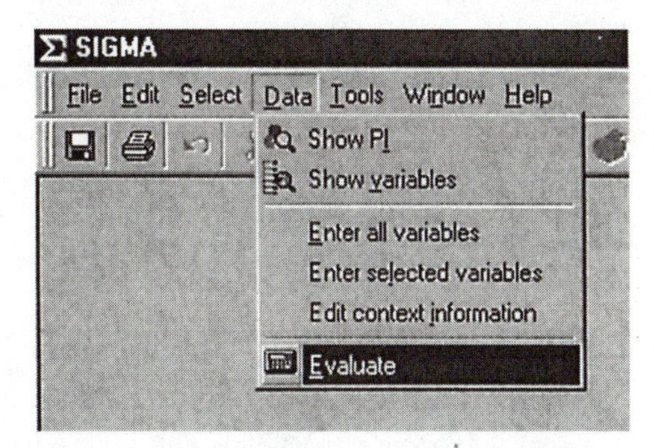

FIGURE 7.21 PI evaluation.

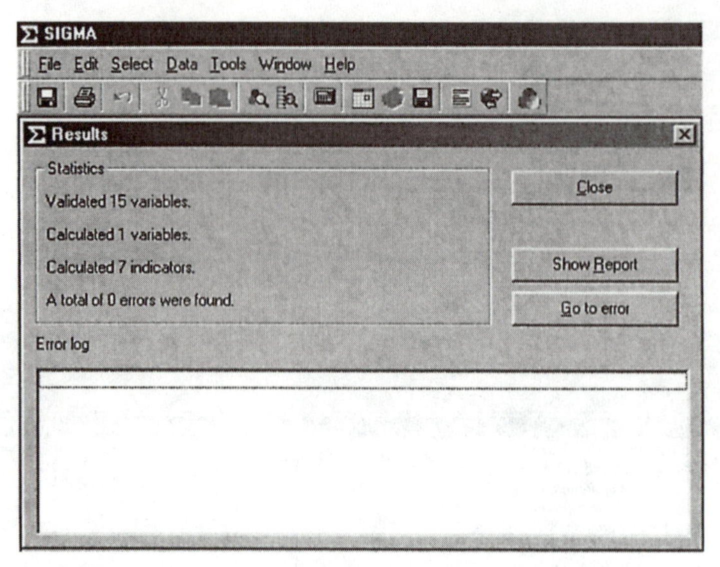

FIGURE 7.22 Statistics and error log file.

SIGMA - [PI Report]

File Edit Select Data Tools Window Help

	A	B	C
1	**Context Information**	**Value**	**Units**
2	Water Undertaking identification	Case Study - personnel	
3	Geographical scope	local	
4	Type of activity	No other activity	
5	Type of assets ownership	public	
6	Type of operations	public	
7	Number of water supply systems	3	No
8	Total personnel	102	No
9	Management and support outsourcing	5	%
10	Financial and commercial outsourcing	10	%
11	Customer service outsourcing	0	%
12	Planning and design outsourcing	50	%
13	Construction outsourcing	90	%
14	Operation and maintenance outsourcing	20	%
15	Laboratory services outsourcing	50	%

FIGURE 7.23 Report: context information.

The managers recognized that the indicator "inefficiency of use of water resources" was appropriate to represent the water resources perspective. They also agreed that the indicator "nonrevenue water by volume" should be selected to represent the financial viewpoint. However, they were surprised that the network efficiency was not included as an operational indicator. In fact, the network efficiency, expressing the water losses as a percentage of the input volume, is the operational indicator of

SIGMA - [PI Report]

File Edit Select Data Tools Window Help

	A	B	C	D	E
1	Code	Variable	Value	Units	Confidence
2	B1	Total personnel	102	No	A1
3	B2	Management and support personnel	16	No	A1
4	B3	Administration personnel	5,5	No	A1
5	B4	Strategic planning personnel	3	No	A1
6	B5	Legal affairs personnel	1,5	No	A1
7	B6	Human resources personnel	5	No	A1
8	B7	Public relations personnel	1	No	A1
9	B8	Quality management personnel	0	No	A1
10	B9	Other supporting activities personnel	1	No	A1
11	B10	Financial and commercial personnel	12	No	A1
12	B11	Customer service personnel	16	No	A1
13	B12	Technical services personnel	58	No	A1
14	B13	Planning and construction personnel	10	No	A2
15	B14	Operations and maintenance personnel	48	No	A2
16	C32	Service connections	32000	No	B2

FIGURE 7.24 Report: variables.

SIGMA - [PI Report]

File Edit Select Data Tools Window Help

	A	B	C	D	E
1	Code	Indicator	Value	Units	Confidence
2	Pe1	Employees per connection	3,19	No./1000 connections	B2
3	Pe2	Management and support personnel	0,5	No./1000 connections	B2
4	Pe3	Financial and commercial personnel	0,38	No./1000 connections	B2
5	Pe4	Customer service personnel	0,5	No./1000 connections	B2
6	Pe5	Technical services personnel	1,81	No./1000 connections	B2
7	Pe6	Planning & construction personnel	0,31	No./1000 connections	B2
8	Pe7	Operations & maintenance personnel	1,5	No./1000 connections	B2

FIGURE 7.25 Report: performance indicators.

water losses most commonly used worldwide. Its adoption is attractive because it is apparently simple to assess and interpret. However, it has the major advantage of being strongly dependent on the consumption, a scaling factor that leads to misinterpretations, as discussed further on. Conversely, this indicator has the shortcoming of not taking into account any of the relevant factors that influence real losses, such as the total length of the mains and of the service connections, the service pressure, the density of service connections, and the location of domestic meters. These are the reasons why the IWA supports the recommendation of the National Committees of Germany, South Africa, and the United Kingdom, as well as of the British regulator OFWAT, which states that the network efficiency is not an adequate PI for technical use.

To understand how misleading the network efficiency can be, two simple comparisons are suggested:

1. Let us take one network and analyze the water losses in 2 subsequent years. Because of the occurrence of a very wet summer during the first year monitored, the total consumption was significantly above average. The volume of water losses was approximately the same in both cases.

However, the observation of the network efficiency, that consequently increased, might lead to the false conclusion that the situation had improved in terms of losses.

2. Let us take two networks exactly equal in terms of material, age, failure rates, diameters, average service pressure, and density of service connections—therefore with the same amount of water losses. The former supplies single-floor buildings, while the later supplies an area of multistore apartment blocks, with a higher consumption per connection and per pipe length. From the operational point of view it is important to find an indicator that provides equivalent values for both situations. The network efficiency does not, and therefore is inappropriate.

The indicator "total water losses" and particularly the indicator "real losses per service connection," as defined in Tables 7.13 and 7.14, are more robust operational indicators. Being aware of these reasons, managers decided to select these two indicators.

Revisiting their objectives, managers verified that only one had been left uncovered: the measure of the performance regarding the minimization of the unmetered consumption. Several options were considered:

- Number of unmetered authorized consumers/total number of authorized consumers
- Unmetered consumption/total input volume
- Unmetered billed consumption/total metered billed consumption
- Unmetered authorized consumption/total authorized consumption

The first indicator was abandoned because metering prioritization should take into account the consumption, and not only the number of consumers. The second indicator was eliminated as well because the other two seemed comparatively better.

The decision between these two latter indicators requires a further clarification of the objectives. If the objective is to minimize the percentage of bills based on estimates, the focus should be put on the billed consumption. If the objective is more general and aims to improve the confidence of the water balance figures, in order to have a proper monitoring of the water losses, the focus should be on the authorized consumption. Managers chose the fourth indicator: "unmetered authorized consumption/total authorized consumption."

7.8.5 Assessment of the Water Balance

Step 0. Define the system (or part of the system) to audit and define the reference dates (1 year) (Fig. 7.26)

Step 1. Define *system input volume* and enter in column A (see Fig. 7.27).

- ***a.*** Water abstracted: $78.5 \times 24 \times 365 = 687{,}660$ m^3
- ***b.*** Imported treated water: $55.0 \times 24 \times 365 = 481{,}800$ m^3
- **c.** Imported treated water: 0 m^3
- ***d.*** *System input volume:* 687,660 m^3 + 481,800 m^3 + 0 = 1,169,460 m^3

Step 2.

- ***a.*** Define *billed metered consumption* (the type of consumers can be different from this example).
  - **(1)** Direct supply:
    - **(*a*)** Domestic consumption: 428,145 m^3
    - **(*b*)** Industrial consumption: 75,555 m^3
    - **(*c*)** Public consumption: 50,370 m^3
    - **(*d*)** Total direct consumption: 554,070 m^3
  - **(2)** Exported water: $16{,}126 \times 24 \times 365 = 141{,}264$ m^3
  - **(3)** *Billed metered consumption:* 554,070 m^3 + 141,264 m^3 = 695,334 m^3
- ***b.*** Define *billed unmetered consumption* in column D.
  - **(1)** Unmetered consumers (12%, 150 L per capita per day): 75,555 m^3

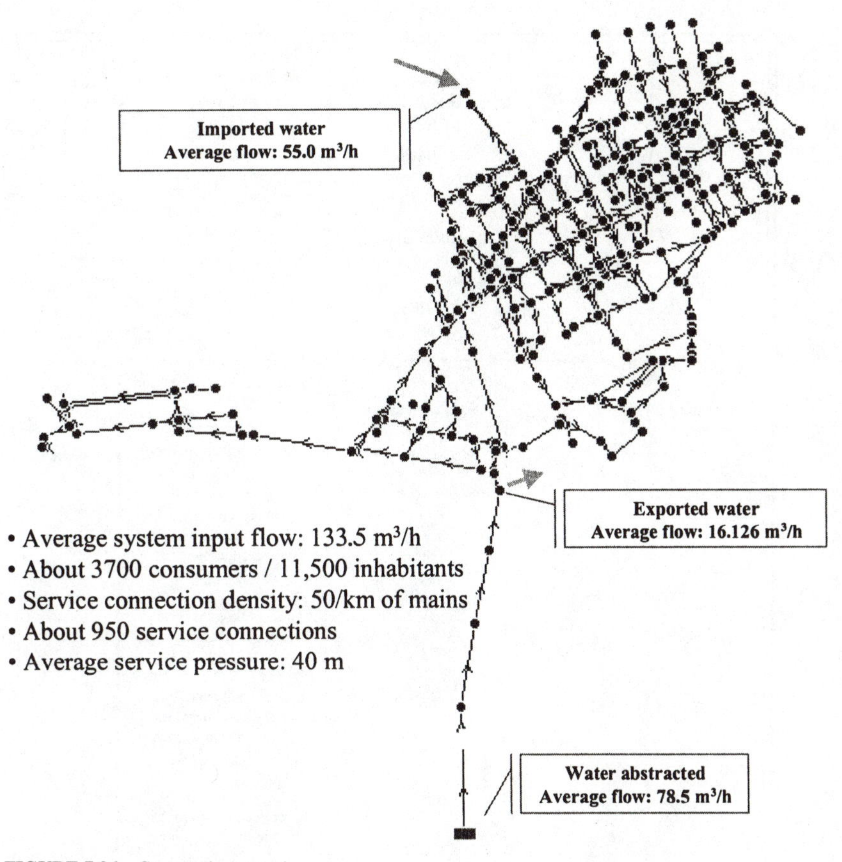

FIGURE 7.26 Case study network.

(**2**) Landscape irrigation: 540 m^3

(***a***) *Example:* By estimating the average daily irrigation time, the average flow consumption per irrigation sprinkler, the approximate number of irrigation sprinklers, and the number of months per year of irrigation.

(**3**) Street cleaning: 375 m^3

(***a***) *Example:* By estimating the average number of vehicles used per day, the number of times each one is refilled per day, their water storage capacity, and the average number of days per year they are used.

(**4**) *Billed unmetered consumption:* 75,555 m^3 + 540 m^3 + 375 m^3 = 76,470 m^3

c. Enter total in *billed authorized consumption* (column C) and *revenue water* (column E).

(**1**) *Billed authorized consumption* (= *revenue water*): 771,804 m^3 + 76,470 m^3 = 771,804 m^3

Step 3. Calculate the volume of (column E) as *system input volume* (column A) minus *revenue water* (column E).

a. *Nonrevenue water:* 1,169,460 m^3 − 771,804 m^3 = 397,656 m^3

Step 4.

a. Define *unbilled metered consumption* in column D.

(**1**) Undertaking's own consumption: 450 m^3

(**2**) Public consumption: 12,500 m^3

<table>
<tr><th>A</th><th>B</th><th>C</th><th>D</th><th>E</th></tr>
<tr><td rowspan="10">System input volume
1,169.460 m^3</td><td rowspan="4">Authorized consumption
855,754 m^3</td><td rowspan="2">Billed authorized consumption
771,804 m^3</td><td>Billed metered consumption (including water exported)
695,334 m^3</td><td rowspan="2">Revenue water
771,804 m^3</td></tr>
<tr><td>Billed unmetered consumption
76,470 m^3</td></tr>
<tr><td rowspan="2">Unbilled authorized consumption
530 m^3</td><td>Unbilled metered consumption
12,950 m^3</td><td rowspan="8">Nonrevenue water
397,656 m^3</td></tr>
<tr><td>Unbilled unmetered consumption
71,000 m^3</td></tr>
<tr><td rowspan="6">Water losses
313,706 m^3</td><td rowspan="2">Apparent losses
115,416 m^3</td><td>Unauthorized consumption
15,000 m^3</td></tr>
<tr><td>Metering inaccuracies
100,416 m^3</td></tr>
<tr><td rowspan="4">Real losses
198,290 m^3</td><td>Real losses on raw water mains and at the treatment works (if applicable)
[m^3/year]</td></tr>
<tr><td>Leakage on transmission and/or distribution mains
[m^3/year]</td></tr>
<tr><td>Leakage and overflows at transmission and/or distribution storage tanks
[m^3/year]</td></tr>
<tr><td>Leakage on service connections up to the measurement point
[m^3/year]</td></tr>
</table>

FIGURE 7.27 Components of water balance for the selected case study.

(3) Other: 0 m^3
(4) *Unbilled metered consumption:* 450 m^3 + 12,500 m^3 + 0 m^3 = 12,950 m^3

b. Define *unbilled unmetered consumption* in column D.
(1) Landscape irrigation: 20,000 m^3
(2) Street cleaning: 43,000 m^3
(3) Fire fighting: 5,000 m^3
(4) Pipe flushing and storage tank cleaning: 3,000 m^3
(5) Other: 0 m^3
(6) *Unbilled unmetered consumption:* 20,000 m^3 + 43,000 m^3 + 5,000 m^3 + 3,000 m^3 = 71,000 m^3

c. Assess *unbilled authorized consumption* and transfer total to column C.
(1) *Unbilled authorized consumption:* 12,950 m^3 + 71,000 m^3 = 83,950 m^3

Step 5. Add volumes of *billed authorized consumption* and *unbilled authorized consumption* in column C; enter sum as *authorized consumption* (column B).
a. Authorized consumption: 771,804 m^3 + 83,950 m^3 = 855,754 m^3

Step 6. Calculate *water losses* (column B) as the difference between *system input volume* (column A) and *authorized consumption* (column B).
a. Water losses. 1,169,460 m^3 − 855,754 m^3 = 313,706 m^3

Step 7.

a. Assess components of *unauthorized consumption* (column D) by best means available.
  (1) Unauthorized temporary use of fire and irrigation hydrants
  (2) Unauthorized connections
  (3) *Unauthorized consumption:* unknown (10,000 to 20,000 m^3)

b. Assess components of *metering inaccuracies* (column D) by best means available.
  (1) Meter reading and registration errors (10%): 10% × (695,334 + 12,950) = 70,828 m^3
  (2) Inaccuracies of unmetered authorized consumption (20%): 20% × (76,940 + 71,000) = 29,588 m^3
  (3) *Metering inaccuracies:* 70,828 m^3 + 29,588 m^3 = 100,416 m^3 (these values can be positive or negative)

c. Add *unauthorized consumption* and *metering inaccuracies* and enter sum in *apparent losses* (column C).
  (1) *Apparent losses:* 15,000 + 100,416 = 115,416 m^3

Step 8. Calculate *real losses* (column C) as *water losses* (column B) minus *apparent losses* (column C).
  a. Real losses: 313,706 m^3 − 115,416 m^3 = 198,290 m^3

Step 9. Assess components of *real losses* (column D) by best means available (night flow analysis, burst frequency/flow rate/duration calculations, modeling, etc.), add these and cross-check with volume of *real losses* in column C.

7.8.6 Assessment of the Indicators

Water Resources Indicators

$$\text{Inefficiency of use of water resources (\%)} = \frac{\text{volume of real losses}}{\text{system input volume} \times 100}$$

$$\frac{198{,}290}{1{,}169{,}460} \times 100 = 17.0\ \%$$

Financial Indicator

$$\text{Nonrevenue indicators (by volume) (\%)} = \frac{\text{volume of nonrevenue water}}{\text{system input volume}} \times 100$$

$$\left(\frac{397{,}656}{1{,}169{,}460}\right) \times 100 = 34.0\%$$

Operational Indicators

$$\text{Water losses (m}^3\text{ per connection per year)} = \frac{\text{volume of water losses}}{\text{number of service connections}}$$

$$= 330\ \text{m}^3 \text{ per connection per year}$$

Real losses (L per connection per day)

$$= \frac{\text{real losses} \times 1000}{\text{number of service connections} \times 365 \times T/100}$$

where T is the percent of the year when the system is pressurized.

$$\frac{198{,}290 \times 1000}{(950 \times 365 \times 100/100)} = 572 \text{ L per connection per day}$$

Unmetered consumption (%) =

$$\frac{\text{Billed unmetered consumption} + \text{unbilled unmetered consumption}}{\text{authorized consumption (\%)}}$$

$$\frac{76{,}470 + 71{,}000}{855{,}754} = 17.2\%$$

7.8.7 Results Interpretation

The results show that the situation of this network in terms of water losses is problematic. In financial terms, the efficiency is low, with 34 percent being nonrevenue water. Real losses of 572 L per connection per day are high, even taking into account that the average service pressure (40 m) is above average. The percentage of unmetered consumption, 17 percent, shows that the consumption assessed from estimates is still significant, although it demonstrates that the undertaking already manages to measure most of the authorized consumption. In a real situation, the interpretation would need to be completed with the analysis of the explanatory factors and the identification of the main aspects where there is room for improvement.

7.9 FINAL REMARKS

The systematic use of performance indicators, when preceded by an adequate planning and implementation and followed by an in-depth comparative analysis of the results and by the improvements derived from it, has a major leverage effect on the increase of efficiency and productivity of an organization.

The implementation of this type of management tool requires the active participation and engagement of the top managers in every stage of the process. The establishment of a team led by a senior staff member and representatives of the main departments of the organization that have clear responsibilities in the process is also a key factor for its success.

The selection of the PI system and of the specific PIs to adopt is another step of major importance. The use of standardized PI systems, such as the IWA's, is recommended as a generally better alternative to ad hoc systems developed expressly to a specific objective. A subset of relevant PIs shall be selected from the listing, and whenever necessary customized (e.g., changing metric units into imperial units) or expanding in some cases (e.g., splitting the PI into more detailed ones).

Any PI value is meaningless if the confidence of the input data is not known. Therefore, it is recommended that the reliability and accuracy of each variable required by the PIs be carefully assessed according to preset criteria.

All the team members shall in one way or another participate in the process of interpretation of the PI values, assisting in the identification and implementation of the procedures where there is room for improvement. This phase is critical to keep the staff motivated.

The continuous use of PIs allows the comparison of the actual results with management objectives and with the results of previous years. This type of analysis is in general more important than the comparison with other undertakings.

One could hardly choose better concluding remarks than the words of Finn Johansen, Director of Oslo Water and Sewage Works (Johansen and Helland, 2000):

> There is today a strong call for efficiency, productivity, quality and good service. As a director of municipally-owned water and wastewater undertaking, I experience these demands from the politicians and administrators in the City Hall and from the users of our services. And in addition there are require-

ments and demands from the State regulators and policy-making Departments. Last, but not least, we have the constant interest from the media and from environmental organizations and other pressure groups....

Let me make it clear that I don't experience the present call for efficiency and effectiveness as a threat, but rather as a stimulant for our organization to become even better.

Benchmarking routines can be implemented internally in our organizations—comparing our results this year with last year's—or externally—comparing our performance with other undertakings with similar size and conditions, and preferably with undertakings that are better than us in certain areas or activities....

The most important benchmarking is done within each undertaking, when we compare the current year's PI with previous years' indicators. This is a relevant motivational factor for employees in our different divisions, who will use all of PI used by the whole group plus several more specific division indicators....

The big question for us has been to find good performance indicators for benchmarking purposes.

Oslo is one of the 70 undertakings participating in the IWA PI field-testing project.

7.10 ACKNOWLEDGMENTS

The author expresses her acknowledgments to Wolfram Hirner, Jaime Melo Baptista, and Renato Parena, her coauthors of *Performance Indicators for Water Supply Services* in IWA Manual of Best Practices Series; and to Francisco Cubillo, Enrique Cabrera Jr., and Jaime Melo Baptista, members of the coordinating team of the ongoing field-testing phase.

A word of appreciation is also due to the 70 organizations who are participating in the field-testing of the IWA PI system and to everyone who contributed by hosting meetings, providing comments and suggestions, or assisting with the background work.

On behalf of the International Water Association, the author acknowledges her employer, the National Laboratory of Civil Engineering, for all the support provided to this project.

7.11 REFERENCES

Adamsson, J., *The Swedish Experience and Viewpoints,* IWSA Workshop on Performance Indicators for Transmission and Distribution Systems, Lisbon, Portugal, 1997.

Adamsson, J., et al., *Performance Benchmarking among 6 Cities in Scandinavia,* working report, 2000.

Alegre, H., J. M. Baptista, and A. L. Faria, *A General Framework of Performance Indicators in the Scope of Water Supply,* IWSA Workshop on Performance Indicators for Transmission and Distribution Systems, Lisbon, Portugal, 1997.

Alegre, H., W. Hirner, J. M. Baptista, and R. Parena, *Performance Indicators for Water Supply Services,* Manual of Best Practice Series, IWA Publishing, London, 2000.

Ancarani, A., "The Evaluation of the Quality of the Public Services Provided by Public/Private Firms: The Water Service Firms Case Study," Ph.D. Thesis in Economical and Managerial Engineering, Rome, December 1999.

AWA, "Australian Non Major Urban Water Utilities," *Performance Monitoring Report 1998–1999.* Australian Water Association, 2000.

AWWARF, "Distribution System Performance Evaluation," *Report of the American Water Works Research Foundation,* 1995.

AWWARF, "Performance Benchmarking for Water Utilities," *Report of the American Water Works Research Foundation,* 1996.

Cabrera, Jr., E., "Design of System for the Management Assessment of Urban Water Supply Systems," Ph.D. thesis, Universidade Politécnica de Valência, Valência, Spain, 2001 (in Spanish).

Cardoso, A., R. Matos, H. Alegre, and S. T. Coelho, "Performance Assessment of Water Supply and Wastewater Systems," *Seminar on Performance Indicators for Urban Drainage,* University of Hertfordshire, Reino Unido, December 6, 2000.

Cardoso, M. A., J. S. Matos, R. S. Matos, and S. T. Coelho, "A New Approach for Diagnosis and Rehabilitation of Sewerage Systems through the Development of Performance Indicators," *Proceedings of the 8th International Conference on Urban Storm Drainage,* Sydney, Australia, September 1999.

Coelho, T., *Performance Indicators in Water Supply—A Systems Approach,* Water Engineering and Management Systems Series, Research Studies Press, John Wiley and Sons, UK, 1997.

Deb, A., and L. Cesario, "Water Distribution Systems Performance Assessment," *IWSA Workshop on Performance Indicators for Transmission and Distribution Systems,* Lisbon, Portugal, 1997. Project funded by the AWWA Research Foundation.

Faria, A. L., and H. Alegre, "Paving the Way to Excellence in Water Supply Systems: a National Framework for Levels-of-Service Assessment Based on Consumer Satisfaction," The Maarten Schalekamp Award—1995, *AQUA,* 45(1):1–12, 1996.

Guérin-Schneider, L., "Regulation of the Water Services in France and the Performance Indicators. Instrumentation and Organization," Ph.D. thesis, ENGREF—École National du Génie Rural, des Eaux et Forêts, Paris, 2001 (in French).

Johansen, F., and B. Helland, "Benchmarking Experience in Nordic Water and Wastewater Utilities," in *H2Objecttivo 2000—XI Edition,* Turin, Italy, May 2000, pp. 3–5.

MWA, *Performance Indicators for Water Supply,* A proposal from the Malaysian Water Association for the consideration of member countries of ASPAC, 1996.

MWA, *Malaysia Water Industry Guide 2001,* Water Supply Branch of the Malaysian Water Association, 2001.

Matos, R., M. A. Cardoso, and P. Duarte, "Preliminary Listing of Definitions, Context Information and Performance Indicators for Wastewater Services," Working document, *Workshop on Performance Indicators for Wastewater Services,* International Water Association, World Congress 2001, Berlin, Germany, 2001.

McIntosh, A. C., and C. Yñoguez, C., *Second Water Utilities Data Book—Asian and Pacific Region,* Asian Development Bank's Regional Technical Assistance No. 5694, Manila, Philippines, 1997.

Merkel, W., "Performance Assessment in the Water Industry," *International Report,* International Water Association, World Congress 2001, Berlin, Germany, 2001.

OFWAT, "Levels-of-Service for the Water Industry in England & Wales," *1999–2000 Report,* Office of Water Services, United Kingdom, 2000.

Paralez, L., "Status of Performance Assessment in Water and Wastewater Utility Systems in the United States," *National Report on Performance Assessment in the Water Industry,* International Water Association, World Congress 2001, Berlin, Germany, 2001, 11 pp.

van der Willigen, F., "Dutch Experience and Viewpoints on Performance Indicators," *IWSA Workshop on Performance Indicators for Transmission and Distribution Systems,* Lisbon, Portugal, 1997.

Westerhoff, G., D. Gale, P. Reiter, S. Haskins, and J. Gilbert, in J. B. Mannion (ed.), *The Changing Water Utility: Creative Approaches to Effectiveness and Efficiency,* American Water Works Association, 1998.

Yeppes, G., and A. Dianderas, *Water & Wastewater Utilities Indicators,* Water and Sanitation Division, The World Bank, 1996.

7.12 ENDNOTES

1. In England and Wales most domestic water consumption is not metered.
2. In 1999 the former International Water Services Association (IWSA) and the former International Association of Water Quality (IAWQ) merged, giving place to the IWA.
3. Text extracted from *Performance Indicators for Water Supply Services* in the IWA Manual of Best Practices Series.
4. Helena Alegre (LNEC, Portugal); Francisco Cubillo (Canal de Isabel II, Spain); Enrique Cabrera Jr. (ITA, Spain); Jaime Melo Baptista (LNEC, Portugal); Patrícia Duarte (LNEC, Portugal).

CHAPTER 8
CLIMATE CHANGE EFFECTS AND WATER MANAGEMENT OPTIONS

Larry W. Mays
Department of Civil and Environmental Engineering
Arizona State University
Tempe, Arizona

8.1 INTRODUCTION

The Pacific Institute has compiled a comprehensive bibliography of peer-reviewed literature dealing with climate change and its effects on water resources and water systems. As of December 2001 the bibliography included over 920 citations. The address for this bibliography is www.pacinst.org/CCBib.html. The International Panel on Climate Change (IPCC) has been active in compiling information concerning future climate change (IPCC, 1990, 1996a, 1996b, 1996c, 1998, 2001a, 2001b, 2001c). These reports can be obtained from the Web address www.ipcc.ch.

Other reports have been published that contain our state-of-the-art knowledge on the topic of global climate change and water resources. Gleick (1993) edited the book *Water Crisis: A Guide to the World's Fresh Water Resources,* in which he and several authors presented summaries on several freshwater resources topics related to our resources on earth. In particular, this work presented in one place many tables of information and data related to water resources around the world. The Conference on Climate and Water held in Helsinki, Finland in September 1989 and the Second International Conference on Climate and Water held in Espoo, Finland in August 1998 provide other sources of information on this topic, particularly for the European region. The December 1999 and April 2000 issues of the *Journal of the American Water Resources Association* were devoted to the topic of water resources and climate change. Gleick (2000) and several authors of the National Water Assessment Group presented a report for the U.S. Global Change Group.

8.1.1 The Climate System

The *climate system* (refer to Fig. 8.1) as defined by the IPCC (2001a) is "an interactive system consisting of five major components: the atmosphere, the hydrosphere, the cryosphere, the land surface, and the biosphere, forced or influenced by various external forcing mechanisms, the most important of which is the Sun." The effect of human activities on the climate system is considered as external forcing. The sun provides the ultimate source of energy that drives the climate system.

The *atmosphere,* which is the most unstable and rapidly changing part of the system, is composed (earth's dry atmosphere) mainly of nitrogen (78.1 percent), oxygen (20.9 percent), and argon (0.93 percent). The atmospheric distribution of ozone in the lowest part of the atmosphere, the troposphere, and the lower stratosphere acts as greenhouse gas. (There are five basic layers in the atmosphere: troposphere, stratosphere, mesosphere, thermosphere, and exosphere.) In the higher part of the stratosphere there is a natural layer of high ozone concentration. This ozone absorbs ultraviolet radiation, playing an essential role in the stratosphere's radiation balance, and at the same time filters out this potentially damaging form of radiation.

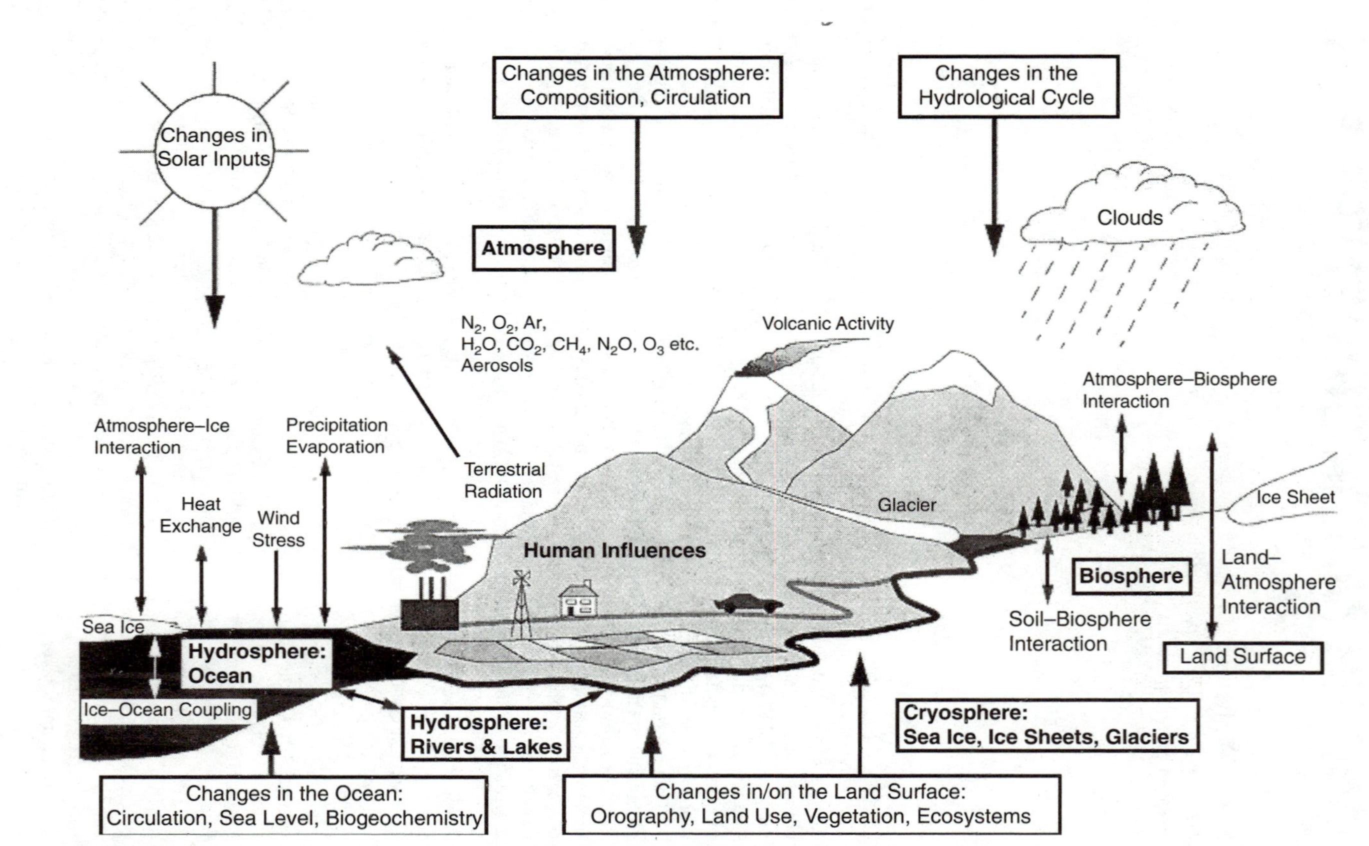

FIGURE 8.1 Schematic view of the components of the global climate system (bold), showing their processes and interactions (thin arrows) and some aspects that may change (bold arrows). (*Source: IPCC, 2001a.*)

The *hydrosphere* comprises all liquid surface water and subterranean water, both fresh water, including rivers, lakes, and aquifers, and saline water of the oceans and seas. The *cryosphere,* which includes the ice sheets of Greenland and Antarctia, continental glaciers and snow fields, sea ice, and permafrost, derives its importance to the climate system from its reflectivity (*albedo*) for solar radiation, its low thermal conductivity, and its large thermal inertia. The cryosphere has a critical role in driving the deep-water circulation. Ice sheets store a large volume of water so that variations in their volume are a potential source of sea-level variations. *Marine* and *terrestrial biospheres* have major impacts on the atmosphere's composition through *biota,* influencing the uptake and release of greenhouse gases, and through the *photosynthetic process,* in which both marine and terrestrial plants (especially forests) store significant amounts of carbon from carbon dioxide. Interaction processes (physical, chemical, and biological) occur among the various components of the climate system on a wide range of space and time scales. This makes the climate system, as shown Fig. 8.1, extremely complex.

The mean energy balance for the earth is illustrated in Fig. 8.2, which shows on the left-hand side what happens with the incoming solar radiation and on the right-hand side how the atmosphere emits the outgoing infrared radiation. A *stable climate* must have a balance between the incoming radiation and the outgoing radiation emitted by the climate system. On average the climate system must radiate 235 W/m^2 back into space. Climate variations may result not only from the radiative forcing but also from internal interactions among components of the climate system; therefore, a distinction is made.

8.1.2 Definition of Climate Change

The earth's temperature is affected by numerous causes, including (1) the incoming solar radiation that is absorbed by the atmosphere and the earth's surface, (2) the characteristics (emissivity) of the matter that absorbs the radiation, and (3) the fact that part of the longwave radiation emitted by the surface is absorbed by the atmosphere, and is then reemitted as longwave radiation either in the upward or downward direction.

The *greenhouse effect* is caused by the net change of the internal radiation balance of the atmosphere due to the continued increased emission of greenhouse gases, resulting in both the atmosphere and the earth's surface being warmer (Mitchell, 1989; Mitchell et al., 1995). Magnitude of the greenhouse effect is dependent on the composition of the atmosphere, the most important factors of which are the concentrations of water vapor and carbon dioxide, and less importantly on certain trace gases such as methane. There is mounting evidence that global warming is under way (Houghton et al., 1990; Levine, 1992; Mitchell et al., 1995; Santer et al., 1996; Tett et al., 1996; Harris and Chapman, 1997; Kaufmann and Stern, 1997). Scientists agree that a global temperature rise of 1 to 2°C is likely by 2050; however, a precise knowledge of the global-average temperature change is of little predictive value for water resources. On a global scale the warming would result in increased evaporation and increased precipitation. What is more important is how the temperature will change regionally and seasonally, affecting precipitation and runoff, which is by no means clear. The articles by Karl et al. (1995) and Karl and Quayle (1988) on climate change may be of interest to the reader.

In general, the hydrologic effects are likely to influence water storage patterns throughout the hydrologic cycle and impact the exchange among aquifers, streams, rivers, and lakes. Chalecki and Gleick (1999) provided a bibliography of the impacts of climate change on water resources in the United States. Climate change can have a significant effect on the organisms living in these aquatic systems (see Fisher and Grimm, 1996; Firth and Fisher, 1992). Other literature concerning lakes, ecosystems, and other related topics includes articles by Burkett and Kusler (2000), Cohon (1987), and Hostetler and Small (1999).

8.1.3 Climate Change Prediction

Climate change predictions are based mostly on computer simulations using *general circulation models* (GCMs) of the atmosphere. These models solve the conservation equations (in discrete space and time) that describe the geophysical fluid dynamics of the atmosphere. GCMs have the same general structure as *numerical weather prediction models* (NWPMs) that are used for weather prediction; however, they have a much larger spatial distribution and are run for much longer time periods. Scenarios of changes in atmospheric variables (such as temperature, precipitation, and evapotranspiration) that effect the surface and subsurface hydrology are determined using methods that range from prescribing hypothetical

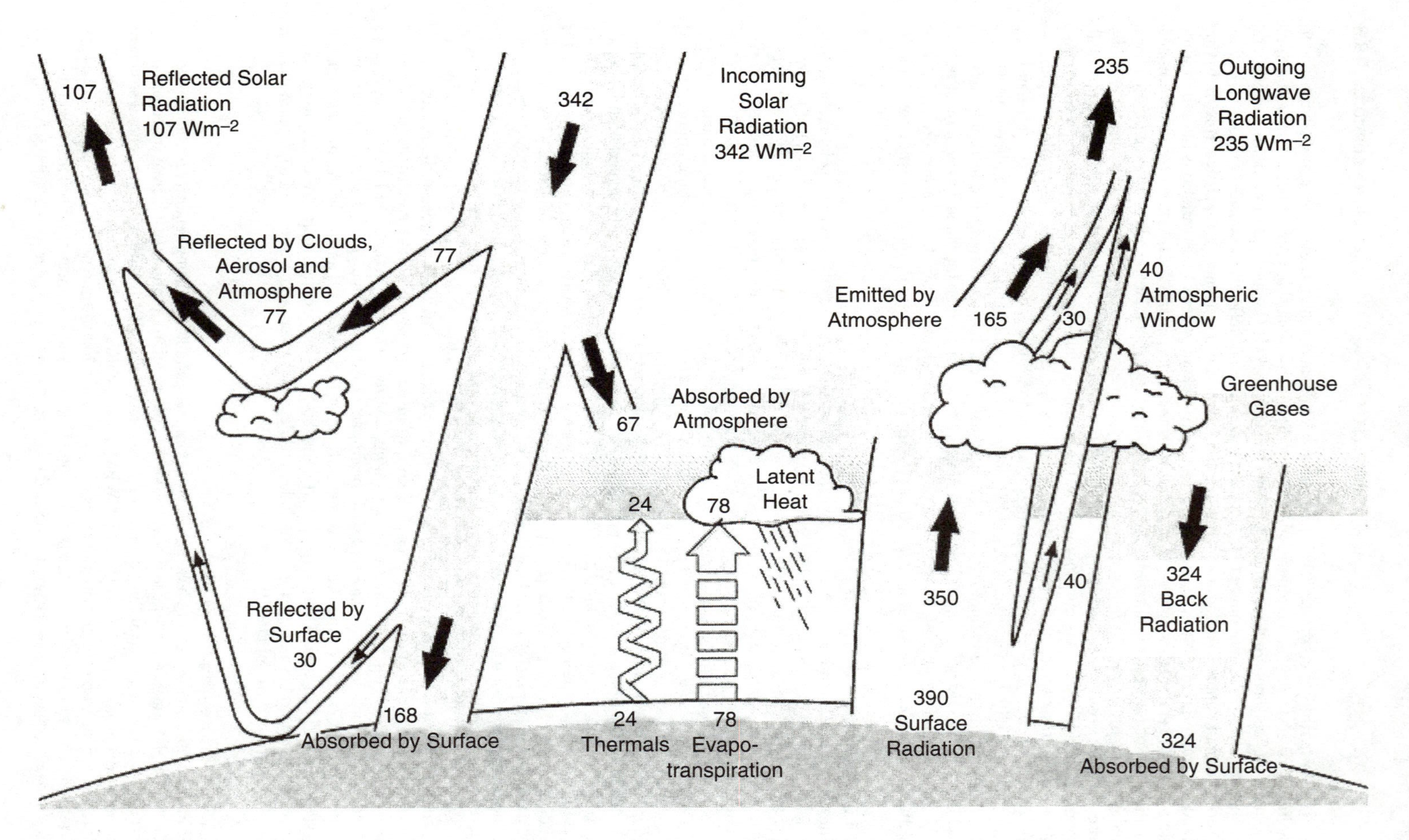

FIGURE 8.2 The earth's annual and global mean energy balance. Of the incoming solar radiation, 49 percent (168 W/m^2) is absorbed by the surface. That heat is returned to the atmosphere as sensible heat, evapotranspiration (latent heat), and thermal infared radiation. Most of this radiation is absorbed by the atmosphere, which in turn emits radiation both up and down. The radiation lost to space comes from cloud tops and atmospheric regions much colder than the surface. This causes a greenhouse effect. [*Source:* Kiehl and Trenberth (1997), as presented in IPCC (2001a).]

climate conditions to the use of GCMs. Lettenmaier et al. (1996) discussed the advantages and disadvantages of using three methods: historical and spatial analogs, output from GCMs, and prescribed climate change scenarios. They further discussed the various types of climate change scenarios based on GCM simulations that include scenarios based directly on GCM simulations, ratio and difference methods, scenarios based on GCM atmospheric circulation patterns, stochastic downscaling models, and nested models. They felt that perhaps the most useful scenarios of future climatic conditions are those developed by using a combination of methods.

Figure 8.3 shows the results of models that were used to make projections of atmospheric concentrations of greenhouse gases and aerosols, and future climate, based on emissions from the IPCC Special Report on Emission Scenarios (SRES) as discussed in the IPCC (2001a) report. A great deal of uncertainty lies in the estimate of climate change, which is not focused on in this chapter. Refer to Dickinson (1989) and the IPCC reports for more information on this topic.

8.1.4 Droughts

Droughts can be classified into different types: *meteorological droughts,* which refer to lack of precipitation; *agricultural droughts,* which refer to lack of soil moisture; and *hydrological droughts,* which refer to reduced streamflow or groundwater levels. *Drought analysis* is used to characterize the magnitude, duration, and severity of the respective type of drought in a region of interest. Figure 8.4 illustrates the progression of droughts and their impacts over a standard 30-year climatic period.

8.2 CLIMATE CHANGE EFFECTS

8.2.1 Hydrologic Effects

Evaporation and Transpiration. The two main factors influencing evaporation from an open water surface are the supply of energy to provide the latent heat of vaporization and the ability to transport the vapor away from the evaporative source (see Chow et al., 1988). Solar radiation is the main source of heat energy, and the ability to transport vapor away from the evaporative surface depends on the wind velocity over the surface and the specific humidity gradient in the air above it. As temperatures rise, the availability of energy for evaporation increases and the atmospheric demand for water in the hydrologic cycle increases.

Evaporation from the land surface includes evaporation directly from the soil and vegetation surface, and transpiration through plant leaves, in which water is extracted by the plant's roots, transported upward through its stem, and diffused into the atmosphere through tiny openings in the leaves called *stomata* (see Chow et al., 1988). The processes of evaporation from the land surface and transpiration from vegetation collectively are termed *evapotranspiration,* which is influenced by the two factors described previously for open water evaporation, and a third factor, the supply of moisture at the evaporative surface. Potential evapotranspiration is that which occurs from a well-vegetated surface when moisture supply is not limiting. The actual evapotranspiration drops below its potential level as the soil dries out.

Soil Moisture. A major part of the hydrologic cycle is subsurface water flow, which can be discussed as three major processes: infiltration of surface water into the soil to become soil moisture, subsurface (unsaturated) flow through the soil, and groundwater (saturated) flow through soil or rock strata. Climate changes that alter the precipitation patterns and the evapotranspiration regime will directly affect soil moisture storage, runoff processes, and groundwater recharge dynamics (see Rind et al., 1990; Vinnikov et al., 1996). Decreases in precipitation may significantly decrease soil moisture. In regions where precipitation increases along with increases in evaporation as a result of higher temperatures, or the timing of the precipitation or runoff changes, the soil moisture on the average or over certain periods may decrease. In regions where precipitation increases significantly, soil moisture will probably increase and in some locations may significantly increase.

Sensitivity studies by Gregory et al. (1997) reported reduced soil-moisture conditions in midlatitude summers in the Northern Hemisphere as temperature and evaporation rise and winter snow cover

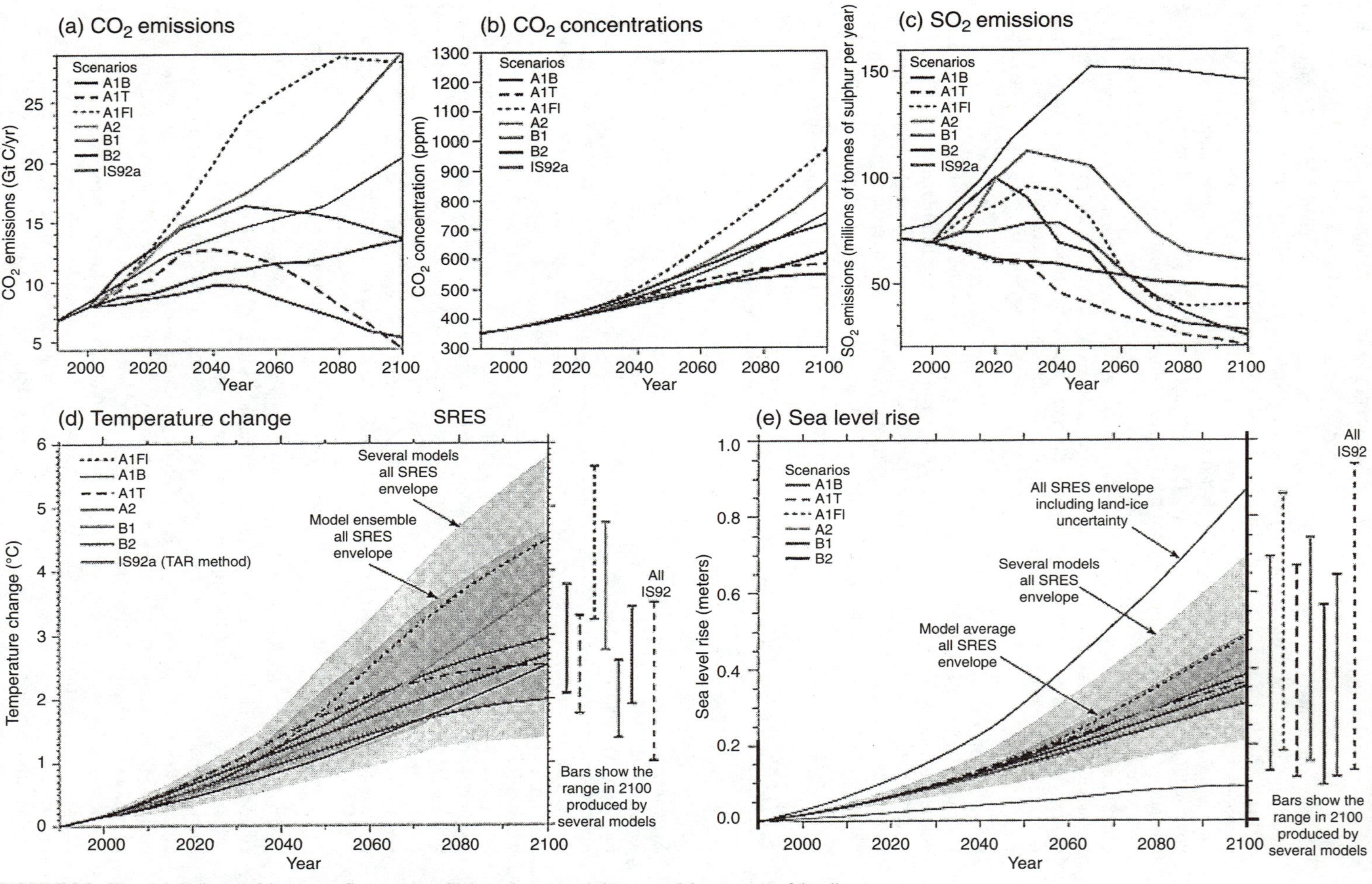

FIGURE 8.3 The global climate of the twenty-first century will depend on natural changes and the response of the climate system to human activities. (*IPCC, 2001a.*)

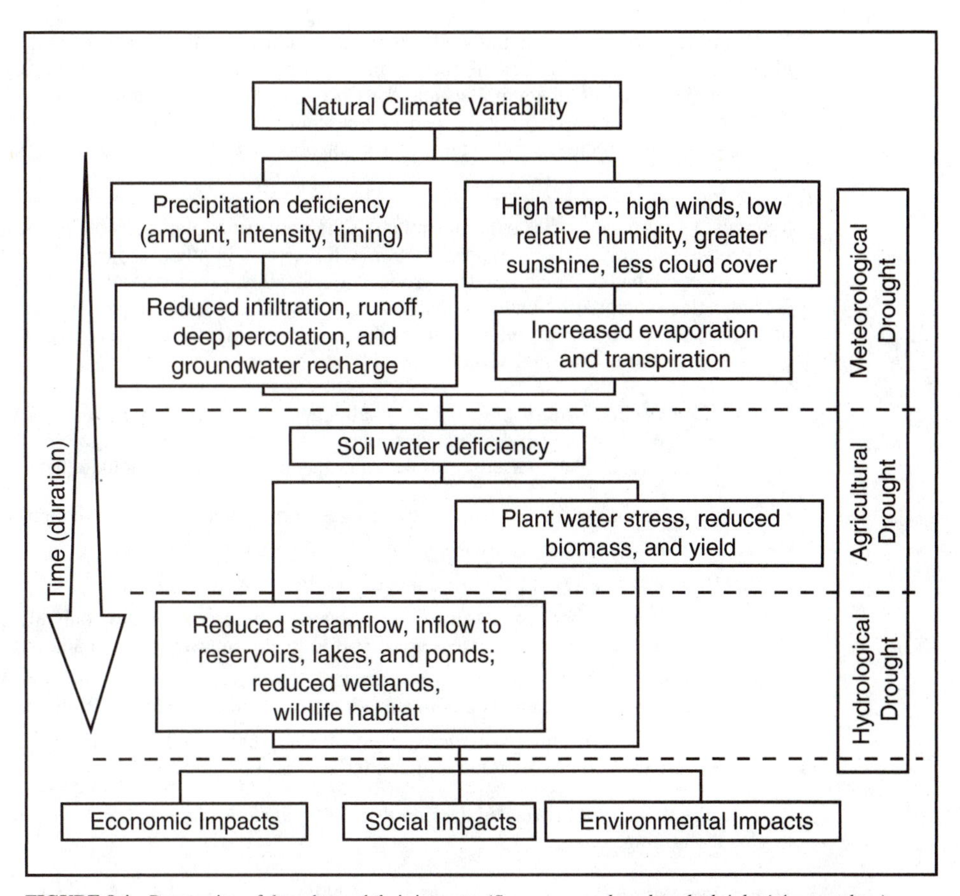

FIGURE 8.4 Progression of droughts and their impacts. (*Source:* www.drought.unl.edu/whatis/concept.htm.)

and spring runoff decline. Wetherald and Manabe (1999) investigated the temporal spatial variation of soil moisture in a coupled ocean–atmosphere model, with results that showed both summer soil moisture dryness and winter wetness in middle and high latitudes of North America and southern Europe. The study results showed a large percentage reduction of soil moisture in summer and a soil moisture decrease for nearly the entire year in response to warming.

Snowfall and Snowmelt. Temperature increases are expected to be greater in the higher-latitude regions because of the dynamics of the atmosphere and feedbacks among ice, albedo, and radiation. Research has established that higher temperatures will lead to dramatic changes in snowfall and snowmelt dynamics in watersheds with substantial snow. Higher temperatures will have several major effects, including (Gleick, 2000):

- Increase in the ratio of rain to snow
- Delay in the onset of the snow season
- Accelerated rate of spring snowmelt
- Shortening of the overall snowfall season
- More rapid and earlier seasonal runoff
- Significant changes in the distribution of permafrost and mass balance of glaciers

One of the most persistent and well-established findings on the impacts of climate change is that higher temperatures will affect the timing and magnitude of runoff in watersheds with substantial snow dynamics. Both climate models and theoretical studies of snow dynamics have long projected that higher temperatures will lead to a decrease in the extent of snow cover in the Northern Hemisphere. More recently, field studies have corroborated these findings.

Storm Frequency and Intensity. In general, global climate change will result in a wetter world. Many model estimates indicate that for much of the northern midlatitude continents, summer rainfall will decrease slightly and winter precipitation will increase. In other studies precipitation changes in midlatitudes are highly variable and uncertain. Unfortunately, general circulation models poorly reproduce detailed precipitation patterns. As pointed out by Gleick (2000), potential changes in rainfall intensity and variability are difficult to evaluate because intense convective storms tend to occur over regions smaller than global models are able to resolve.

Runoff: Floods and Droughts. A major challenge for the hydrologic community in the study of climate change is to establish the linkage between local-scale and global-scale processes.

Wigley and Jones (1985) used a simple water balance model to conclude that

- Changes in runoff are more sensitive to changes in precipitation than to changes in evaporation.
- Relative change in runoff is always greater than that in precipitation.
- Runoff is most sensitive to climate changes in arid and semiarid regions.
- Relative change in runoff exceeds the relative change in evapotranspiration only in regions where the *runoff ratio w* (ratio of long-term average runoff to long-term average precipitation) is less than 0.5.
- Overall it is expected that there will be very large increases in average runoff in response to predicted warming, unless there is a compensating large increase in land evapotranspiration.
- Higher surface temperatures will increase global precipitation in the range of 3 to 11 percent, probably leading to even greater relative increases in runoff.

Loaiciga et al. (1996) reviewed the findings and limitations of predictions of hydrologic responses to global warming.

Streamflow Variability. Karl and Riebsame (1989) used historical data from the United States with a water balance model to study the sensitivity of streamflow to changes in temperature and precipitation. Their study concluded that a temperature change of 1 to 2°C typically had little effect on streamflow. On the other hand, a relative change in precipitation magnified to a one- to six-fold change in relative streamflows.

Schaake (1990) and Nemec and Schaake (1982) showed that on an average basis (e.g., using mean annual precipitation, evaporation, and runoff) that the *elasticity of runoff* with respect to precipitation is greater than one, and is larger relative to the *long-term runoff ratio,* which is less in arid than in humid climates.

Because streamflow is climatically determined and consequently highly variable, it is instructive to look at historical data on streamflow variability in order to develop a context for detecting and predicting the effects of climate change on the hydrologic cycle. Dingman (2002) illustrates that there is a much higher relative variability of streamflow in arid regions (upper and lower plains and the southwestern United States) than in humid regions (northeastern and northwestern regions). Also evident is that the time series of river discharge shows considerable synchronism over the United States.

Long records of flow in large rivers such as the Nile show evidence of the climate-related persistence of streamflow variability. The Nile River has the longest streamflow record in the world, with information available from 622 to 1520 and from 1700 to the present. Riehl and Meitin (1979) discussed three contrasting patterns in the record: (1) the period from 622 to about 950 had periods of high flow alternated with periods of low flow, with these cycles lasting from 50 to 90 years with moderate amplitude; (2) the period from 950 to 1225 had no major trends; and (3) the period from 1700 to present had alternating periods of high and low flow, having cycles of 100 to 180 years with much higher amplitudes than during the period from 622 to 950. These periods of variability appear to be linked to global climatic fluctuations. Eltahir (1996) found that the Nile River flows were influenced by the El Niño—Southern Oscillation (ENSO) cycle. Richey et al. (1989) studied the Amazon River record (1903–1985) and found no indication of climate or land-use change over that period; however, they did find a 2- to 3-year period of declining flow following the warm phase of the ENSO cycle. They found that periods of high flow were coincident with the ENSO cold phase.

Groundwater. The effects of climate change on groundwater sustainability include those mentioned by Alley et al. (1999):

- Changes in groundwater recharge resulting from changes in average precipitation and temperature or in seasonal distribution of precipitation
- More severe and longer-lasting droughts
- Changes in evapotranspiration resulting from changes in vegetation
- Possible increased demands of groundwater as a backup source of water supply

Surficial aquifers are likely to be part of the groundwater system, which is most sensitive to climate. These aquifers supply much of the flow to streams, lakes, wetlands, and springs. Because groundwater systems tend to respond more slowly to short-term variability in climate conditions than do surface water systems, the assessment of groundwater resources and related model simulations are based on average conditions, such as annual recharge and/or average annual discharge to streams. The use of average conditions may underestimate the importance of droughts (Alley et al., 1999).

The impacts of climate change on (1) specific groundwater basins, (2) the general groundwater recharge characteristics, and (3) groundwater quality have received little attention in the literature. Vaccaro (1992) addressed the climate sensitivity of groundwater recharge, finding that a warmer climate (doubling CO_2) resulted in a relatively small sensitivity to recharge and depended on land use. Sandstrom (1995) studied a semiarid basin in Africa and concluded that a 15 percent reduction in rainfall could lead to a 45 percent reduction in groundwater recharge. Sharma (1989) and Green et al. (1997) reported similar sensitivities of the effects of climate change on groundwater in Australia. Panagoulia and Dimou (1996) studied the effect of climate change on groundwater–streamflow interactions in a mountainous basin in central Greece. They realized large impacts in the spring and summer months as a result of temperature-induced changes in snowfall and snowmelt patterns. Oberdorfer (1996) looked at the impacts of climate change on groundwater discharge to the ocean using a simple water balance model to study the effect of changes in recharge rates and sea level on groundwater resources and flows in a California coastal watershed. The impacts of sea-level rise on groundwater will include increased intrusion of salt water into coastal aquifers.

Water Resource System Effects. Studies that have looked at the impact of climate change on water resource management include those by Lettenmaier et al. (1999), Kirshen and Fennessey (1993), Lettenmaier and Sheer (1991), Nash and Gleick (1991), and Nemec and Schaake (1982).

A broad assessment was performed by Lettenmaier et al. (1999) of the sensitivity to climate change of six major water systems in the United States: Columbia River basin, Missouri River basin, Savannah River basin, Apalachicola-Chattahooche-Flint (ACF) River basin, Boston water supply system, and Tacoma water supply system. The investigators evaluated *system reliability* (percentage of time that the system operates without failure), the *system resiliency* (ability of a system to recover from failure), and *system vulnerability* (average severity of failure). Some thoughts or conclusions from this study were

- The most important factors determining climate sensitivity of system performance were changes in runoff, even when direct effects of climate change on water demands were affected.
- Sensitivities depended on the purposes for which water was needed and the priority given to those uses.

- Higher temperatures increased system use in many basins, but these increases tended to be modest. Effects of higher temperatures on system reliability were similar.
- Influence of long-term demand on system performance had a greater impact than did climate change when long-term withdrawals were projected to substantially grow.

Regional assessments can focus on three major questions (IPCC, 1996b; Miles et al., 2000):

1. How sensitive is the region to climate variability?
2. How adaptable is the region to climate variability and change?
3. How vulnerable is the region to climate variability and change?

Sensitivity is the degree to which a system will respond to a change in climatic conditions (IPCC, 1996b). *Adaptability* refers to the degree to which adjustments in systems' practices, processes, or structures to projected or actual changes of climate are possible (IPCC, 1996b). *Vulnerability* defines the extent to which climate change may damage or harm a system (IPCC, 1996b). Vulnerability depends not only on a system's sensitivity but also on its ability to adapt to new climatic conditions. Major (1998) discussed the role of risk management methods for climate change and water resources.

8.2.2 River Basin/Regional Runoff Effects

Numerous studies of the effect of climate change on various river basins have been reported in the literature. These include those by Ayers et al. (1993), Cohon (1991), Georgakakos et al. (1998), Hamlet and Lettenmaier (1999), Hay et al. (2000), Hotchkiss et al. (2000), Hurd et al. (1999), Kaczmarek et al. (1996a, 1996b), Kirshen and Fennessey (1993), Lettenmaier and Gan (1990), Leung and Wigmosta (1999), McCabe and Ayers (1989), Miles et al. (2000), Nash and Gleick (1991), Stone et al. (2001), and many others.

Stone et al. (2001) studied the impacts of climate change on the Missouri River basin water yield (Fig. 8.5). They considered the effect of doubling the atmospheric carbon dioxide on water yield using a Regional Climate Model (RegCM) and the Soil and Water Assessment (SWAT) model. Results of this analysis include the following (Stone et al., 2001):

- Precipitation and temperature both increase more in the northern part of the basin, thus resulting in the greatest changes in the predicted water yields.
- Water yields (modeled) for the spring, summer, and fall seasons showed more distinct spatial trends than water yields for the winter months. SWAT simulation results exhibited dramatic increases (70 percent and much more in some subbasins) in water yields across the northern and northwestern portions of the Missouri River basin in the spring, summer, and fall.
- Southeastern portions of the basin displayed an overall decrease (most decreased by less than 20 percent, but some decreased by more than 80 percent) in water yields during these three seasons. During the spring and summer the decreased water yields in the southeastern portions are sufficient to lower the total water yield of the entire basin by 10 to 20 percent.
- Water balance summaries for the principal tributaries indicate the large increase of runoff in the northern subbasins and decreased runoff in the southern subbasins.
- Six main stem reservoirs are operated in the upper river basin and control 55 percent of the area upstream of the reservoirs with the storage capacity of three times the mean annual runoff from the area. Total main stem reservoir storage changed dramatically from the baseline conditions, doubling the average annual releases, enough additional water under doubling CO_2 conditions to provide 24 cm of additional water annually for all irrigated areas in the Missouri River basin.

Hamlet and Lettenmaier (1999) and Miles et al. (2000) studied the impact of climate change on the hydrology and water resources of the Columbia River basin. The Columbia River system and the

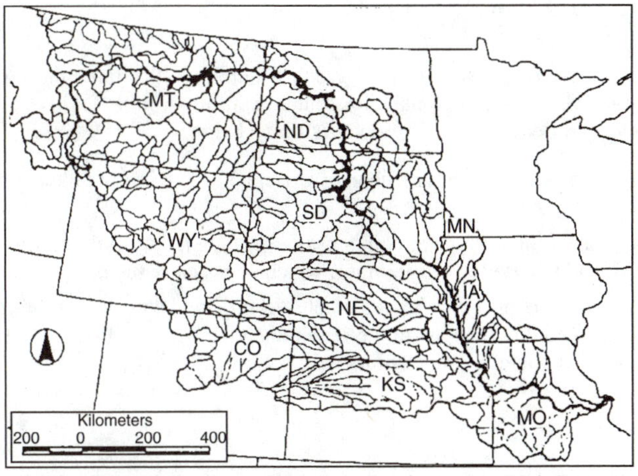

(a) The Missouri River Basin Separated Into the 310 USGS Eight-Digit Subbasins.

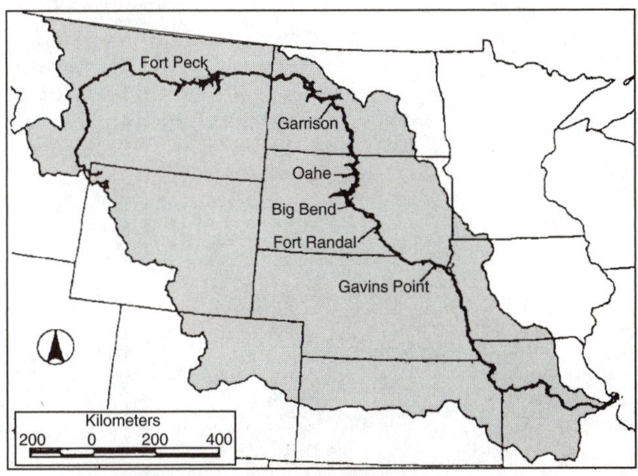

(b) The Six Missouri River Main Stem Reservoirs.

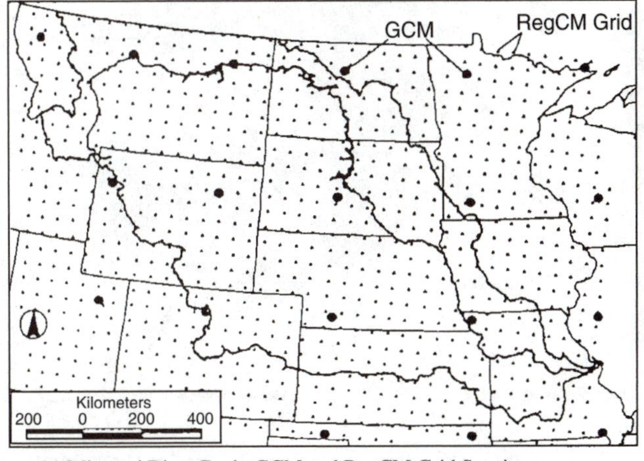

(c) Missouri River Basin GCM and RegCM Grid Spacing.

FIGURE 8.5 Missouri River basin. (*Stone et al., 2001.*)

major dam locations are shown in Fig. 8.6. Figure 8.7 illustrates the dominant impact pathway through which changes in regional climate are manifested in the basin. Some of the major findings from Hamlet and Lettenmaier (1999) are outlined below:

- Hydrologic response to climate change will tend to shift flow volumes from the summer to the winter.
- Management of the system to compensate for this effect would require that water be stored in the winter and released in the summer, which is basically the opposite to what is done now.
- Energy production, however, has its greatest economic value during the winter months. The result would be potential reductions in the required spring/summer flood storage and a shift in primary hydropower production to the summer.

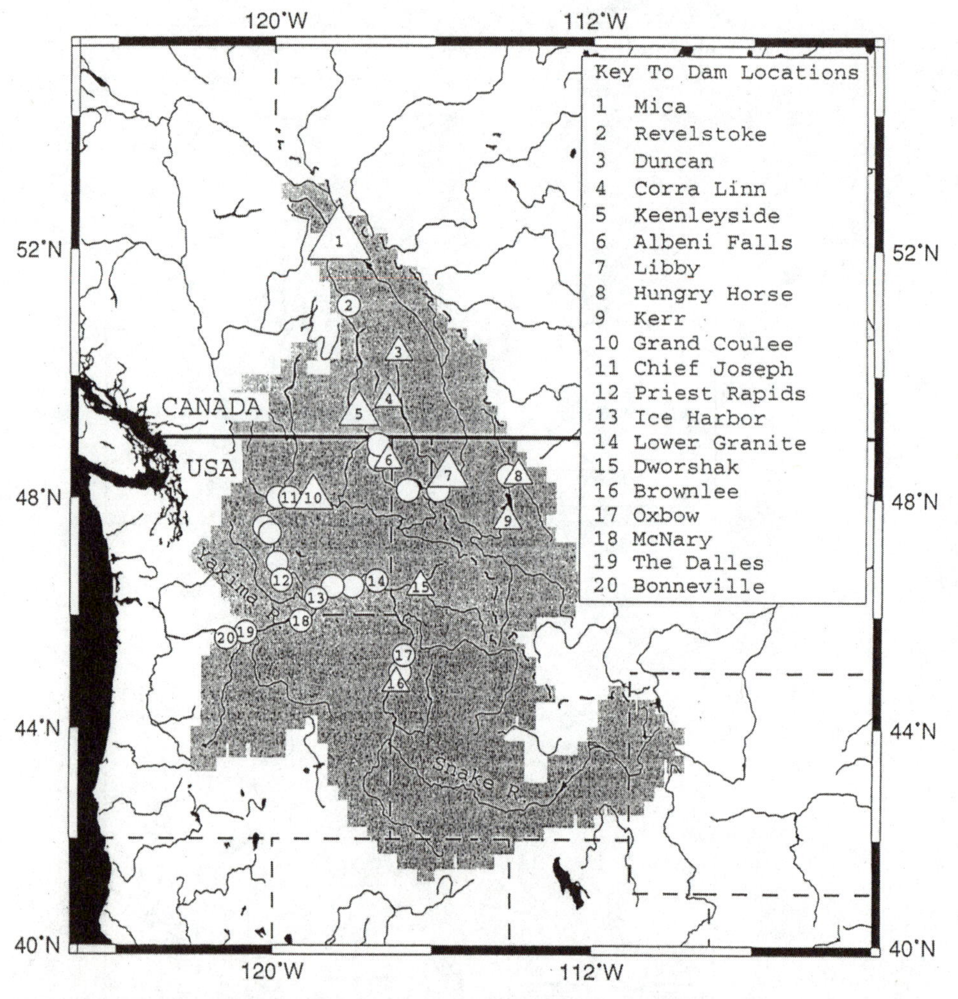

FIGURE 8.6 The Columbia River basin and major dams. The gray shaded area defines the drainage basin upstream of the Dalles Dam in Oregon. Triangles represent major storage reservoirs (size of icon represents relative storage capacity), and circles are major run-of-river projects. (*Hamlet and Lettenmaier, 1999.*)

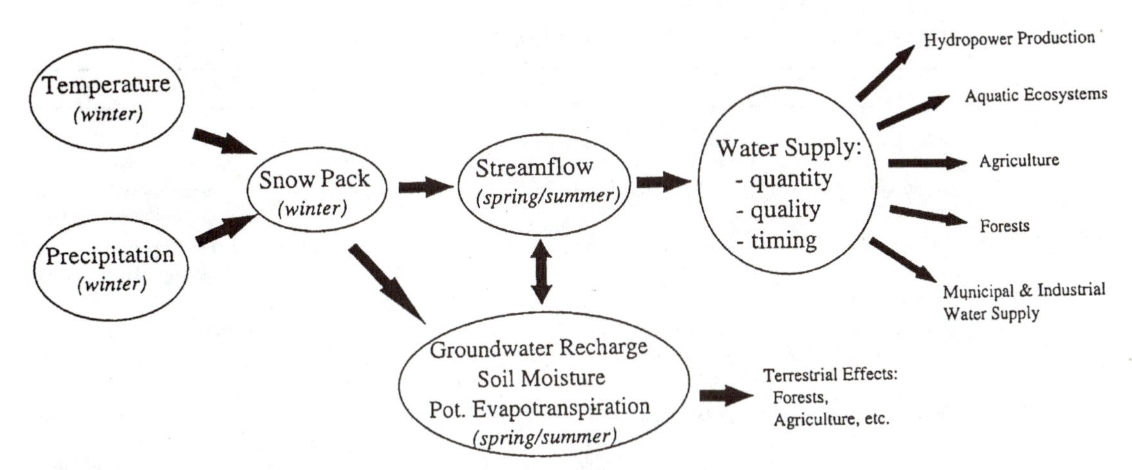

FIGURE 8.7 The dominant impact of the pathway through which changes in regional climate are manifested in the Pacific Northwest. (*Miles et al., 2000.*)

Summary of Regional Runoff Effects Found in the United States

Arid and Semiarid Western Regions. Relatively modest changes in precipitation can have proportionately larger impacts on runoff, and higher temperatures result in higher evaporation rates, reduced streamflows, and increased frequency of droughts.

Cold and Cool Temperate Zones. Where a large portion of annual runoff comes from snowmelt, the major effect of warming is the change in timing of streamflow, both timing and intensity of the peaks. A declining proportion of the total precipitation occurs as snow with rising temperatures, and the remaining snow melts sooner and faster in the spring, resulting in more winter runoff. In some basins there may be increases in spring peak runoff, and in other basins runoff volumes may significantly shift to winter months.

Warmer Climates. In the southern region of the United States, where seasonal cycles of rainfall and evaporation are predominant, runoff is affected much more significantly by total precipitation with seasonal cycles of rainfall and precipitation.

8.3 WATER MANAGEMENT OPTIONS

Table 8.1 summarizes some supply-side and demand-side adaptive options for various water-use sectors.

8.4 UNCERTAINTIES

As pointed out by many investigators, the limitations of state-of-the-art climate models are the primary sources of uncertainty in the experiments that study the hydrologic and water resources impact of climate change. Future improvement to the climate models, hopefully resulting in more accurate regional predictions, should greatly improve the types of experiments to more accurately define the hydrologic and hydraulic impacts of climate change. Future precipitation and temperature are the primary drivers for determining future hydrologic response. Because of the uncertainties of the predictions of the future precipitation and temperatures, the hydrologic responses of various river basins are uncertain, resulting in uncertainties of our future hydraulic resources.

TABLE 8.1 Supply-Side and Demand-Side Adaptive Options for Municipal Water Supply

Supply-Side		Demand-Side	
Option	Comments	Option	Comments
Increase reservoir capacity	Expensive; potential environmental impact	Incentives to use less (e.g., through pricing)	Possibly limited opportunity; needs institutional framework
Extract more from rivers or groundwater	Potential environmental impact	Legally enforceable water standards (e.g., for appliances)	Potential political impact; usually cost-inefficient
Alter system operating rules	Possibly limited opportunity	Increase use of gray water turbines; encourage energy efficiency	Potentially expensive
Interbasin transfer	Expensive; potential environmental impact; may not be feasible	Reduce leakage	Potentially expensive to reduce to very low levels, especially in old systems
Desalination	Expensive; potential environmental impact	Development of non-water-based sanitation systems	Possibly too technically advanced for wide application

Source: IPCC (2001b).

8.5 REFERENCES

Alley, W. M., T. E. Reilly, and O. L. Franke, *Sustainability of Ground-Water Resources,* U.S. Geological Survey Circular 1186, U.S. Geological Survey, Denver Colorado, 1999.

Ayers, M. A., D. M. Wolock, G. J. McCabe, L. E. Hay, and G. D. Tasker, *Sensitivity of Water Resources in the Delaware River Basin to Climate Variability and Change,* U.S. Geological Survey Open-File Report 02-52, 1993.

Burkett, V., and J. Kusler, "Climate Change: Potential Impacts and Interactions in Wetlands of the United States," *Journal of the American Water Resources Association,* 36:313–320, 2000.

Chalecki, L. H., and P. H. Gleick, "A Comprehensive Bibliography of the Impacts of Climate Change and Variability on Water Resources of the United States," *Journal of the American Water Resources Association,* 35:1657–1665, 1999.

Chow, V. T., D. R. Maidment, and L. W. Mays, *Applied Hydrology,* McGraw-Hill, New York, 1988.

Cohon, S. J., "Influences of Past and Future Climates on the Great Lakes Region of North America," *Water International,* 12:163–169, 1987.

Cohon, S. J., "Possible Impacts of Climatic Warming Scenarios on Water Resources in the Saskatchewan River Sub-Basin, Canada," *Climatic Change,* 19:291–317, 1991.

Dickinson, R. E., "Uncertainties of Estimates of Climate Change: A Review," *Climate Change,* 15:5–14, 1989.

Dingman, S. L., *Physical Hydrology,* 2d ed., Prentice-Hall, Upper Saddle River, N.J., 2002.

Eltahir, E. A. B., "El Niño and the Natural Variability of the Flow of the Nile River," *Water Resources Research,* 32:131–137, 1996.

Falkenmark, M., and G. Lindh, "Water and Economic Development," in P. H. Gleick (ed.), *Water in Crisis: A Guide to the World's Fresh Water Resources,* Oxford University Press, Oxford, U.K., 1993.

Firth, P., and S. G. Fisher (eds.), *Climate Change and Freshwater Ecosystems,* Springer-Verlag, New York, 1992.

Fisher, S. G., and N. B. Grimm, "Ecological Effects of Global Climate Change on Freshwater Ecosystems with Emphasis on Streams and Rivers," in L. W. Mays (ed.), *Water Resources Handbook,* McGraw-Hill, New York, 1996.

Georgakakos, A., H. Yao, M. Mullusky, and K. Georgakakos, "Impacts of Climate Variability on the Operational Forecast and Management of the Upper Des Moines River Basin," *Water Resources Research,* 34:799–821, 1998.

Gleick, P. H. (ed.), *Water in Crisis: A Guide to the World's Fresh Water Resources,* Oxford University Press, Oxford, U.K., 1993.

Gleick, P. H. (lead author), *Water: The Potential Consequences of Climate Variability and Change for the Water Resources of the United States,* a report of the National Water Assessment Group, for the U.S. Global Change Research Program, Pacific Institute for Studies in Development, Environment, and Security, Oakland, Calif., September 2000.

Green, T. R., B. C. Bates, P. M. Fleming, S. P. Charles, and M. Taniguchi, "Simulated Impacts of Climate Change on Groundwater Recharge in the Subtropics of Queensland, Australia," *Subsurface Hydrological Responses to Land Cover and Land Use Changes,* Kluwer Academic Publishers, Norwell, Mass., pp. 187–204, 1997.

Gregory, J. M., J. F. B. Mitchell, and A. J. Brady, "Summer Drought in Northern Midlatitudes in a Time-Dependent CO_2 Climate Experiment," *Journal of Climate,* 10:662–686, 1997.

Hamlet, A. F., and D. P. Lettenmaier, "Effects of Climate Change on Hydrology and Water Resources Objectives in the Columbia River Basin," *Journal of the American Water Resources Association,* 35:1597–1624, 1999.

Harris, R. N., and D. S. Chapman, "Borehole Temperatures and a Baseline for 20th-Century Global Warming Estimates," *Science,* 275:1618–1621, 1997.

Hay, L. E., R. L. Wilby, and G. H. Leavesley, "A Comparison of Delta Change and Downscaled GCM Scenarios for Three Mountainous Basins in the United States," *Journal of the American Water Resources Association,* 36:387–398, 2000.

Hostetler, S. W., and E. E. Small, "Response of North American Freshwater Lakes to Simulated Future Climates," *Journal of the American Water Resources Association,* 35:1625–1638, 1999.

Hotchkiss, R. H., S. F. Jorgensen, M. C. Stone, and T. A. Fontaines, "Regulated River Modeling for Climate Impacts Assessments: The Missouri River," *Journal of the American Water Resources Association,* 36:375–386, 2000.

Houghton, J. T., G. J. Jenkins, and J. J. Ephraums (eds.), *Climate Change: The IPCC Scientific Assessment,* Cambridge University Press, Cambridge, U.K., 1990.

Hurd, B., N. Leary, R. Jones, and J. Smith, "Relative Regional Vulnerability of Water Resources to Climate Change," *Journal of the American Water Resources Association,* 35:1399–1410, 1999.

Intergovernmental Panel on Climate Change (IPCC), *Climate Change: The IPCC Scientific Assessment,* J. T. Houghton, G. J. Jenkins, and J. J. Ephrauns (eds.), Cambridge University Press, Cambridge, U.K., 1990.

Intergovernmental Panel on *Climate Change (IPCC), Climate Change 1995: The Science of Climate Change: Contribution of Working Group I to the Second Assessment Report of the Intergovernmental Panel on Climate Change,* Cambridge University Press, New York, 1996a.

Intergovernmental Panel on Climate Change (IPCC), *Climate Change 1995: Impacts, Adaptations, and Mitigation of Climate Change: Scientific-Technical Analysis: Contribution of Working Group II to the Second Assessment Report of the Intergovernmental Panel on Climate Change,* Cambridge University Press, New York, 1996b.

Intergovernmental Panel on Climate Change (IPCC), "Hydrology and Freshwater Ecology," in *Climate Change 1995: Impacts, Adaptations, and Mitigation of Climate Change: Contribution of Working Group II to the Second Assessment Report of the Intergovernmental Panel on Climate Change,* Cambridge University Press, New York, 1996c.

Intergovernmental Panel on Climate Change (IPCC), *Regional Impacts of Climate Change: An Assessment of Vulnerability,* Cambridge University Press, New York, 1998.

Intergovernmental Panel on Climate Change (IPCC), *Climate Change 2001: The Scientific Basis,* www.ipcc.ch, 2001a.

Intergovernmental Panel on Climate Change (IPCC), *Climate Change 2001: Impacts, Adaptation, and Vulnerability,* www.ipcc.ch, 2001b.

Intergovernmental Panel on Climate Change (IPCC), *Climate Change 2001: Mitigation,* www.ipcc.ch, 2001c.

Kaczmarek, Z., N. W. Arnell, and E. Z. Stakhiv, "Water Resources Management," Chap. 10 in Intergovernmental Panel on Climate Change (IPCC), *Second Assessment Report,* Cambridge University Press, 1996a.

Kaczmarek, Z., J. J. Napiorkowski, and K. M. Strzepek, "Climate Change Impact on the Water Supply System in the Warta River Catchment," *Water Resources Development,* 12(2):165–180, 1996b.

Karl, T. R., and R. G. Quayle, "Climate Change in Fact and Theory: Are We Collecting the Facts?" *Climate Change,* 13:5–17, 1988.

Karl, T. R., and W. E. Riebsame, "The Impact of Decadal Fluctuations in Mean Precipitation and Temperature on Runoff: A Sensitivity Study over the United States," *Climate Change,* 15:423–447, 1989.

Karl, T. R., R. W. Knight, and N. Plummer, "Trends in High Frequency Climate Variability in the Twentieth Century," *Nature,* 377:217–220, 1995.

Kaufmann, R. K., and D. I. Stern, "Evidence for Human Influence on Climate from Hemispheric Temperature Relations," *Nature,* 388:39–44, 1997.

Kiehl, J.T., and K.E. Trenberth, "Earth's Annual Global Mean Energy Budget," *Bulletin American Meteorological Society,* 78:197–208, 1997.

Kirshen, P. H., and N. M. Fennessey, *Potential Impacts of Climate Change upon the Water Supply of the Boston Metropolitan Area,* U.S. Environmental Protection Agency, 1993.

Lettenmaier, D. P., and T. Y. Gan, "Hydrologic Sensitivities of the Sacramento–San Joaquin River Basin, California, to Global Warming," *Water Resources Research,* 26:69–86, 1990.

Lettenmaier, D. P., and D. P. Sheer, "Climatic Sensitivity of California Water Resources," *Journal of Water Resources Planning and Management,* 117(1):108–125, 1991.

Lettenmaier, D. P., G. McCabe, and E. Z. Stakhiv, "Global Climate Change: Effect on Hydrologic Cycle," in L. W. Mays (ed.), *Water Resources Handbook,* McGraw-Hill, New York, 1996.

Lettenmaier, D. P., A. W. Wood, R. N. Palmer, E. F. Wood, and E. Z. Stakhiv, "Water Resources Implications of Global Warming: A U.S. Regional Perspective," *Climate Change,* 43(3):537–579, 1999.

Leung, L. R., and M. S. Wigmosta, "Potential Climate Change Impacts on Mountain Watersheds in the Pacific Northwest," *Journal of the American Water Resources Association,* 36(6):1463–1471, 1999.

Levine, J. S., "Global Climate Change," in P. Firth and S. Fisher (eds.), *Climate Change and Freshwater Ecosystems,* Springer-Verlag, New York, 1992.

Loaiciga, H. A. et al., "Global Warming and the Hydrologic Cycle," *Journal of Hydrology,* 174:83–127, 1996.

Major, D. C., "Climate Change and Water Resources: The Role of Risk Management Methods," *Water Resources Update,* 112:47–50, 1998.

McCabe, G. J., and M. A. Ayers, "Hydrologic Effects of Climate Change in the Delaware River Basin," *Water Resources Bulletin,* 25(6):1231–1242, 1989.

Miles, E. L., A. K. Snover, A. F. Hamlet, B. Callahan, and D. Fluharty, "Pacific Northwest Regional Assessment: Impacts of the Climate Variability and Climate Change on the Water Resources of the Columbia River Basin," *Journal of the American Water Resources Association,* 36(2):399–420, April 2000.

Mitchell, J. F. B., "The `Greenhouse' Effect and Climate Change," *Reviews of Geophysics,* 27:115–139, 1989.

Mitchell, J. F. B., T. C. Johns, J. M. Gregory, and S. F. B. Tett, "Climate Response to Increasing Levels of Greenhouse Gases and Sulphate Aerosols," *Nature,* 376:501–504, 1995.

Nash, L. L., and P. H. Gleick, "Sensitivity of Streamflow in the Colorado Basin to Climatic Changes," *Journal of Hydrology,* 125:221–241, 1991.

Nemec, J., and J. C. Schaake, "Sensitivity of Water Resource Systems to Climate Variation," *Journal of Hydrological Sciences,* 27:327–343, 1982.

Oberdorfer, J. A., "Numerical Modeling of Coastal Discharge: Predicting the Effects of Climate Change," in R. W. Buddemeier (ed.), *Groundwater Discharge in the Coastal Zone: Proceedings of an International Symposium,* Moscow, Russia, pp. 85–91, July 1996.

Panagoulia, D., and G. Dimou, Sensitivities of Groundwater-Streamflow Interactions to Global Climate Change, *Hydrological Sciences Journal,* 41:781–796, 1996.

Richey, J. E., C. Nobre, and C. Deser, "Amazon River Discharge and Climate Variability: 1903 to 1985," *Science,* 246:101–103, 1989.

Riehl, H., and J. Meitin, "Discharge of the Nile River: A Barometer of Short-Period Climatic Fluction," *Science,* 206:1178–1179, 1979.

Rind, D. R., R. Goldberg, J. Hansen, C. Rosensweig, and R. Ruedy, "Potential Evapotranspiration and the Likelihood of Future Drought," *Journal of Geophysical Review,* 95:9983–10,004, 1990.

Sandstrom, K., "Modeling the Effects of Rainfall Variability on Groundwater Recharge in Semi-Arid Tanzania," *Nordic Hydrology,* 26:313–320, 1995.

Santer, B. et al., "A Search for Human Influences on the Thermal Structure of the Atmosphere," *Nature,* 382:39–46, 1996.

Schaake, J. C., "From Climate to Flow," in P. E. Wagner (ed.), *Climate Change and U.S. Water Resources,* Wiley, New York, 1990.

Sharma, M. L., "Impact of Climate Change on Groundwater Recharge," *Proceedings of the Conference on Climate and Water,* Valton Painatuskeskus, Helsinki, Finland, vol. I:511–520, 1989.

Stone, M. C., R. H. Hotchkiss, C. M. Hubbard, T. A. Fontaine, L. O. Mearns, and J. G. Arnold, "Impacts of Climate Change on Missouri River Basin Water Yield," *Journal of the American Water Resources Association,* 37(5):1119–1129, 2001.

Tett, F. B., J. F. B. Mitchell, D. E. Parker, and M. R. Allen, "Human Influence on the Atmospheric Vertical Structure: Detection and Observations," *Science,* 274:1170–1173, 1996.

Vaccaro, J. J., "Sensitivity of Groundwater Recharge Estimates to Climate Variability and Change, Columbia Plateau, Washington," *Journal of Geophysical Research,* 97(D3):2821–2833, 1992.

Vinnikov, K. Y., A. Robock, N. A. Speranskaya, and C. A. Schlosser, "Scales of Temporal and Spatial Variability of Midlatitude Soil Moisture," *Journal of Geophysical Research,* 101:7163–7174, 1996.

Wetherald, R. T., and S. Manabe, "Detectability of Summer Dryness Caused by Greenhouse Warming," *Climate Change,* 43(3):495–522, 1999.

Wigley, T. M. L., and P. D. Jones, "Influence of Precipitation Changes and Direct CO_2 Effects on Streamflow," *Nature,* 314:149–151, 1985.

8.6 FURTHER READING

Alexandrov, V., "A Strategy Evaluation of Irrigation Management of Maize Crop under Climate Change in Bulgaria," *Proceedings of the Second International Conference on Climate and Water,* R. Lemmela and N. Helenius (eds.), Espoo, Finland, August 17–20, vol. 3, pp. 1545–1555, 1998.

American Water Works Association (WWA), "Climate Change and Water Resources, Committee Report of the AWWA Public Advisory Forum," *Journal of the American Water Works Association,* 89(11): 107–110, 1997.

Arnell, N. W., *Global Warming, River Flows and Water Resources,* Wiley, Chichester, United Kingdom, 1996.

Arnell, N. W., "The Effect of Climate Change on Hydrological Regimes in Europe: A Continental Perspective," *Global Environmental Change,* 9:5–23, 1999.

Arnold, J. G., R. Srinivasan, R. S. Muttiah, and J. R. Williams, "Large Area Hydrologic Modeling and Assessment, Part I: Model Development," *Journal of American Water Resources Association,* 34(1):73–89, 1998.

Arnold, J. G., R. Srinivasan, R. S. Muttiah, and P. M Allen, "Continental Scale Simulation of the Hydrologic Balance," *Journal of American Water Resources Association,* 35:1037–1051, 1999.

Dvorak, V., J. Hladny, and L. Kasparek, "Climate Change Hydrology and Water Resources Impact and Adaptation for Selected River Basins in the Czech Republic," *Climate Change,* 36:93–106, 1997.

Eagleson, P. S., "The Emergence of Global-Scale Hydrology," *Water Resources Research,* 22:6S–14S, 1986.

Gleick, P. H., "Climate Change, Hydrology, and Water Resources," *Reviews of Geophysics,* 7(3):329–344, 1989.

Grabs, W., *Impact of Climate Change on Hydrological Regimes and Water Resources Management in the Rhine,* Cologne, Germany, 1997.

Hayden, B. P., "Climate Change and Extratropical Storminess in the United States: An Assessment," *Journal of the American Water Resources Association,* 35:1387–1398, 1999.

Herrington, P., *Climate Change and the Demand for Water,* Her Majesty's Stationary Office, London, United Kingdom, 1996.

Klein, R. J. T., J. Aston, E. N. Buckley, M. Capobianco, N. Mizutani, R. J. Nicholls, P. D. Nunn, and S. Ragoonaden, in Metz et al. (eds.), *Coastal Adaptation Technologies: IPCC Special Report on Methodological and Technological Issues in Technology Transfer,* Cambridge University Press, New York, 2000.

Kos, Z., "Sensitivity of Irrigation and Water Resources Systems to Climate Change," *Journal of Water Management,* 41(4–5):247–269, 1993.

Mander, U., and A. Kull, "Impacts of Climatic Fluctuations and Land Use Change on Water Budget and Nutrient Runoff: The Porijogi," *Proceedings of the Second International Conference on Climate and Water,* R. Lemmela and N. Helenius (eds.), Espoo, Finland, August 17–20, 2:884–896, 1998.

Saelthun, N. R. et al., "Climate Change Impacts on Runoff and Hydropower in the Nordic Countries," *TemaNord,* 552:170, 1998.

Vorosmarty, C. V., C. A. Federer, and A. L. Schloss, "Potential Evaporation Functions Compared on U.S. Watersheds: Possible Implications for Global-Scale Water Balance and Terrestrial Ecosystem Modeling," *Journal of Hydrology,* 207:147–169, 1998.

Vorosmarty, C. V., P. Green, J. Salisbury, and R. B. Lammers, "The Vulnerability of Global Water Resources: Major Impacts from Climate Change or Human Development?" *Science,* 289:284–288, 2000.

Vorosmarty, C., L. Hinzman, B. Peterson, D. Bromwich, L. Hamilton, J. Morison, V. Romanovsky, M. Sturm, and R. Webb, "Artic-CHAMP: A Program to Study Artic Hydrology and Its Role in Global Change," EOS, *Transactions, American Geophysical Union,* 83(22), May 28, 2002.

Walsh, J. et al., *Enhancing NASA's Contributions to Polar Science: A Review of Polar Geophysical Data Sets,* Commission on Geosciences, Environment, and Resources, National Research Council, National Academy of Science Press, Washington, D.C., 2001.

CHAPTER 9
WATER SUPPLY SAFETY AND SECURITY: AN INTRODUCTION

Larry W. Mays
Department of Civil and Environmental Engineering
Arizona State University
Tempe, Arizona

> It is a doctrine of war not to assume the enemy will not come, but rather to rely on one's readiness to meet him; not to presume that he will not attack, but rather to make one's self invincible
>
> SUN TZU

9.1 THE WATER SUPPLY SYSTEM: A BRIEF DESCRIPTION

The events of September 11, 2001, have significantly changed the approach to management of water utilities. Previously the consideration of terrorist threat to our nation's drinking water supply was minimal. Now we have an intensified approach to the consideration of terrorist threat. The objective of this chapter is to provide an introduction to the very costly process of developing water security measures for our nation's water utilities.

Figure 9.1 illustrates a typical municipal water utility showing the water distribution system as a part of this overall water utility. In some locations where excellent quality groundwater is available, water treatment may include only chlorination. Other handbooks on the subject of water supply and water distribution systems include those by Mays (1989, 2000, 2002, 2004).

Water distribution systems are composed of three major components: pumping stations, distribution storage, and distribution piping. These components may be further divided into subcomponents, which in turn can be divided into sub-subcomponents. For example, the pumping station component consists of structural, electrical, piping, and pumping unit subcomponents. The pumping unit can be divided further into sub-subcomponents: pump, driver, controls, and power transmission. The exact definition of components, subcomponents, and sub-subcomponents depends on the level of detail of the required analysis and to a somewhat greater extent the level of detail of available data. In fact, the concept of component, subcomponent, sub-subcomponent merely defines a hierarchy of building blocks used to construct the water distribution system. Figure 9.2 shows the hierarchical relationship of system, components, subcomponents, and sub-subcomponents for a water distribution system.

A water distribution system operates as a system of independent components. The hydraulics of each component is relatively straightforward; however, these components depend directly on each other and as a result affect the performance of other components. The purpose of design and analysis

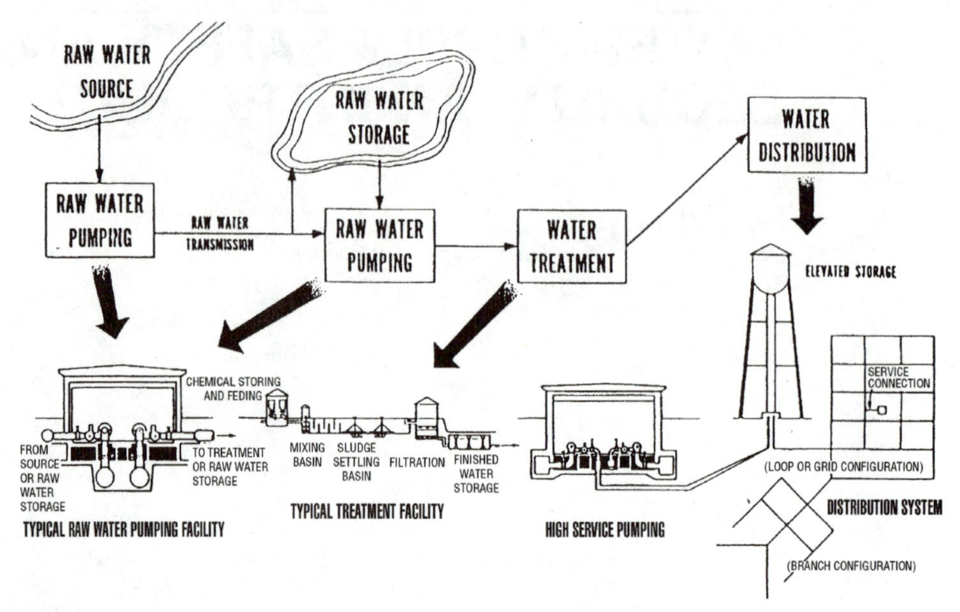

FIGURE 9.1 A typical water distribution system.

is to determine how the systems perform hydraulically under various demands and operation conditions. These analyses are used for the following situations:

- Design of a new distribution system
- Modification and expansion of an existing system
- Analysis of system malfunction such as pipe breaks, leakage, valve failure, and pump failure
- Evaluation of system reliability
- Preparation for maintenance
- System performance and operation optimization

9.2 WHY WATER SUPPLY SYSTEMS?

Distribution system of pipelines, pipes, pumps, storage tanks, and the appurtenances such as various types of valves, meters, and other components offers the greatest opportunity for terrorism because it is extensive, relatively unprotected and accessible, and often isolated. The physical destruction of a water distribution system's assets or the disruption of water supply may be more likely than contamination. A likely avenue for such an act of terrorism is a bomb carried by car or truck, similar to the recent events listed in Table 9.1. Truck or car bombs require less preparation, skill, or labor than did the complex attacks such as those of September 11, 2001.

9.2.1 The Threats

Cyber threats

- Physical disruption of SCADA (supervisory control and data acquisition) network
- Attacks on central control system to create simultaneous failures
- Electronic attacks using worms and viruses

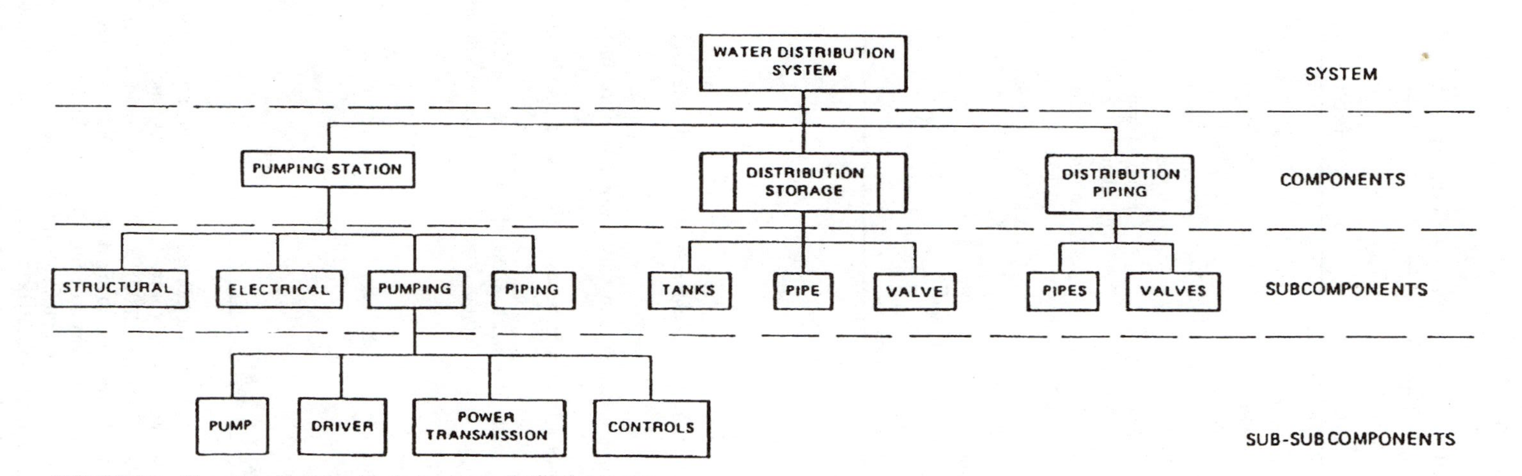

FIGURE 9.2 Hierarchy of building blocks in a water distribution system.

TABLE 9.1 Recent Terrorist Attacks against American Targets Using Car-Bomb Technologies

Date	Target/location	Delivery/material	TNT equivalent, lb	Reference
April 1983	U.S. Embassy, Beirut, Lebanon	Van	2,000	www.beirut-memorial.org
October 1983	U.S. Marine Barracks, Beirut, Lebanon	Truck, TNT with gas enhancement	12,000	www.usmc.mil
February 1993	World Trade Center, New York, USA	Van, urea nitrate and hydrogen gas	2,000	www.interpol.int
April 1995	Murrah Federal Bldg., Oklahoma City, USA	Truck, ammonium nitrate fuel oil	5,000	U.S. Senate documents
June 1996	Khobar Towers, Dhahran, Saudi Arabia	Tanker truck, plastic explosive	20,000	www.fbi.gov
August 1998	U.S. Embassy, Nairobi, Kenya	Truck, TNT, possibly Semtex	1,000	News reports, U.S. Senate documents
August 1998	U.S. Embassy, Dar es Salaam, Tanzania	Truck	1,000	U.S. Senate documents
October 2000	Destroyer *USS Cole,* Aden Harbor, Yemen	Small watercraft, possibly C-4	440	www.al-bab.com news.bbc.co.uk

Source: Peplow et al. (2003).

- Network flooding
- Jamming
- Disguising data to neutralize chlorine or add no disinfectant, allowing addition of microbes

Physical threats

- Physical destruction of system's assets or disruption of water supply is more likely than contamination.
- Loss of water pressure compromises firefighting capabilities and could lead to possible bacterial buildup in the system.
- Potential for creating a water hammer effect by opening and closing major control valves and turning pumps on and off too quickly, which could result in simultaneous main breaks.

Chemical threats
See Table 9.2.

Biological threats
See Table 9.3.

9.3 PRIOR TO SEPTEMBER 11, 2001

Prior to September 11, 2001, the literature contained numerous publications concerning the threat of terrorist attacks to our water supply infrastructure. These included reports by Burrows and Renner (1999), DeNileon (2001), Dickey (2000), Foran and Brosnan (2000), Grayman et al. (2001), Haimes et al. (1998), Hickman (1999), and many others. The topic was receiving some attention, but basically the water utility industry was not implementing mitigation measures to such threats.

President Bush's Commission on Critical Infrastructure Protection (PCCIP) was established by President Clinton in 1996. The PCCIP determined that the water infrastructure is highly vulnerable to a range of potential attacks and convened a public-private partnership called the *Water Sector Critical*

TABLE 9.2 Summary of Chemicals Effective in Drinking Water

Chemical agents [milligrams per liter (mg/l) unless otherwise noted]	Acute concentration* 0.5L	Recommended Guidelines 5 L/day	Recommended Guidelines 15 L/day
Chemical warfare agents:			
Hydrogen cyanide[§]	25	6.0	2.0
Tabun (GA, microgram/liter (μg/l))	50	70.0	22.5
Sarin (GB, μg/l)	50	13.8	4.6
Soman (GD, μg/l)	50	6.0	2.0
VX (μg/l)	50	7.5	2.5
Lewisite (arsenic fraction)	100–130	80.0	27.0
Sulfer Mustard (μg/l)		140.0	47.0
3-quinuclidinyl benzilate (BZ, μg/l)		7.0	2.3
lysergic acid diethylamide (LSD)	0.050		
Industrial chemical poisons:			
Cyanides	25	6.0	2.0
Arsenic	100–130	80.0	27.0
Fluoride	3000		
Cadmium	15		
Mercury	75–300		
Dieldrin	5000		
Sodium fluoroacetate[‡]		None Provided	
Parathion[‡]		None Provided	

*Major John Garland, *Water Vulnerability Assessments* (Armstrong Laboratory, AL-TR-1991-0049), April 1991, 8–9. The author assumes acute effects (death or debilitation) after consumption of 0.5 L.

†National Research Council, Committee on Toxicology, *Guidelines for Chemical Warfare Agents in Military Field Drinking Water*, 1995, 10. Listed doses are "safe."

‡W. Dickinson Burrows, J. A. Valcik, and Alan Seitzinger, "Natural and Terrorist Threats to Drinking Water Systems," presented at the American Defense Preparedness Association 23rd Environmental Symposium and Exhibition, 7–10 April 1997, New Orleans, LA. The authors consider the organophospate nerve agent VX, the two hallucinogens BZ and LSD, sodium cyanide, fluoroacetate, and parathion as potential threat agents. They do not provide acute concentrations or lethal doses.

§Hydrogen cyanide (blood agent), the nerve agents Tabun, Sarin, Soman, and VX, the blistering agents Lewisite and sulfur mustard, and the hallucinogen BZ are potential drinking water poisons. Garland focuses on LSD (a hallucinogen), nerve agents (VX is listed as most toxic), arsenic (Lewisite), and cyanide (hydrogen cyanide). Burrows, et al., list BZ, LSD, and VX. These agents, however, are not the only chemicals a saboteur might use in drinking water.

Source: As presented in Hackman (1999).

Infrastructure Advisory Group. According to the President's Commission on Critical Infrastructure Protection (1997), three attributes are crucial to water supply users:

- There must be adequate quantities of water on demand.
- It must be delivered at sufficient pressure.
- It must be safe to use.

The first two are influenced by physical damage, and the third attribute (water quality) is susceptible to physical damage as well as the introduction of microorganisms, toxins, chemicals, or radioactive materials. Actions (terrorist activities) that affect any one of these three attributes can be debilitating for the water supply system.

TABLE 9.3 Potential Threat of Biological Weapons Agents: Summary

Agent	Type	Weaponized	Water threat	Stable in water	Chlorine* tolerance
Anthrax	Bacteria	Yes	Yes	2 years (spores)	Spores resistant
Brucellosis	Bacteria	Yes	Probable	20–72 days	Unknown
C. perfringens	Bacteria	Probable	Probable	Common in sewage	Resistant
Tularemia	Bacteria	Yes	Yes	Up to 90 days	Inactivated, 1 ppm, 5 min
Glanders	Bacteria	Probable	Unlikely	Up to 30 days	Unknown
Meliodosis	Bacteria	Possible	Unlikely	Unknown	Unknown
Shigellosis	Bacteria	Unknown	Yes	2–3 days	Inactivated, 0.05 ppm, 10 min
Cholera	Bacteria	Unknown	Yes	"Survives well"	"Easily killed"
Salmonella	Bacteria	Unknown	Yes	8 days, fresh water	Inactivated
Plague	Bacteria	Probable	Yes	16 days	Unknown
Q fever	Rickettsia	Yes	Possible	Unknown	Unknown
Typhus	Rickettsia†	Probable	Unlikely	Unknown	Unknown
Psittacosis	Rickettsia-like	Possible	Possible	18–24 h, sea water	Unknown
Encephalomyelitis	Virus	Probable	Unlikely	Unknown	Unknown
Hemorrhagic fever	Virus	Probable	Unlikely	Unknown	Unknown
Variola	Virus	Possible	Possible	Unknown	Unknown
Hepatitis A	Virus	Unknown	Yes	Unknown	Inactivated, 0.4 ppm, 30 min
Cryptosporidiosis	Protozoan‡	Unknown	Yes	Stable for days or more	Oocysts resistant
Botulinum toxins	Biotoxin§	Yes	Yes	Stable	Inactivated, 6 ppm, 20 min
T-2 mycotoxin	Biotoxin	Probable	Yes	Stable	Resistant
Aflatoxin	Biotoxin	Yes	Yes	Probably stable	Probably tolerant
Ricin	Biotoxin	Yes	Yes	Unknown	Resistant at 10 ppm
Staph enterotoxins	Biotoxin	Probable	Yes	Probably stable	Unknown
Microcystins	Biotoxin	Possible	Yes	Probably stable	Resistant at 100 ppm
Anatoxin A	Biotoxin	Unknown	Probable	Inactivated in days	Unknown
Tetrodotoxin	Biotoxin	Possible	Yes	Unknown	Inactivated, 50 ppm
Saxitoxin	Biotoxin	Possible	Yes	Stable	Resistant at 10 ppm

*Ambient temperature, <1 part per million (ppm) free available chlorine, 30 min, or as indicated.
†Parasites that are pathogens for humans and animals.
‡Consisting of one cell or of a colony of like or similar cells.
§Toxic to humans.
Source: Burrows and Renner (1999).

9.4 RESPONSE TO SEPTEMBER 11, 2001

Within a very short time after September 11, 2001, we began to see a concerted effort at all levels of government to begin addressing issues related to the threat of terrorist activities on our nation's water supply. Reports began appearing, including those by Bailey (2001), Blomgren (2002), Copeland and Cody (2002), Haas (2002), and many others. We saw a number of acts passed, such as the Security and Bioterrorism Preparedness and Response Act and the Homeland Security Act, which addressed the nation's water supply. These acts resulted in agencies such as the U.S. Environmental Protection Agency (USEPA) developing new protocols to address their new responsibilities under these acts.

9.4.1 Public Health, Security and Bioterrorism Preparedness and Response Act ("Bioterrorism Act"), (PL 107-188), June 2002

This act requires every community water system that serves a population of greater than 3300 persons to

- Conduct a vulnerability assessment
- Certify and submit a copy of the assessment to the USEPA administer

- Prepare or revise an emergency response plan that incorporates the results of the vulnerability assessment
- Certify to the USEPA administrator, within 6 months of completing the vulnerability assessment, that the system has completed or updated their emergency response plan

Table 9.4 lists the key provisions of the security-related amendments.

9.4.2 Homeland Security Act (PL 107-296), November 25, 2002

This act

- Directs the greatest reorganization of the federal government in decades by consolidating a host of security-related agencies into a single cabinet-level department to be headed by Tom Ridge, head of the White House Office of Homeland Security.
- Creates four major directorates to be led by White House–appointed undersecretaries. These are the directorates of:
  1. Information Analysis and Infrastructure Protection (IAIP) (most directly affects USEPA and the drinking water community)
  2. Science and Technology
  3. Border and Transportation Security
  4. Emergency Preparedness and Response

The law grants IAIP access to all pertinent information, including infrastructure vulnerabilities, and directs all federal agencies to "promptly provide" IAIP with all information they have on terrorism threats and infrastructure vulnerabilities. IAIP is responsible for overseeing transferred functions of the NIPC, the CIAO, and the Energy Department's National Infrastructure Simulation and Analysis Center and Energy Assurance Office, and the General Services Administration's Federal Computer Incident Response Center. IAIP will administer the Homeland Security Advisory System, which is the government's voice for public advisories about homeland threats as well as specific warnings and counterterrorism advice to state and local governments, the private sector, and the public.

TABLE 9.4 Security-Related Amendments to Bioterroism Act*

1. Requires community water systems serving populations greater than 3300 to conduct vulnerability assessments and submit them to USEPA
2. Requires specific elements to be included in a vulnerability assessment
3. Requires each system that completes a vulnerability assessment to revise an emergency response plan and coordinate (to the extent possible) with local emergency planning committees
4. Identifies specific completion dates for both vulnerability assessments and emergency response plans
5. USEPA to develop security protocols as may be necessary to protect the copies of vulnerability assessments in its possession
6. USEPA to provide guidance to community water systems serving populations of 3300 or less on how to conduct vulnerability assessments, prepare emergency response plans, and address threats
7. USEPA to provide baseline information to community water systems regarding types of probable terrorist or other intentional threats
8. USEPA to review current and future methods to prevent, detect, and respond to the intentional introduction of chemical, biological or radiological contaminants into community water systems and their respective source waters
9. USEPA to review methods and means by which terrorists or other individuals or groups could disrupt the supply of safe drinking water
10. USEPA to authorize funds to support these activities

*In June 2002 President Bush signed PL 107-188, the Public Health, Security, and Bioterrorism Preparedness and Response Act ("Bioterrorism Act"), which includes provisions to help safeguard the nation's public drinking water systems against terrorist and other intentional acts. Key provisions of the new security-related amendments are summarized in this table.

9.4.3 USEPA's Protocol

On October 2, 2002, the U.S. Environmental Protection Agency announced its Strategic Plan for Homeland Security (www.epa.gov/epahome/headline 100202.htm). The goals of the plan are separated into four distinct mission areas: critical infrastructure protection; preparedness, response, and recovery; communication and information; and protection of USEPA personnel and infrastructure. USEPA's strategic plan lays out goals, tactics, and results for each of these areas.

USEPA has developed a compilation of water infrastructure security website links and tools located at www.epa.gov/safewater/security/index.html. Table 9.5 lists the USEPA's strategic objectives to address drinking water system and wastewater utility security needs to meet the requirements of the Bioterrorism Act for public drinking water security.

9.5 VULNERABILITY ASSESSMENT

9.5.1 Definition of Vulnerability Assessment

PL 107-188 [Public Health, Security, and Bioterrorism Preparedness and Response Act (Bioterrorism Act of 2002)] requires community water systems (CWSs) serving populations greater than 3300 to conduct vulnerability assessments and submit them to USEPA. The dates of compliance are listed in Table 9.6.

The common elements of vulnerability assessments are

- Characterization of the water system, including its mission and objectives
- Identification and prioritization of adverse consequences to avoid
- Determination of critical assets that might be subject to malevolent acts that could result in undesired consequences
- Assessment of the likelihood (qualitative probability) of such malevolent acts from adversaries
- Evaluation of existing countermeasures
- Analysis of current risk and development of a prioritized plan for risk reduction

Obviously the complexity of vulnerability assessments will vary according to the design and operation of the water system. The relative points to consider for each of the basic elements listed above are presented in App. 9.A.

9.5.2 Security Tools

- RAM-W^{SM}: Risk Assessment Methodology for Water Utilities developed in cooperation with the Energy Department's Sandia National Laboratories with funding from USEPA.

TABLE 9.5 USEPA Objectives to Ensure Safe Drinking Water*

Providing tools and guidance to drinking water systems and wastewater utilities
Providing training and technical assistance including "train-the-trainer" programs
Providing financial assistance to undertake vulnerability assessments and emergency response plans as funds are made available
Build and maintain reliable communication processes
Build and maintain reliable information systems
Improve knowledge of potential threats, methods to detect attacks, and effectiveness of security enhancements in the water sector
Improve networking among groups involved in security-related matters: water, emergency response, laboratory, environmental, intelligence, and law enforcement communities

*USEPA has developed several strategic objectives to address drinking water system and wastewater utility security needs and also meet requirements set forth in the Bioterrorism Act for public drinking water security. These strategic objectives are summarized in this table.

TABLE 9.6 Dates of Compliance for Community Water Systems to Submit Vulnerability Assessments to USEPA

A Systems serving populations* of	B Submit VA and VA certification† prior to	C Certify ERP within 6 months of VA but no later than‡
≥100,000 persons	March 31, 2003	September 30, 2003
50,000–99,999 persons	December 31, 2003	June 30, 2004
3301–49,999 persons	June 30, 2004	December 31, 2004

*See also text for discussion of determination of system population size.
†Compliance with these deadlines is determined by the date of the postmark or the date the courier places on the mailing label of the submission.
‡VA certifications submitted to USEPA earlier than the dates shown in column B means that the CWS must submit an ERP certification earlier than the dates shown in column C.

- VSAT: Vulnerability Self-Assessment Tool developed by the Association of Metropolitan Sewerage Agencies (AMSA) in collaboration with two consulting firms with USEPA funding.
- National Rural Water Association (NRWA) and Association of State Drinking Water Administrators (ASDWA) with USEPA assistance: "Security Self-Assessment Guide for Small Systems Serving between 3,300 and 10,000." This self-assessment is provided in App. 9.B and www.asdwa.org.

9.5.3 Security Vulnerability Assessment Using VSAT

The Vulnerability Self Assessment Methodology (www.vsatusers.net) as developed in the VSAT software is depicted in Fig. 9.3. The methodology is based on a qualitative risk assessment approach shown in the figure, adopting a broad-based approach that assesses vulnerability, prepares for extreme events and responds should they occur, and restores back to normal business conditions thereafter. The self-assessment framework can be divided into dimensions. The first examines the utility assets including the physical plant, people, knowledge base, information technology (IT platform), and the customers. The second dimension of the framework recognizes that there is a process over time beginning with early assessment and planning activities, followed by later response actions as a result of an extreme event and eventually business recovery activities that occur after the event. VSAT is a tool that can help utilities identify vulnerabilities and evaluate the potential vulnerabilities, along with documentation of the decision process, rationale employed, and relative ranking of risks.

9.5.4 Security Vulnerability Assessment for Small Drinking Water Systems

The Association of State Drinking Water Administrators and the National Rural Waster Association (2002) published the *Security Vulnerability Self-Assessment Guide for Small Drinking Water Systems Serving between 3300 and 10,000* [people]. This self-assessment guide (App. 9.B) is meant to encourage smaller water system administrators, local officials, and water system owners to review their vulnerabilities. The intent, however, is not to take the place of a comprehensive review by security experts. Completion of the documents does meet the requirement for conducting a vulnerability assessment as directed under Public Health Security and Bioterrorism Preparedness and Response Act of 2002. The goal of the vulnerability assessment is to develop a system-specific list of priorities intended to reduce risks to threats of attack.

The self-assessment guide is designed for use by water system personnel to perform the vulnerability assessment. The assessment "should include, but not be limited to a review of pipes and constructed conveyances, physical barriers, water collection, pretreatment, treatment, storage and distribution facilities, electronic, computer or other automated systems which are utilized by the public water system, the use, storage, or handling of various chemicals, and the operation and maintenance of such system." The self-assessment should be conducted on all components of the water system: wellhead or surface water intake, treatment, plant, storage tank(s), pumps, distribution system, and other important components of the water system.

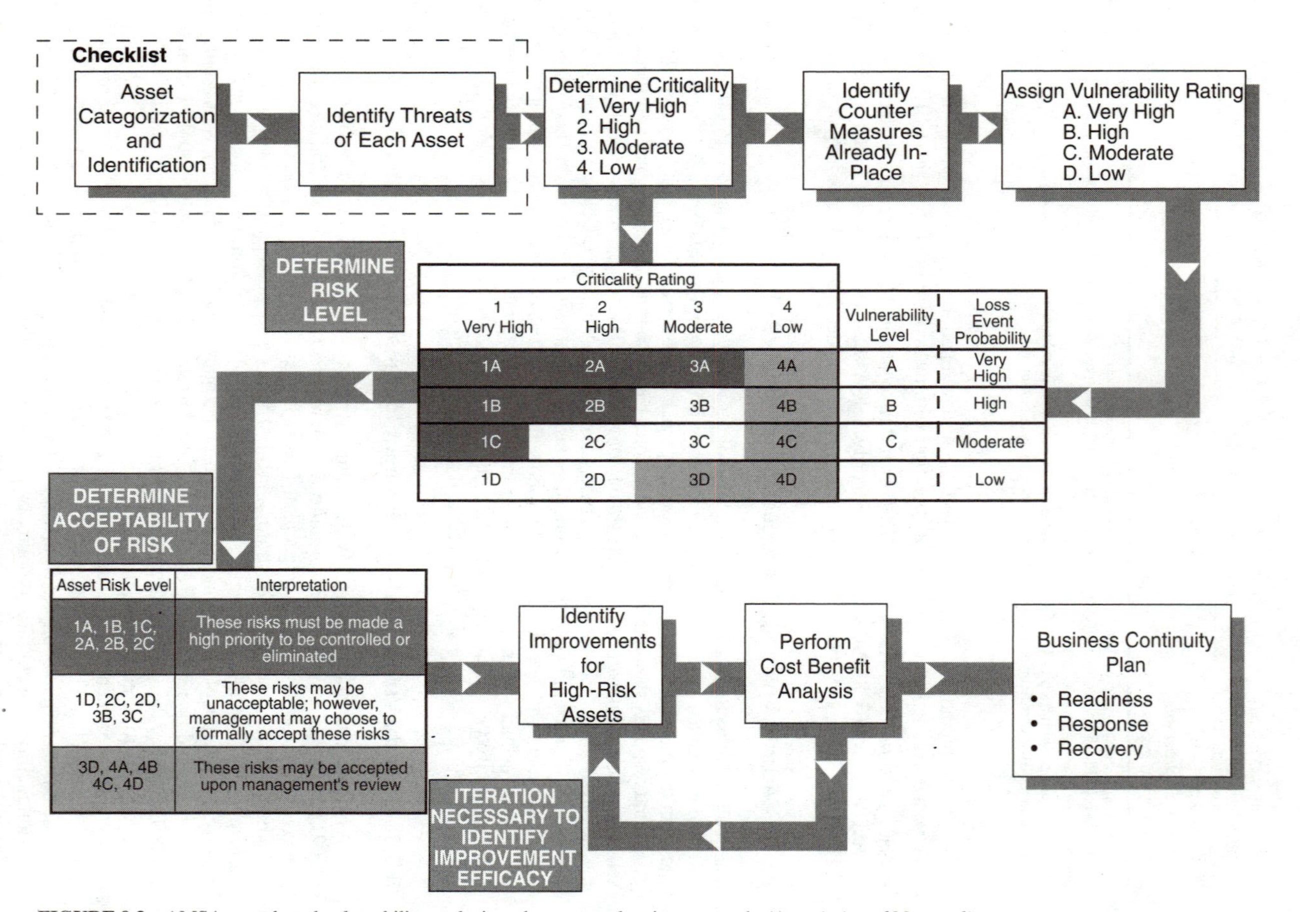

FIGURE 9.3 AMSA asset-based vulnerability analysis and response planning approach. *(Association of Metropolitan Sewage Agencies, 2002)*

The self-assessment has a simple design; the final product is the list of priority actions based on the most likely threats to the water system. The inventory of small water system critical components and the security vulnerability self-assessment general questions are listed in App. 9.B. Once the questions have been answered, the prioritization of actions is developed. A certificate of completion must be completed and sent to the state drinking water primacy agency.

9.6 EMERGENCY RESPONSE PLANNING: RESPONSE, RECOVERY, AND REMEDIATION GUIDANCE

Water utilities need to have an emergency operations or a response plan that is coordinated with state and local emergency response organizations, regulatory authorities, and local government officials. Emergency response planning primarily needs to be a local responsibility. USEPA (April 2002) developed a water utility response, recovery, and remediation guidance for human-made and/or technological emergencies, as a result of their responsibilities under Presidential Decision Directive (PDD) 63. This guidance (presented in App. 9.C) was developed for five different incident types:

- Contamination event: articulated threat with unspecified material
- Contamination threat at a major event
- Notification from health officials of potential water contamination
- Intrusion through supervisory control and data acquisition (SCADA)
- Significant structural damage resulting from a intentional act

Appendix 9.C presents the response actions, recovery actions (recovery notifications and appropriate utility elements), and remediation actions in tabular form for each incident type. Each category contains a section on notifications and utility actions. This guidance can be used to supplement existing water utility emergency operations plans developed to prepare for and respond to natural disasters and emergencies. Even though the guidance is oriented toward the response, recovery, and remediation actions for the five incident types listed above, it could be utilized for other threatened or actual intentional acts.

9.7 INFORMATION SHARING (WWW.WATERISAC.ORG)

WaterISAC is an information service developed to provide America's drinking water and wastewater systems with a secure Web-based environment for early warning of potential threats and a source of knowledge about water system security. The WaterISAC is open to all U.S. drinking water and wastewater systems.

9.8 RELIABILITY ASSESSMENT

There is a need for the use of reliability assessment in the reliability analysis of water supply and water distribution systems. The ideas of using event/fault-tree analysis for this purpose were first introduced in Mays (1989). Unfortunately, these methods have not been implemented by utilities for reliability assessment. The purpose of this section is to reintroduce some very valuable methodologies that can be utilized in the reliability computations for vulnerability assessment. This type of analysis goes above and beyond the type of qualitative vulnerability assessments discussed earlier in this chapter.

Event/fault-tree calculations have been used in the chemical-petroleum and nuclear industries for some time (Fullwood and Hall, 1988; Peplow et al., 2003; Sulfredge et al., 2003); Software tools used for event/fault-tree analysis typically rely on cut-set approaches. These packages, designed for reliable systems, take advantage of low failure probabilities to use several approximations that greatly speed up the calculations. One example is the Visual Interactive Site Analysis Code (VISAC) developed at Oak Ridge National Laboratory (ORNL), which uses a geometric model of a facility coupled to an event/fault-tree model of plant systems to evaluate the vulnerability of a nuclear power plant as well as other infrastructure targets. The event/fault-tree models associated with facility vulnerability calculations typically involve systems with high component failure probabilities resulting from an attack scenario. Such methodologies could be employed for water distribution systems. VISAC is a Java-based graphical user interface (GUI) that can analyze a variety of accidents and incidents at nuclear or industrial facilities ranging from simple component sabotage to an attack with military or terrorist weapons. A list of damaged components from a scenario is then propagated through a set of event/fault trees to determine the overall facility kill probability, the probability of an accompanying radiological release, and the expected facility downtime.

9.8.1 State Enumeration Method

The *state enumeration method* lists all possible mutually exclusive states of the system components that define the state of the entire system. In general, for a system containing M components, each of which can be classified into N operating states, there will be N^M possible states for the entire system. For example, if the state of each of the M components is classified into failed and operating states, the system has 2^M possible states.

Once all the possible system states are enumerated, the states that result in successful system operation are identified and the probability of the occurrence of each successful state is computed. The last step is to sum all the successful state probabilities, which yield the system reliability.

Consider the simple water distribution network in Fig. 9.4. The tree diagram, such as Fig. 9.5, is called an *event tree,* and the analysis involving the construction of an event tree is referred to as *event-tree analysis.* As can be seen, an event tree simulates not only the topology of a system but, more importantly, the sequential or chronological operation of the system.

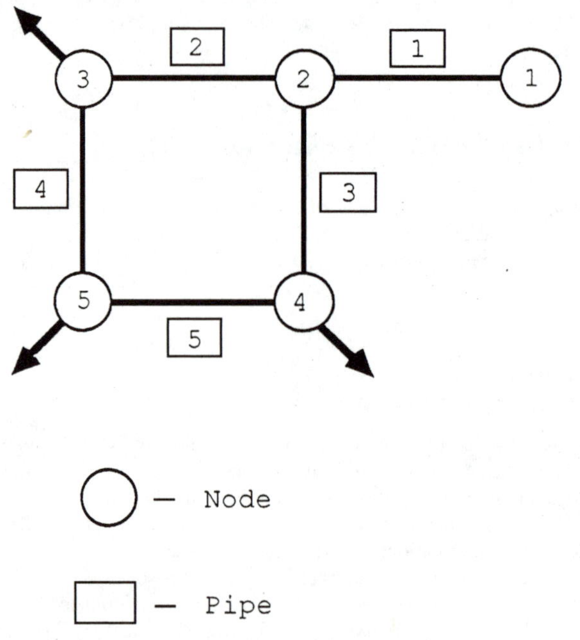

FIGURE 9.4 Simple example water distribution network.

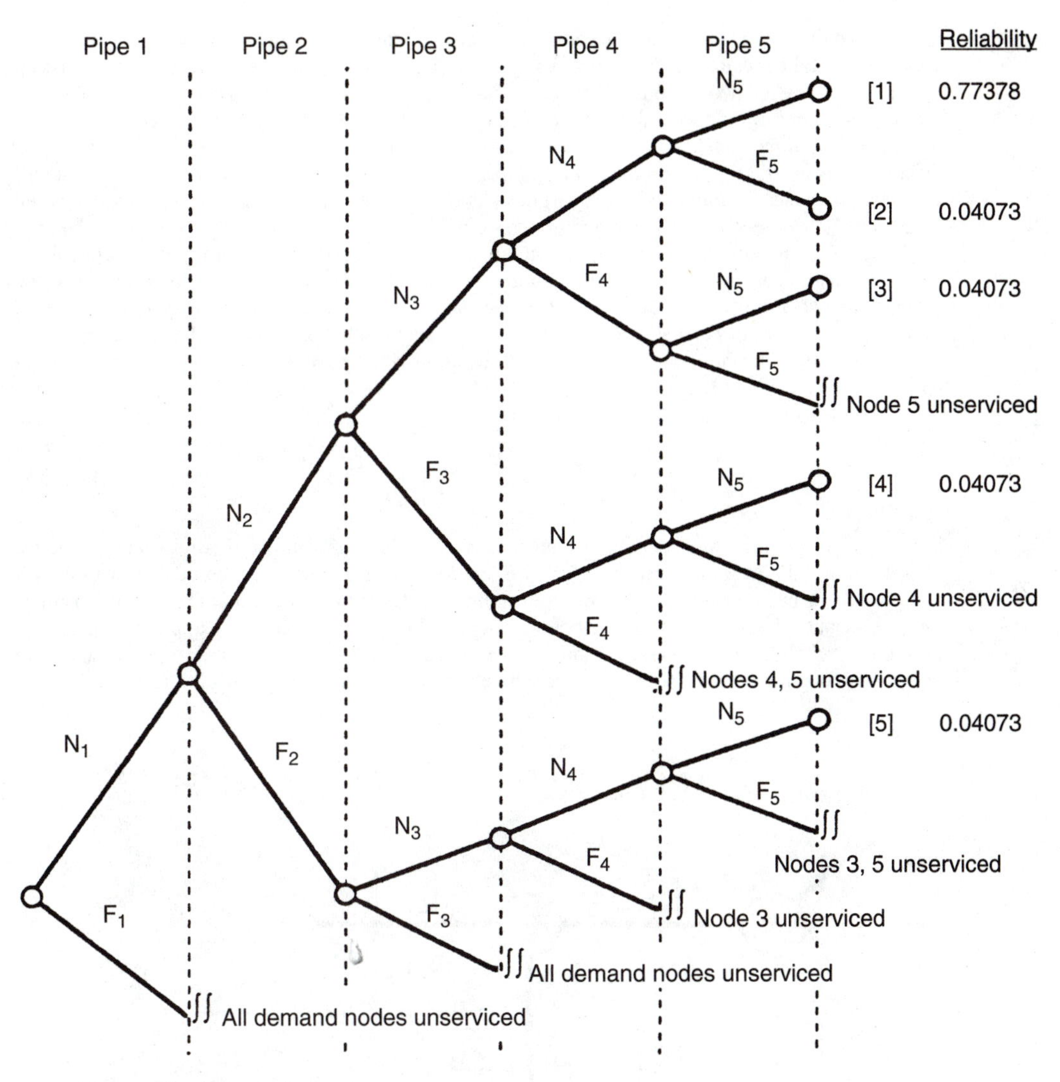

N_i = Nonfailure of pipe i
F_i = Failure of pipe i
∬ = Branch associated with unserviceability of one or more demand nodes
[i] = The ith branch indicating system reliability

FIGURE 9.5 Example event tree for state enumeration method.

9.8.2 Path Enumeration Method

This is a very powerful method for system reliability evaluation. A *path* is defined as a set of components or modes of operation which lead to a certain outcome of the system. In system reliability analysis, the system outcomes of interest are those of failed state or operational state. A *minimum path* is one in which no component is traversed more than once in going along the path. Under this methodological category, the *tie-set analysis* and *cut-set analysis* are the two well-known techniques.

Cut-Set Analysis. The *cut-set* is defined as a set of system components or modes of operation which, when failed, cause the failure of the system. The cut-set analysis is powerful for evaluating system reliability for two reasons: (1) it can be easily programmed on digital computers for fast and efficient solutions of any general system configuration, especially in the form of a network; and (2) the cut-sets are directly related to the modes of system failure and therefore identify the distinct and discrete ways in which a system may fail. For example, in a water distribution system, a cut-set will be the set of system components such as pipe sections, pumps, and storage facilities which, when failed jointly, would disrupt the service to certain users.

The cut-set method utilizes the minimum cut-sets for calculating the system failure probability. The *minimum cut-set* is a set of system components which, when all failed, causes failure of the system, but when any one component of the set does not fail, does not cause system failure. A minimum cut-set implies that all components of the cut-set must be in the failure state to cause system failure. Therefore, the components or modes of operation involved in the minimum cut-set are effectively connected in parallel, and each minimum cut-set is connected in series. Consequently, the failure probability of a system can be expressed as

$$p_{f,\mathrm{sys}} = P\left[\bigcup_{i=1}^{I} C_i\right] = P\left[\bigcup_{i=1}^{I}\left(\bigcap_{j=1}^{J_i} F_{ij}\right)\right] \tag{9.1}$$

in which C_i is the ith of the total I minimum cut-sets, J_i is the total number of components or modes of operation in the ith minimum cut-set, and F_{ij} represents the failure event associated with the jth components or mode of operation in the ith minimum cut-set. In the case that the number of minimum cut-sets I is large, computing the bounds for probability of a union can be applied. The bounds on failure probability of the system should be examined for their closeness to ensure that adequate precision is obtained. Figure 9.6 shows the cut-sets for the simple water distribution system.

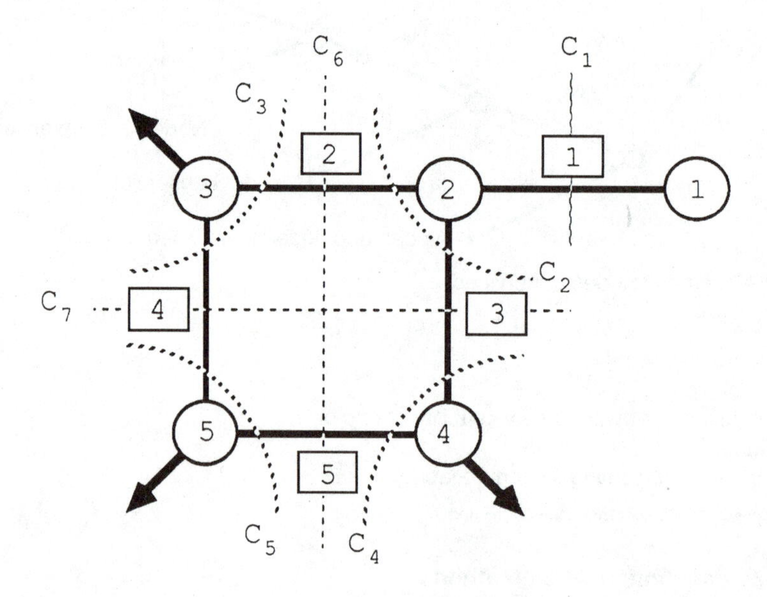

FIGURE 9.6 Cut-sets for the example water distribution network.

Tie-Set Analysis. As the complement of the cut-set, a *tie-set* is a minimal path of the system in which system components or modes of the operation are arranged in series. Consequently, a tie-set fails if any of its components or modes of operation fail. All tie-sets are effectively connected in parallel; that is, the system will be in the operating state if any of its tie-sets are functioning. Therefore, the system reliability can be expressed as

$$p_{s,\mathrm{sys}} = P\left[\bigcup_{i=1}^{I} T_i\right] = P\left[\bigcup_{i=1}^{I}\left(\bigcap_{j=1}^{J_i} F'_{ij}\right)\right] \tag{9.2}$$

in which T_i is the ith tie-set of all I tie-sets, J_i is the total number of components or modes of operation in the ith tie-set, and F_{ij}' represents the nonfailure state of the jth component in the ith tie-set. Again, when the number of tie-sets is large, computation of exact system reliability by Eq. (9.2) could be computationally cumbersome. In such condition, bounds for system reliability could be computed.

The main disadvantage of the tie-set method is that failure modes are not directly identified. Direct identification of failure modes is sometimes essential if a limited amount of a resource is available to focus on a few dominant failure modes. Figure 9.7 shows the tie-sets for the simple water distribution system.

In summary, the path enumeration method involves the following steps (Henley and Gandhi, 1975):

1. Find all minimum paths. In general, this has to be done using a computer when the number of components is large and the system configuration is complex.
2. Find all required unions of the paths.
3. Give each path union a reliability expression in terms of module reliability.
4. Compute the system reliability in terms of module reliabilities.

9.8.3 Conditional Probability Approach

The approach starts with a selection of key components and modes of operation whose states (operational or failure) would decompose the entire system into simple series and/or parallel subsystems for which the reliability or failure probability of subsystems can be easily evaluated. Then, the reliability of the entire system is obtained by combining those of the sub-systems using conditional probability rules as

$$p_{s,\mathrm{sys}} = p_{s|F_i'} \times p_{s,i} + p_{s|F_i} \times p_{f,i} \tag{9.3}$$

in which $p_{s|Fi'}$ and $p_{s|Fi}$ are the conditional system reliability given that the ith component is operational, F_i', and failed, F_i, respectively, and $p_{s,i}$ and $p_{f,i}$ are the reliability and failure probabilities of the ith component, respectively.

Except for very simple and small systems, a nested conditional probability operation is inevitable. Efficient evaluation of system reliability of a complex system hinges entirely on a proper selection of key components, which generally is a difficult task when the scale of the system is large. Furthermore, the method cannot be easily adapted to computerization for problem solving. Figure 9.8 illustrates the conditional probability method for the simple water distribution system.

9.8.4 Fault-Tree Analysis

The major objective of fault-tree construction is to represent the system condition, which may cause system failure, in a symbolic manner. In other words, the fault tree consists of sequences of events that lead to system failure. *Fault-tree analysis,* unlike event-tree analysis, is a backward analysis, which begins with a system failure and traces backward, searching for possible causes of the failure. Fault-tree analysis was initiated at Bell Telephone Laboratories and Boeing Aircraft Company

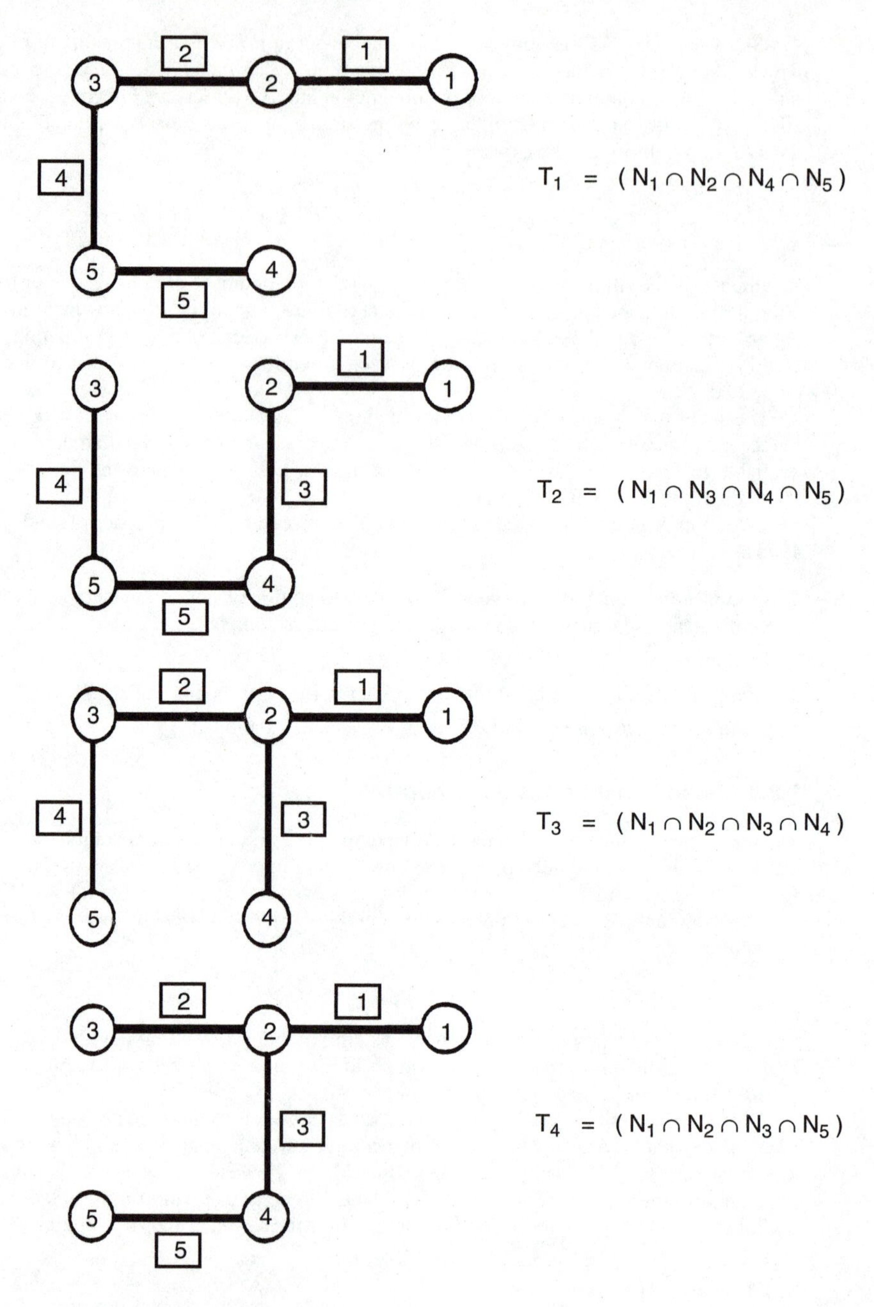

T_i = the ith tie-set
N_k = Nonfailure state of pipe section k

FIGURE 9.7 The four minimum tie-sets (or paths) for the example network.

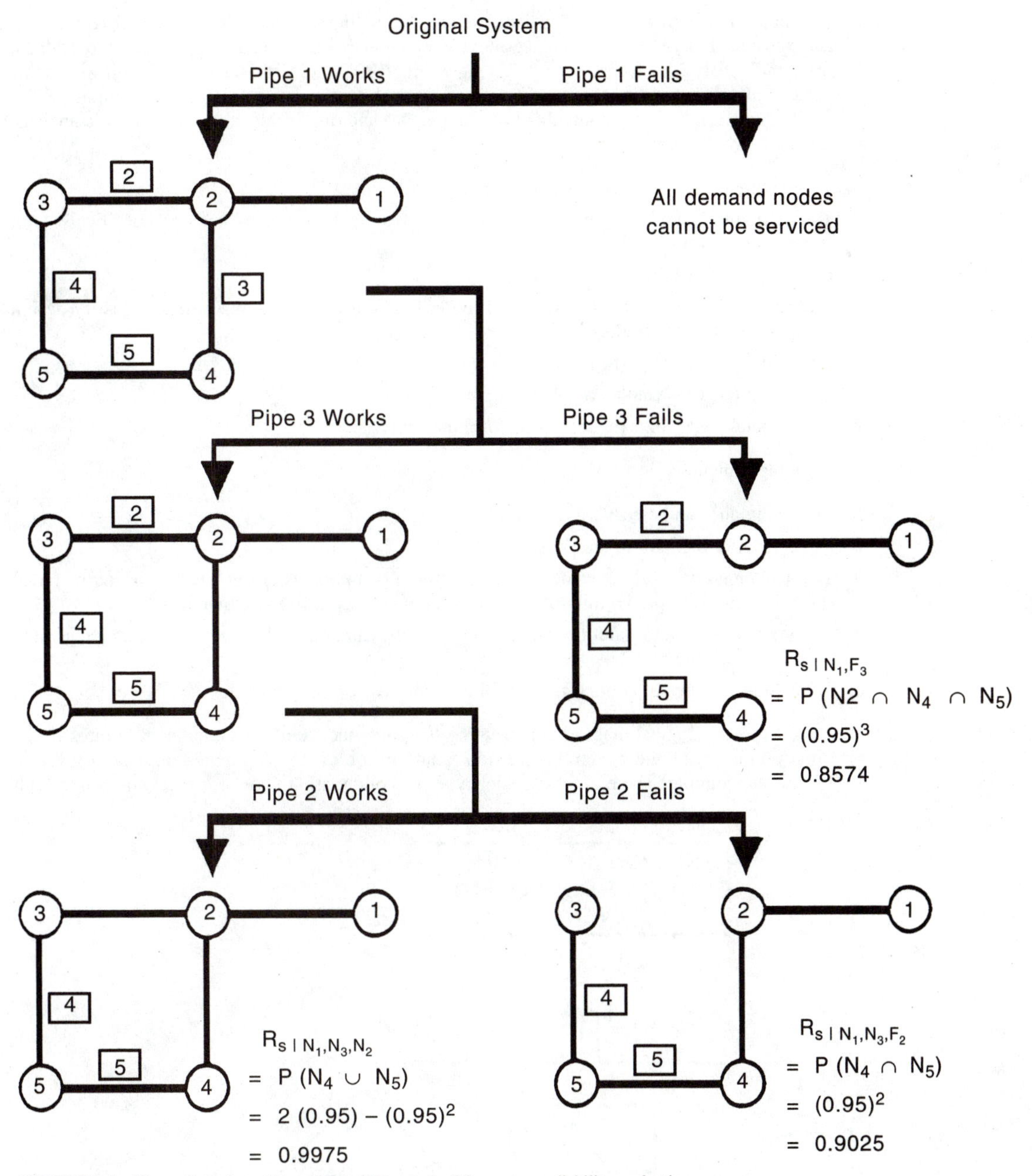

FIGURE 9.8 Illustration of conditional probability method for system reliability evaluation.

(Barlow et al., 1975). Since then it has been used for evaluating the reliability of many different engineering systems. In hydrosystem engineering design, fault-tree analysis has been applied to evaluate the risk and reliability of earth dams (Cheng et al., 1993), underground water control systems (Bogardi et al., 1987), and water-retaining structures such as dikes and sluice gates (Vrijling, 1993).

Dhillon and Singh (1981) pointed out the advantages and disadvantages of the fault-tree analysis technique. Advantages include

1. This technique provides insight into the system behavior.
2. This technique requires engineers to understand the system thoroughly and deal specifically with one particular failure at a time.
3. It helps ferret out failures deductively.
4. It provides a visible and instructive tool to designers, users, and management to justify design changes and tradeoff studies.
5. It provides options to perform quantitative or qualitative reliability analysis.
6. This technique can handle complex systems.
7. Commercial codes are available to perform the analysis.

Disadvantages include

1. This technique can be costly and time-consuming.
2. Results can be difficult to check.
3. This technique normally considers that the system components are in either working or failed state; therefore, the partial failure states of components are difficult to handle.
4. Analytical solutions for fault trees containing standbys and repairable components are difficult to obtain for the general case.
5. To include all types of common-cause failure requires considerable effort.

A fault tree is a logical diagram representing the consequence of the component failures (basic or primary failures) on the system failure (top failure or top event). A simple fault tree is given in Fig. 9.9 as an example. There are two major types of combination nodes (or gates) used in a fault

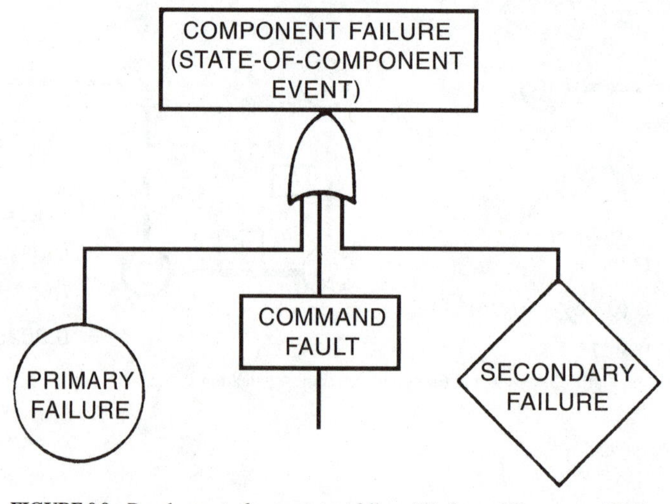

FIGURE 9.9 Development of a component failure. (*Henley and Kumamoto, 1981.*)

tree. The *AND node* implies that the output event occurs only if all the input events occur simultaneously, corresponding to the intersection operation in probability theory. The *OR node* indicates that the output event occurs if any one or more of the input events occur, that is, if there is a union. System reliability $p_{s,sys}(t)$ is the probability that the top event does not occur over the time interval $(0,t)$.

Before constructing a fault tree, engineers must thoroughly understand the system and its intended use. One must determine the higher-order functional events and continue the fault-event analysis to determine their logical relationships with lower-level events. Once this is accomplished, the fault tree can be constructed. A brief description of fault-tree construction is presented below. The basic concepts of fault-tree analysis are presented in Henley and Kumamoto (1981) and Dhillon and Singh (1981).

There are actually two types of building blocks: gate symbols (Table 9.7) and event symbols (Table 9.8). *Gate symbols* connect events according to their causal relation such that they may have one or more input events but only one output event. A *fault event,* denoted by a rectangular box, results from a combination of more basic faults acting through logic gates. A circle denotes a basic component failure that represents the limit of resolution of a fault tree. A diamond represents a fault event whose causes have not been fully developed. For more complete descriptions of other types of gate and event symbols, readers are referred to Henley and Kumamoto (1981).

Henley and Kumamoto (1981) present heuristic guidelines for constructing fault trees, which are summarized in Table 9.9, and are listed below:

1. Replace abstract events by less abstract events.
2. Classify an event into more elementary events.
3. Identify distinct causes for an event.
4. Couple trigger event with "no protection actions."
5. Find cooperative causes for an event.
6. Pinpoint component failure events.
7. Develop component failure using Fig. 9.9.

Another example of a fault-tree construction is given for the pumping system in Fig. 9.10. In this pumping system, the tank is filled in 10 min and empties in 50 min, having a cycle time of 60 min. After the switch is closed, the time is set to open the contacts in 10 min. If the mechanism fails, the horn sounds and the operator opens the switch to prevent pressure tank rupture. The fault tree for the pumping system is shown in Fig. 9.11.

Evaluation of Fault Trees. The basic steps used to evaluate fault trees include

1. Construct the fault tree.
2. Determine the minimal cut-sets.
3. Develop primary event information.
4. Develop cut-set information.
5. Develop top-event information.

To evaluate the fault tree, one should always start from the minimal cut-sets, which, in essence, are *critical paths.* Basically, the fault-tree evaluation comprises two distinct processes: (1) determination of the logical combination of events that cause top-event failure expressed in the minimal cut-sets, and (2) numerical evaluation of the expression.

Cut-sets, as discussed previously, are collections of basic events such that if all these basic events occur, then the top event is guaranteed to occur. The tie-set is a dual concept to the cut-set in that it is a collection of basic events in which none of the events in the tie-set occur, and in which the top event is guaranteed not to occur. As one could imagine, a large system has an enormous number of failure modes. A minimal cut-set is one that if any basic event is removed from the set, the remaining events collectively are no longer a cut-set. By the use of minimum cut-sets, the number of cut-sets and basic events are reduced in order to simplify the analysis.

TABLE 9.8 Gate Symbols in Fault-Tree Analysis

	GATE SYMBOL	GATE NAME	CAUSAL RELATION
1		AND GATE	OUTPUT EVENT OCCURS IF ALL INPUT EVENTS OCCUR SIMULTANEOUSLY.
2		OR GATE	OUTPUT EVENT OCCURS IF ANY ONE OF THE INPUT EVENTS OCCURS.
3	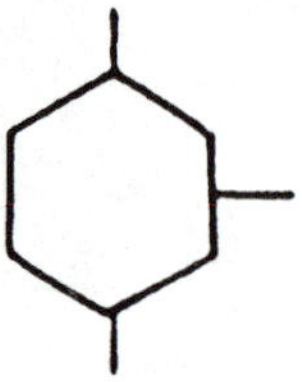	INHIBIT GATE	INPUT PRODUCES OUTPUT WHEN CONDITIONAL EVENT OCCURS.
4		PRIORITY AND GATE	OUTPUT EVENT OCCURS IF ALL INPUT EVENTS OCCUR IN THE ORDER FROM LEFT TO RIGHT.
5		EXCLUSIVE OR GATE	OUTPUT EVENT OCCURS IF ONE, BUT NOT BOTH, OF THE INPUT EVENTS OCCUR.
6		m OUT OF n GATE (VOTING OR SAMPLE GATE)	OUTPUT EVENT OCCURS IF m OUT OF n INPUT EVENTS OCCUR.

Source: Henley and Kumamoto (1981).

TABLE 9.9 Event Symbols in Fault-Tree Analysis

	EVENT SYMBOL	MEANING OF SYMBOLS
1	CIRCLE	BASIC EVENT WITH SUFFICIENT DATA
2	DIAMOND	UNDEVELOPED EVENT
3	RECTANGLE	EVENT REPRESENTED BY A GATE
4	OVAL	CONDITIONAL EVENT USED WITH INHIBIT GATE
5	HOUSE	HOUSE EVENT. EITHER OCCURRING OR NOT OCCURRING
6	TRIANGLES	TRANSFER SYMBOL

Source: Henley and Kumamoto (1981).

The system availability $A_{sys}(t)$ is the probability that the top event does not occur at time *t*, which is the probability of the systems operating successfully when the top event is an OR combination of all system hazards. System unavailability $U_{sys}(t)$, on the other hand, is the probability that the top event occurs at time *t*, which is either the probability of system failure or the probability of a particular system hazard at time *t*.

System reliability $p_{s,sys}(t)$ is the probability that the top event does not occur over time interval $(0,t)$. System reliability requires continuation of the nonexistence of the top event, and its value is less than or equal to the availability. On other hand, the system unreliability $p_{f,sys}(t)$ is the probability that the top event occurs before time *t* and is complementary to the system reliability. Also, system unreliability, in general, is greater than or equal to system unavailability. The system failure density $f_{sys}(t)$ can be obtained from the system unreliability.

TABLE 9.10 Heuristic Guidelines for Fault-Tree Construction

	DEVELOPMENT POLICY	CORRESPONDING PART OF FAULT TREE
1	EQUIVALENT BUT LESS ABSTRACT EVENT F	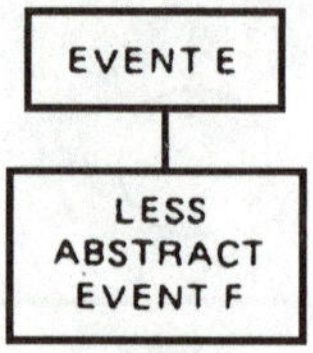 EVENT E; LESS ABSTRACT EVENT F
2	CLASSIFICATION OF EVENT E	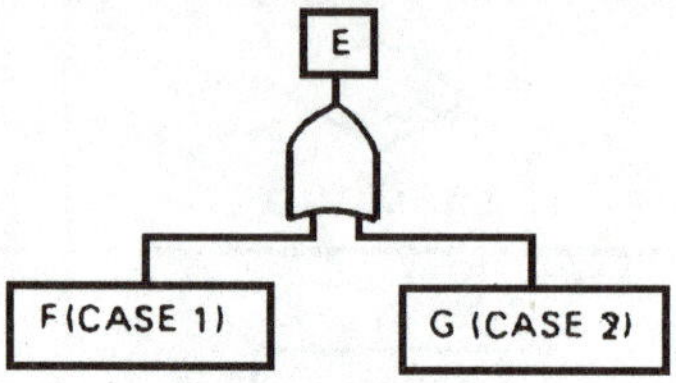 E; F(CASE 1); G (CASE 2)
3	DISTINCT CAUSES FOR EVENT E	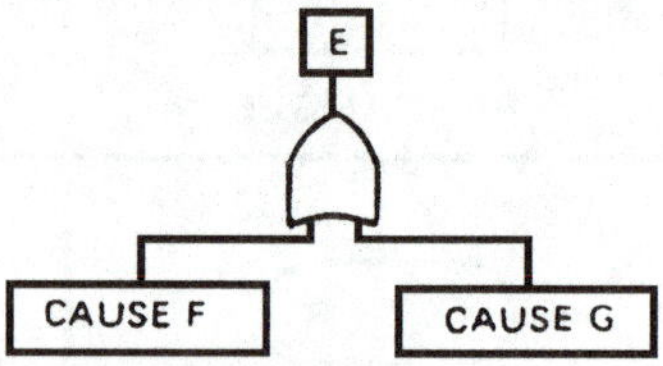 E; CAUSE F; CAUSE G
4	TRIGGER VERSUS NO PROTECTIVE EVENT	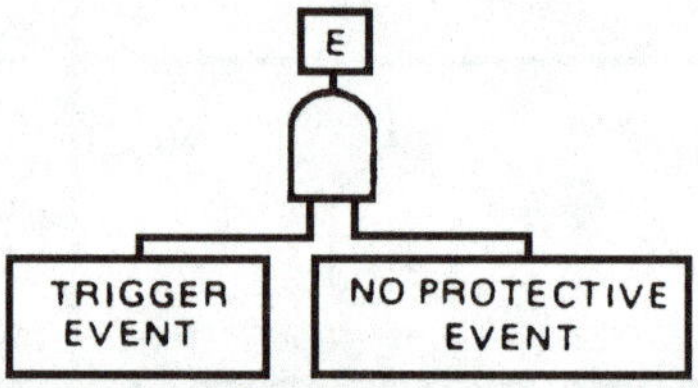 E; TRIGGER EVENT; NO PROTECTIVE EVENT
5	COOPERATIVE CAUSE	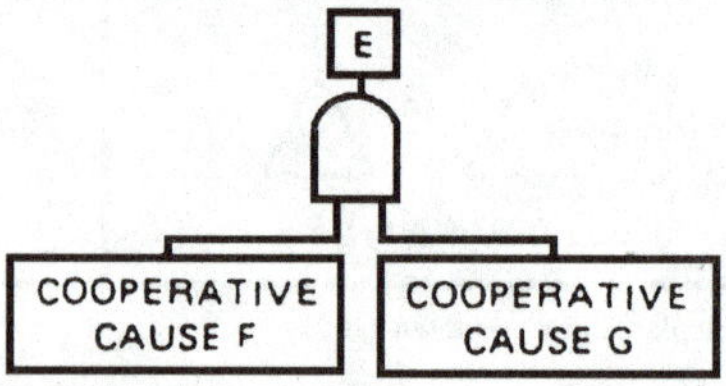 E; COOPERATIVE CAUSE F; COOPERATIVE CAUSE G
6	PINPOINT A COMPONENT FAILURE EVENT	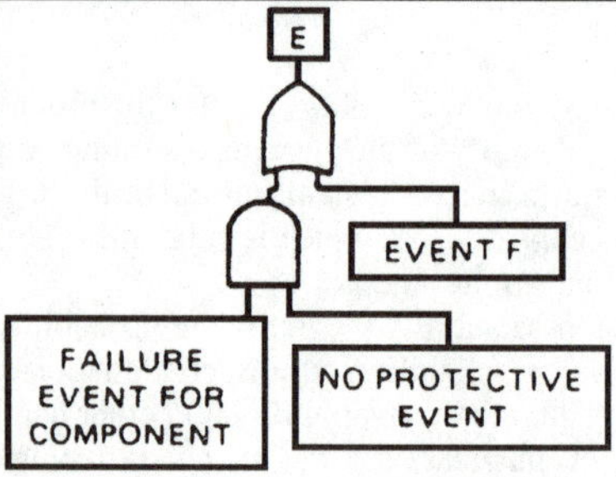 E; EVENT F; FAILURE EVENT FOR COMPONENT; NO PROTECTIVE EVENT

Source: Henley and Kumamoto (1981).

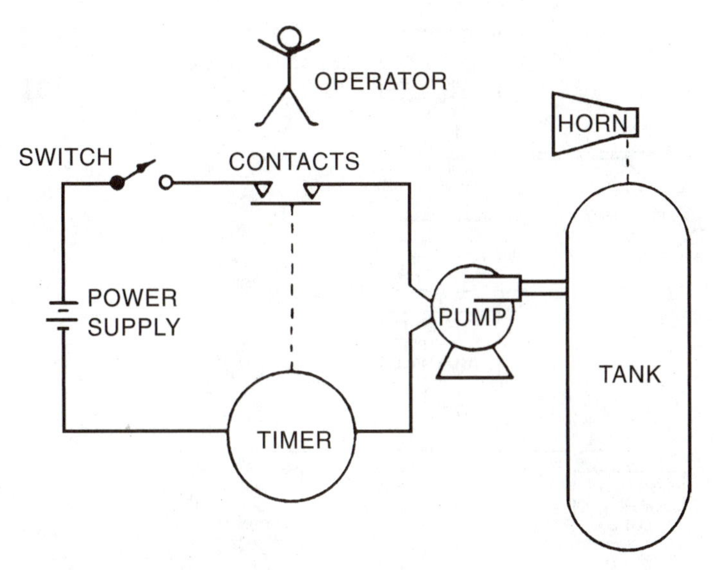

FIGURE 9.10 Schematic diagram for an example pumping system. (*Henley and Kumamoto, 1981.*)

9.9 REFERENCES

Bailey, K. C., *The Biological and Toxin Weapons Threat to the United States,* National Institute for Public Policy, Fairfax, Va., October 2001.

Barlow, R. E., J. B. Fussell, and N. D. Singpurwalla, *Reliability and Fault Tree Analysis: Theoretical and Applied Aspects of System Reliability and Safety Assessment,* SIAM, Philadelphia, 1975.

Blomgren, P., "Utility Managers Need to Protect Water Systems from Cyberterrorism," *U.S. News,* 19:10, October 2002.

Bogardi, I., L. Duckstein, and F. Szidaroviszky, "Reliability Estimation of Underground Water Control Systems under Natural and Sampling Uncertainty," in L. Duckstein and E. J. Plate (eds.), *Engineering Reliability and Risk in Water Resources,* 423–441, Martinus Nijhoff Publishers, Drodrecht, The Netherlands, 1987.

Burrows, W. D., and S. E. Renner, "Biological Warfare Agents as Threats to Potable Water," *Environmental Health Perspectives,* 107(12): 975–984, 1999.

Cheng, S.-T., B. C. Yen, and W. H. Tang, "Stochastic Risk Modeling of Dam Overtopping," in Ben C. Yen and Y.-K. Tung (eds.), *Reliability and Uncertainty Analyses in Hydraulic Design,* 123–132, American Society of Civil Engineers, New York, 1993.

Copeland, C., and B. Cody, *Terrorism and Security Issues Facing the Water Infrastructure Sector,* Order Code RS21026, CRS Report for Congress, Congressional Research Service, Library of Congress, Washington, D.C., June 18, 2002.

DeNileon, G. P., "The Who, Why, and How of Counterterrorism Issues," *Journal of the American Water Works Association,* 93(5): 78–85, 2001.

Dhillon, B. S., and C. Singh, *Engineering Reliability: New Techniques and Applications,* Wiley, New York, 1981.

Dickey, M. E., *Biocruise: A Contemporary Threat,* Counterproliferation Paper no. 7, Future Warfare Series no. 7, USAF Counterproliferation Center, Air War College, Air University, Maxwell Air Force Base, Alabama, www.au.af.mil/au/awc/awcgate/cpc-pubs/dickey.htm, September 2000.

Foran, J. A., and T. M. Brosnan, "Early Warning Systems for Hazardous Biological Agents in Potable Water," *Environmental Health Perspectives,* 108(10): 993–996, 2000.

Fullwood, R. R., and R. E. Hall, *Probabilistic Risk Assessment in Nuclear Power Industry,* Pergamon Press, Oxford, U.K., 1988.

Grayman, W. M., R. A. Deininger, and R. M. Males, *Design of Early Warning and Predictive Source-Water Monitoring Systems,* AWWA Research Foundation and AWWA, 2001.

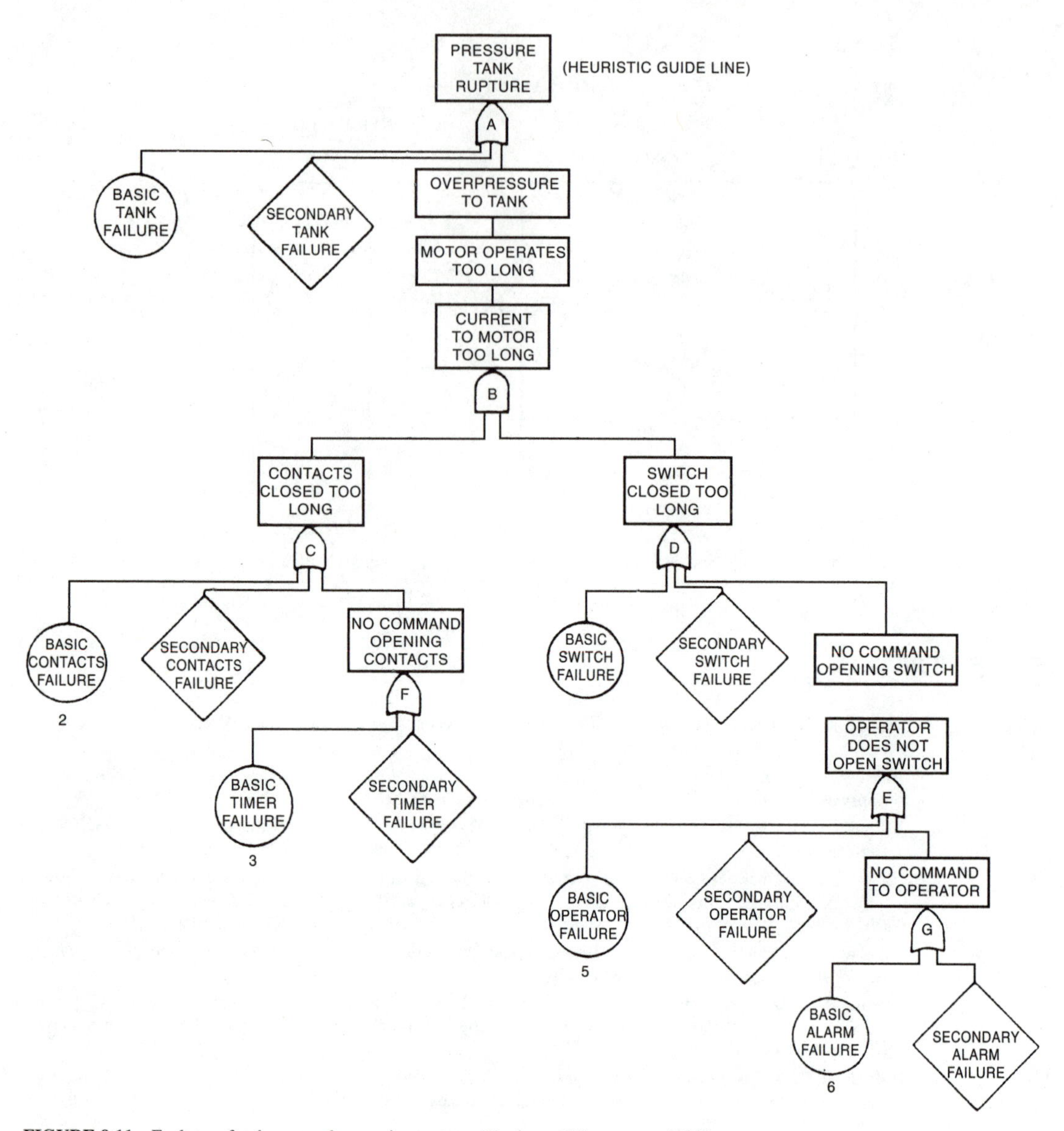

FIGURE 9.11 Fault tree for the example pumping system. (*Henley and Kumamoto, 1981.*)

Haas, C. N., "The Role of Risk Analysis in Understanding Bioterrorism," *Risk Analysis,* 22(2):671–677, 2002.

Haimes, Y. Y., et al., "Reducing Vulnerability of Water Supply Systems to Attack," *Journal of Infrastructure Systems* (ASCE) 4(4):164–177, 1998.

Henley, E. J., and S. L. Gandhi, "Process Reliability Analysis," *American Institute of Chemical Engineering Journal,* 21(4):677–686, 1975.

Henley, E. J., and H. Kumamoto, *Reliability Engineering and Risk Assessment,* Prentice-Hall, Englewood Cliffs, N.J., 1981.

Hickman, D. C., *A Chemical and Biological Warfare Threat: USAF Water Systems at Risk,* Counterproliferation Paper no. 3, Future Warfare Series no. 3, USAF Counterproliferation Center, Air War College, Air University, Maxwell Air Force Base, Alabama, www.au.af.mil/au/awc/awcgate/cpc-pubs/hickman.htm, September 1999.

Mays, L. W. (ed.), *Reliability Analysis of Water Distribution Systems,* American Society of Civil Engineers, New York, 1989.

Mays, L. W. (ed.), *Water Distribution Systems Handbook,* McGraw-Hill, New York, 2000.

Mays, L. W. (ed.), *Urban Water Supply Handbook,* McGraw-Hill, New York, 2002.

Mays, L. W. (ed.), *Water Supply System: Security Handbook,* McGraw-Hill, New York, 2004.

Peplow, D. E., C. D. Sulfredge, R. L. Saunders, R. H. Morris, and T. A. Hann, *Calculating Nuclear Power Plant Vulnerability Using Integrated Geometry and Event/Fault Tree Models,* Oak Ridge National Laboratory, Oak Ridge, Tenn., 2003.

President's Commission on Critical Infrastructure Protection, "Sector Summary Reports," in App. A, *Critical Foundations: Protecting America's Infrastructure,* A-45, http://www.ciao.gov/PCCIP/PCCIP_Report.pdf.

Sulfredge, C. D., R. L. Saunders, D. E. Peplow, and R. H. Morris, *Graphical Expert System for Analyzing Nuclear Facility Vulnerability,* Oak Ridge National Laboratory, Oak Ridge, Tenn., 2003.

U.S. Army Medical Research Institute of Infectious Disease, *USAMRID's Medical Management of Biological Causalities Handbook,* www.usamriid.army.mil/education/bluebook.html, 2001.

U.S. Environmental Protection Agency (USEPA), *Guidance for Water Utility Response, Recovery, and Remediation Actions for Man-Made and/or Technological Emergencies,* http://www.epa.gov/safewater/security/er-guidance.pdf.

U.S. Environmental Protection Agency (USEPA), *Guidance for Water Utility Response, Recovery & Remediation Actions for Man-Made and/or Technological Emergencies,* EPA 810-R-02-001, Office of Water (4601), www.epa.gov/safewater, April 2002.

U.S. Environmental Protection Agency (USEPA), *Water Security Strategy for Systems Serving Populations Less than 100,000/15 MGD or Less,* July 9, 2002a.

U.S. Environmental Protection Agency (USEPA), *Vulnerability Assessment Fact Sheet 12-19,* EPA 816-F-02-025, Office of Water, www.epa.gov/safewater/security/va fact sheet 12-19.pdf, also at www.epa.gov/ogwdw/index.html, November 2002b.

U.S. Environmental Protection Agency (USEPA), *Instructions to Assist Community Water Systems in Complying with the Public Health Security and Bioterrorism Preparedness and Response Act of 2002,* EPA 810-R-02-001, Office of Water, www.epa.gov/safewater/security, January 2003.

U.S. Environmental Protection Agency (USEPA), information at http://www.epa.gov/swercepp/cntr-ter.html.

Vrijling, J. K., "Development in Probabilistic Design of Flood Defenses in the Netherlands," in Ben C. Yen and Y.-K. Tung (eds.), *Reliability and Uncertainty Analyses in Hydraulic Design,* 133–178, American Society of Civil Engineers, New York, 1993.

APPENDIX 9.A WHAT ARE SOME POINTS TO CONSIDER IN A VULNERABILITY ASSESSMENT?

Some points to consider related to the six basic elements are included in the following table. The manner in which the vulnerability assessment is performed is determined by each individual water utility. It will be helpful to remember throughout the assessment process that the ultimate goal is twofold: *to safeguard public health and safety, and to reduce the potential for disruption of a reliable supply of pressurized water.*

Basic element	Points to consider
1. Characterization of the water system, including its mission and objectives. (Answers to system-specific questions may be helpful in characterizing the water system.)	What are the important missions of the system to be assessed? Define the highest priority services provided by the utility. Identify the utility's customers: General public Government Military Industrial Critical care Retail operations Firefighting What are the most important facilities, processes, and assets of the system for achieving the mission objectives and avoiding undesired consequences? Describe the Utility facilities Operating procedures Management practices that are necessary to achieve the mission objectives How the utility operates (e.g., water source, including groundwater and surface water) Treatment processes Storage methods and capacity Chemical use and storage Distribution system In assessing those assets that are critical, consider critical customers, dependence on other infrastructures (e.g., electricity, transportation, other water utilities), contractual obligations, single points of failure (e.g., critical aqueducts, transmission systems, aquifers), chemical hazards and other aspects of the utility's operations, or availability of other utility capabilities that may increase or decrease the criticality of specific facilities, processes and assets.
2. Identification and prioritization of adverse consequences to avoid.	Take into account the impacts that could substantially disrupt the ability of the system to provide a safe and reliable supply of drinking water or otherwise present significant public health concerns to the surrounding community. Water systems should use the vulnerability assessment process to determine how to reduce risks associated with the consequences of significant concern.

	Ranges of consequences or impacts for each of these events should be identified and defined. Factors to be considered in assessing the consequences may include Magnitude of service disruption Economic impact (such as replacement and installation costs for damaged critical assets or loss of revenue due to service outage) Number of illnesses or deaths resulting from an event Impact on public confidence in the water supply Chronic problems arising from specific events Other indicators of the impact of each event as determined by the water utility. Risk reduction recommendations at the conclusion of the vulnerability assessment should strive to prevent or reduce each of these consequences.
3. Determination of critical assets that might be subject to malevolent acts that could result in undesired consequences.	What are the malevolent acts that could reasonably cause undesired consequences? Consider the operation of critical facilities, assets, and/or processes and assess what an adversary could do to disrupt these operations. Such acts may include physical damage to or destruction of critical assets, contamination of water, intentional release of stored chemicals, interruption of electricity, or other infrastructure interdependencies. The Public Health Security and Bioterrorism Preparedness and Response Act of 2002 (PL 107-188) states that a community water system which serves a population of greater than 3300 people must review the vulnerability of its system to a terrorist attack or other intentional acts intended to substantially disrupt the ability of the system to provide a safe and reliable supply of drinking water. The vulnerability assessment shall include, but not be limited to, a review of Pipes and constructed conveyances Physical barriers Water collection, pretreatment and treatment facilities Storage and distribution facilities Electronic, computer or other automated systems which are utilized by the public water system [e.g., supervisory control and data acquisition (SCADA)] The use, storage, or handling of various chemicals The operation and maintenance of such systems
4. Assessment of the likelihood (qualitative probability) of such malevolent acts from adversaries (e.g., terrorists, vandals).	Determine the possible modes of attack that might result in consequences of significant concern based on the critical assets of the water system. The objective of this step of the assessment is to move beyond what is merely possible and determine the likelihood of a particular attack scenario. This is a very difficult task as there is often insufficient information to determine the likelihood of a particular event with any degree of certainty.

Basic element	Points to consider
4. *(Continued)*	The threats (the kind of adversary and the mode of attack) selected for consideration during a vulnerability assessment will dictate, to a great extent, the risk reduction measures that should be designed to counter the threat(s). Some vulnerability assessment methodologies refer to this as a *design basis threat* (DBT), where the threat serves as the basis for the design of countermeasures, as well as the benchmark against which vulnerabilities are assessed. It should be noted that there is no single DBT or threat profile for all water systems in the United States. Differences in geographic location, size of the utility, previous attacks in the local area, and many other factors will influence the threat(s) that water systems should consider in their assessments. Water systems should consult with the local FBI and/or other law enforcement agencies, public officials, and others to determine the threats on which their risk reduction measures should be based. Water systems should also refer to USEPA's *Baseline Threat Information for Vulnerability Assessments of Community Water Systems* to help assess the most likely threats to their system. This document is available to community water systems serving populations greater than 3300 people. If your system has not yet received instructions on how to receive a copy of this document, then contact your regional USEPA office immediately. You will be sent instructions on how to securely access the document via the Water Information Sharing and Analysis Center (ISAC) web site or obtain a hard copy that can be mailed directly to you. Water systems may also want to review their incident reports to better understand past breaches of security.
5. Evaluation of existing countermeasures. (Depending on countermeasures already in place, some critical assets may already be sufficiently protected. This step will aid in identification of the areas of greatest concern, and help to focus priorities for risk reduction.)	What capabilities does the system currently employ for detection, delay and response? Identify and evaluate current detection capabilities such as intrusion detection systems, water quality monitoring, operational alarms, guard post orders, and employee security awareness programs. Identify current delay mechanisms such as locks and key control, fencing, structure integrity of critical assets, and vehicle access checkpoints. Identify existing policies and procedures for evaluation and response to intrusion and system malfunction alarms, adverse water quality indicators, and cyber system intrusions. *It is important to determine the performance characteristics. Poorly operated and maintained security technologies provide little or no protection.* What cyber protection system features does the utility have in place? Assess what protective measures are in place for the SCADA and business-related computer information systems such as Firewalls Modem access Internet and other external connections, including wireless data and voice communications Security policies and protocols *It is important to identify whether vendors have access rights and/or "backdoors" to conduct system diagnostics remotely.* What security policies and procedures exist, and what is the compliance record for them? Identify existing policies and procedures concerning Personnel security Physical security Key and access badge control

	Control of system configuration and operational data Chemical and other vendor deliveries Security training and exercise records
6. Analysis of current risk and development of a prioritized plan for risk reduction.	Information gathered on threat, critical assets, water utility operations, consequences, and existing countermeasures should be analyzed to determine the current level of risk. The utility should then determine whether current risks are acceptable or risk reduction measures should be pursued. Recommended actions should measurably reduce risks by reducing vulnerabilities and/or consequences through improved deterrence, delay, detection, and/or response capabilities or by improving operational policies or procedures. Selection of specific risk reduction actions should be completed prior to considering the cost of the recommended action(s). Utilities should carefully consider both short- and long-term solutions. An analysis of the cost of short- and long-term risk reduction actions may impact which actions the utility chooses to achieve its security goals. Utilities may also want to consider security improvements in light of other planned or needed improvements. Security and general infrastructure may provide significant multiple benefits. For example, improved treatment processes or system redundancies can both reduce vulnerabilities and enhance day-to-day operation. Generally, strategies for reducing vulnerabilities fall into three broad categories: *Sound business practices* affect policies, procedures, and training to improve the overall security-related culture at the drinking water facility. For example, it is important to ensure that rapid communication capabilities exist between public health authorities and local law enforcement and emergency responders. *System upgrades* include changes in operations, equipment, processes, or infrastructure itself that make the system fundamentally safer. *Security upgrades* improve capabilities for detection, delay, or response.

Source: U.S. EPA (2002b)

APPENDIX 9.B SECURITY VULNERABILITY SELF-ASSESSMENT FOR SMALL WATER SYSTEMS

The first 15 questions in this vulnerability self-assessment are general questions designed to apply to all components of your system [wellhead or surface water intake, treatment plant, storage tank(s), pumps, distribution system, and offices]. These are followed by more specific questions that look at individual systems components in great detail.

Question	Answer	Comment	Action needed/taken
1. Do you have a written emergency response plan (ERP)?	Yes ☐ No ☐	Under the provisions of the Public Health Security and Bioterrorism Preparedness and Response Act of 2002, you are required to develop and/or update an ERP within 6 months after completing this assessment. If you do not have an ERP, you can obtain a sample from your state drinking water primacy agency. As a first step in developing your ERP, you should develop your emergency contact list. A plan is vital in case there is an incident that requires immediate response. Your plan should be reviewed at least annually (or more frequently if necessary) to ensure that it is up-to-date and addresses security emergencies including ready access to laboratories capable of analyzing water samples. You should coordinate with your LEPC. You should designate someone to be contacted in case of emergency regardless of the day of the week or time of day. This contact information should be kept up-to-date and made available to all water system personnel and local officials (if applicable). Share this ERP with police, emergency personnel, and your state primacy agency. Posting contact information is a good idea only if authorized personnel are the only ones seeing the information. These signs could pose a security risk if posted for public viewing since it gives people information that could be used against the system.	
2. Have you reviewed USEPA's baseline threat information document?	Yes ☐ No ☐	The USEPA baseline threat document is available through the Water Information Sharing and Analysis Center at www.waterisac.org. It is important you use this document to determine potential threats to your system and to obtain additional security-related information. USEPA should have provided a certified letter to your system that provided instructions on obtaining the threat document.	
3. Is access to the critical components of the water system (i.e., a part of the physical infrastructure of the system that is essential for water flow and/or water quality) restricted to authorized personnel only?	Yes ☐ No ☐	You should restrict or limit access to the critical components of your water system to authorized personnel only. This is the first step in security enhancement for your water system. Consider the following: Issue water system photo-identification cards for employees, and require them to be displayed within the restricted area at all times. Post signs restricting entry to authorized personnel and ensure that assigned staff escort people without proper ID.	

4. Are all critical facilities fenced, including wellhouses and pump pits, and are gates locked where appropriate?	Yes ☐ No ☐	Ideally, all facilities should have a security fence around the perimeter. The fence perimeter should be walked periodically to check for breaches and maintenance needs. All gates should be locked with chains and a tamperproof padlock that at a minimum protects the shank. Other barriers such as concrete "jersey" barriers should be considered to guard certain critical components from accidental or intentional vehicle intrusion.	
5. Are all critical doors, windows, and other points of entry such as tank and roof hatches and vents kept closed and locked?	Yes ☐ No ☐	Lock all building doors and windows, hatches, and vents, gates, and other points of entry to prevent access by unauthorized personnel. Check locks regularly. Deadbolt locks and lock guards provide a high level of security for the cost. A daily check of critical system components enhances security and ensures that an unauthorized entry has not taken place. Doors and hinges to critical facilities should be constructed of heavy-duty reinforced material. Hinges on all outside doors should be located on the inside. To limit access to water systems, all windows should be locked and reinforced with wire mesh or iron bars, and bolted on the inside. Systems should ensure that this type of security meets with the requirements of any fire codes. Alarms can also be installed on windows, doors, and other points of entry.	
6. Is there external lighting around all critical components of your water system?	Yes ☐ No ☐	Adequate lighting of the exterior of water systems' critical components is a good deterrent to unauthorized access and may result in the detection or deterrence of trespassers. Motion detectors that activate switches that turn lights on or trigger alarms also enhance security.	
7. Are warning signs (tampering, unauthorized access, etc.) posted on all critical components of your water system (e.g., well-houses and storage tanks)?	Yes ☐ No ☐	Warning signs are an effective means to deter unauthorized access. "Warning—tampering with this facility is a federal offense" should be posted on all water facilities. These are available from your state rural water association. "Authorized personnel only," "Unauthorized access prohibited," and "Employees only" are examples of other signs that may be useful.	
8. Do you patrol and inspect all source intake, buildings, storage tanks, equipment, and other critical components?	Yes ☐ No ☐	Frequent and random patroling of the water system by utility staff may discourage potential tampering. It may also help identify problems that may have arisen since the previous patrol. All systems are encouraged to initiate personal contact with the local law enforcement to show them the drinking water facility. The tour should include the identification of all critical components with an explanation of why they are important. Systems are encouraged to review, with local law enforcement, the *NRWA/ASDWA Guide for Security Decisions* or similar state document to clarify respective roles and responsibilities in the event of an incident. Also consider asking the local law enforcement to conduct periodic patrols of your water system.	

Question	Answer	Comment	Action needed/taken
9. Is the area around all the critical components of your water system free of objects that may be used for breaking and entering?	Yes ☐ No ☐	When assessing the area around your water system's critical components, look for objects that could be used to gain entry (e.g., large rocks, cement blocks, pieces of wood, ladders, valve keys, and other tools).	
10. Are the entry points to all of your water system easily seen?	Yes ☐ No ☐	You should clear fence lines of all vegetation. Overhanging or nearby trees may also provide easy access. Avoid landscaping that will permit trespassers to hide or conduct unnoticed suspicious activities. Trim trees and shrubs to enhance the visibility of your water system's critical components. If possible, park vehicles and equipment in places where they do not block the view of your water system's critical components.	
11. Do you have an alarm system that will detect unauthorized entry or attempted entry at all critical components?	Yes ☐ No ☐	Consider installing an alarm system that notifies the proper authorities or your water system's designated contact for emergencies when there has been a breach of security. Inexpensive systems are available. An alarm system should be considered whenever possible for tanks, pump houses, and treatment facilities. You should also have an audible alarm at the site as a deterrent and to notify neighbors of a potential threat.	
12. Do you have a key control and accountability policy?	Yes ☐ No ☐	Keep a record of locks and associated keys, and to whom the keys have been assigned. This record will facilitate lock replacement and key management (e.g., after employee turnover or loss of keys). Vehicle and building keys should be kept in a lockbox when not in use. You should have all keys stamped (engraved) "Do not duplicate."	
13. Are entry codes and keys limited to water system personnel only?	Yes ☐ No ☐	Suppliers and personnel from collocated organizations (e.g., organizations using your facility for telecommunications) should be denied access to codes and/or keys. Codes should be changed frequently if possible. Entry into any building should always be under the direct control of water system personnel.	
14. Do you have an updated operations and maintenance manual that includes evaluations of security systems?	Yes ☐ No ☐	Operation and maintenance plans are critical in assuring the ongoing provision of safe and reliable water service. These plans should be updated to incorporate security considerations and the ongoing reliability of security provisions, including security procedures and security-related equipment.	
15. Do you have a neighborhood watch program for your water system?	Yes ☐ No ☐	Watchful neighbors can be very helpful to a security program. Make sure that they know whom to call in the event of an emergency or suspicious activity.	

Water Sources

In addition to the preceding general checklist for your entire water system (questions 1 to 15), you should give special attention to the following issues, presented in separate tables, related to various water system components. Your water sources (surface water intakes or wells) should be secured. Surface water supplies present the greatest challenge. Typically they encompass large land areas. Where areas cannot be secured, steps should be taken to initiate or increase law enforcement patrols. Pay particular attention to surface water intakes. Ask the public to be vigilant and report suspicious activity.

Question	Answer	Comment	Action needed/taken
16. Are your wellheads sealed properly?	Yes ☐ No ☐	A properly sealed wellhead decreases the opportunity for the introduction of contaminants. If you are not sure whether your wellhead is properly sealed, contact your well drilling/maintenance company, your state drinking water primacy agency, your staff rural water association, or other technical assistance providers.	
17. Are well vents and caps screened and securely attached?	Yes ☐ No ☐	Properly installed vents and caps can help prevent the introduction of a contaminant into the water supply. Ensure that vents and caps serve their purpose, and cannot be easily breached or removed.	
18. Are observation/test and abandoned wells properly secured to prevent tampering?	Yes ☐ No ☐	All observation/test and abandoned wells should be properly capped or secured to prevent the introduction of contaminants into the aquifer or water supply. Abandoned wells should be either removed or filled with concrete.	
19. Is your surface water source secured with fences or gates? Do water system personnel visit the source?	Yes ☐ No ☐	Surface water supplies present the greatest challenge to secure. Often, they encompass large land areas. Where areas cannot be secured, steps should be taken to initiate or increase patrols by water utility personnel and law enforcement agents.	

Treatment Plant and Suppliers

Some small systems provide easy access to their water system for suppliers of equipment, chemicals, and other materials for the convenience of both parties. This practice should be discontinued.

Question	Answer	Comment	Action needed/taken
20. Are deliveries of chemicals and other supplies made in the presence of water system personnel?	Yes ☐ No ☐	Establish a policy that an authorized person, designated by the water system, must accompany all deliveries. Verify the credentials of all drivers. This prevents unauthorized personnel from having access to the water system.	
21. Have you discussed with your supplier(s) procedures to ensure the security of their products?	Yes ☐ No ☐	Verify that your suppliers take precautions to ensure that their products are not contaminated. Chain of custody procedures for delivery of chemicals should be reviewed. You should inspect chemicals and other supplies at the time of delivery to verify that they are sealed and in unopened containers. Match all delivered goods with purchase orders to ensure that they were, in fact, ordered by your water system. You should keep a log or journal of deliveries. It should include the driver's name (taken from the driver's photo ID), date, time, material delivered, and the supplier's name.	

Question	Answer	Comment	Action needed/taken
22. Are chemicals, particularly those that are potentially hazardous (e.g., chlorine gas) or flammable, properly stored in a secure area?	Yes ☐ No ☐	All chemicals should be stored in an area designated for their storage only, and the area should be secure and access to the area restricted. Access to chemical storage should be available only to authorized employees. Pay special attention to the storage, handling, and security of chlorine gas because of its potential hazard. You should have tools and equipment on site (such as a fire extinguisher, drysweep, etc.) to take immediate actions when responding to an emergency.	
23. Do you monitor raw and treated water so that you can detect changes in water quality?	Yes ☐ No ☐	Monitoring of raw and treated water can establish a baseline that may allow you to know if there has been a contamination incident. Some parameters for raw water include pH, turbidity, total and fecal coliform, total organic carbon, specific conductivity, ultraviolet adsorption, color, and odor. Routine parameters for finished water and distribution systems include free and total chlorine residual, heterotrophic plate count (HPC), total and fecal coliform, pH, specific conductivity, color, taste, odor, and system pressure. Chlorine demand patterns can help you identify potential problems with your water. A sudden change in demand may be a good indicator of contamination in your system. For those systems that use chlorine, absence of chlorine residual may indicate possible contamination. Chlorine residuals provide protection against bacterial and viral contamination that may enter the water supply.	
24. Are tank ladders, access hatches, and entry points secured?	Yes ☐ No ☐	The use of tamperproof padlocks at entry points (hatches, vents, and ladder enclosures) will reduce the potential for unauthorized entry. If you have towers, consider putting physical barriers on the legs to prevent unauthorized climbing.	
25. Are vents and overflow pipes properly protected with screens and/or grates?	Yes ☐ No ☐	Air vents and overflow pipes are direct conduits to the finished water in storage facilities. Secure all vents and overflow pipes with heavy-duty screens and/or grates.	
26. Can you isolate the storage tank from the rest of the system?	Yes ☐ No ☐	A water system should be able to take its storage tank(s) out of operation or drain its storage tank(s) if there is a contamination problem or structural damage. Install shutoff or bypass valves to allow you to isolate the storage tank in the case of a contamination problem or structural damage. Consider installing a sampling tap on the storage tank outlet to test water in the tank for possible contamination.	

Distribution

Hydrants are highly visible and convenient entry points into the distribution system. Maintaining and monitoring positive pressure in your system is important to provide fire protection and prevent introduction of contaminants.

Question	Answer	Comment	Action needed/taken
27. Do you control the use of hydrants and valves?	Yes ☐ No ☐	Your water systems should have a policy that regulates the authorized use of hydrants for purposes other than fire protection. Require authorization and backflow devices if a hydrant is used for any purpose other than firefighting. Consider designating specific hydrants for use as filling station(s) with proper backflow prevention (e.g., to meet the needs of construction firms). Then, notify local law enforcement officials and the public that these are the only sites designated for this use. Flush hydrants should be kept locked to prevent contaminants from being introduced into the distribution system, and to prevent improper use.	
28. Does your system monitor for, and maintain, positive pressure?	Yes ☐ No ☐	Positive pressure is essential for firefighting and for preventing backsiphonage that may contaminate finished water in the distribution system. Refer to your state primacy agency for minimum drinking water pressure requirements.	
29. Has your system implemented a backflow prevention program?	Yes ☐ No ☐	In addition to maintaining positive pressure, backflow prevention programs provide an added margin of safety by helping to prevent the intentional introduction of contaminants. If you need information on backflow prevention programs, contact your state drinking water primacy agency.	

Personnel

You should add security procedures to your personnel policies.

Question	Answer	Comment	Action needed/taken
30. When hiring personnel, do you request that local police perform a criminal background check, and do you verify employment eligibility (as required by the Immigration and Naturalization Service, Form I-9)?	Yes ☐ No ☐	It is good practice to have all job candidates fill out an employment application. You should verify professional references. Background checks conducted during the hiring process may prevent potential employee-related security issues. If you use contract personnel, check on the personnel practices of all providers to ensure that their hiring practices are consistent with good security practices.	
31. Are your personnel issued photo-identification cards?	Yes ☐ No ☐	For positive identification, all personnel should be issued water system photo-identification cards and be required to display them at all times. Photo identification will also facilitate identification of authorized water system personnel in the event of an emergency.	

Question	Answer	Comment	Action needed/taken
32. When terminating employment, do you require employees to turn in photo IDs, keys, access codes, and other security-related items?	Yes ☐ No ☐	Former or disgruntled employees have knowledge about the operation of your water system, and could have both the intent and the physical capability to harm your system. Requiring employees who will no longer be working at your water system to turn in their IDs, keys, and access codes helps limit these types of security breaches.	
33. Do you use uniforms and vehicles with your water system name prominently displayed?	Yes ☐ No ☐	Requiring personnel to wear uniforms, and requiring that all vehicles prominently display the water system name, helps inform the public when water system staff is working on the system. Any observed activity by personnel without uniforms should be regarded as suspicious. The public should be encouraged to report suspicious activity to law enforcement authorities.	
34. Have water system personnel been advised to report security vulnerability concerns and to report suspicious activity?	Yes ☐ No ☐	Your personnel should be trained and knowledgeable about security issues at your facility, what to look for, and how to report any suspicious event or activity. Periodic meetings of authorized personnel should be held to discuss security issues.	
35. Do your personnel have a checklist to use for threats or suspicious calls or to report suspicious activity?	Yes ☐ No ☐	To properly document suspicious or threatening phone calls or reports of suspicious activity, a simple checklist can be used to record and report all pertinent information. Calls should be reported immediately to appropriate law enforcement officials. Checklists should be available at every telephone. Also consider installing caller ID on your telephone system to keep a record of incoming calls.	

Information Storage, Computers, Controls, and Maps

Security of the system, including computerized controls such as a supervisory control and data acquisition (SCADA) system, goes beyond the physical aspects of operation. It also includes records and critical information that could be used by someone planning to disrupt or contaminate your water system.

Question	Answer	Comment	Action needed/taken
36. Is computer access "password protected"? Is virus protection installed and software upgraded regularly and are your virus definitions updated at least daily? Do you have Internet firewall software installed on your computer? Do you have a plan to back up your computers?	Yes ☐ No ☐	All computer access should be password protected. Passwords should changed every 90 days and (as needed) following employee turnover. be When possible, each individual should have a unique password that is not shared with others. If you have Internet access, a firewall protection program should be installed on your side of the computer and reviewed and updated periodically. Also consider contacting a virus protection company and subscribing to a virus update program to protect your records. Backing up computers regularly will help prevent the loss of data in the event that your computer is damaged or breaks. Backup copies of computer data should be made routinely and stored at a secure off-site location.	

37. Is there information on the Web that can be used to disrupt your system or contaminate your water?	Yes ☐ No ☐	Posting detailed information about your water system on the web site may make the system more vulnerable to attack. Web sites should be examined to determine whether they contain critical information that should be removed. You should do a Web search (using a search engine such as Google, Yahoo!, or Lycos) using key words related to your water supply to find any published data on the Web that is easily accessible by someone who may want to damage your water supply.	
38. Are maps, records, and other information stored in a secure location?	Yes ☐ No ☐	Records, maps, and other information should be stored in a secure location when not in use. Access should be limited to authorized personnel only. You should make backup copies of all data and sensitive documents. These should be stored in a secure off-site location on a regular basis.	
39. Are copies of records, maps, and other sensitive information labeled confidential, and are all copies controlled and returned to the water system?	Yes ☐ No ☐	Sensitive documents (e.g., schematics, maps, and plans and specifications) distributed for construction projects or other uses should be recorded and recovered after use. You should discuss measures to safeguard your documents with bidders for new projects.	
40. Are vehicles locked and secured at all times?	Yes ☐ No ☐	Vehicles are essential to any water system. They typically contain maps and other information about the operation of the water system. Water system personnel should exercise caution to ensure that this information is secure. Water system vehicles should be locked when they are not in use or left unattended. Remove any critical information about the system before parking vehicles for the night. Vehicles also usually contain tools (e.g., valve wrenches) and keys that could be used to access critical components of your water system. These should be secured and accounted for daily.	

Public Relations

You should educate your customers about your system. You should encourage them to be alert and to report any suspicious activity to law enforcement authorities.

Question	Answer	Comment	Action needed/taken
41. Do you have a program to educate and encourage the public to be vigilant and report suspicious activity to assist in the security protection of your water system?	Yes ☐ No ☐	Advise your customers and the public that your system has increased preventive security measures to protect the water supply from vandalism. Ask for their help. Provide customers with your telephone number and the telephone number of the local law enforcement authority so that they can report suspicious activities. The telephone number can be made available through direct mail, billing inserts, notices on community bulletin boards, flyers, and consumer confidence reports.	
42. Does your water system have a procedure to deal with public information requests, and to restrict distribution of sensitive information?	Yes ☐ No ☐	You should have a procedure for personnel to follow when you receive an inquiry about the water system or its operation from the press, customers, or the general public. Your personnel should be advised not to speak to the media on behalf of the water system. Only one person should be designated as the spokesperson for the water system. Only that person should respond to media inquiries. You should establish a process for responding to inquiries from your customers and the general public.	
43. Do you have a procedure in place to receive notification of a suspected outbreak of a disease immediately after discovery by local health agencies?	Yes ☐ No ☐	It is critical to be able to receive information about suspected problems with the water at any time and respond to them quickly. Written procedures should be developed in advance with your state drinking water primacy agency, local health agencies, and your local emergency planning committee and reviewed periodically.	
44. Do you have a procedure in place to advise the community of contamination immediately after discovery?	Yes ☐ No ☐	As soon as possible after a disease outbreak, you should notify testing personnel and your laboratory of the incident. In outbreaks caused by microbial contaminants, it is critical to discover the type of contaminant and its method of transport (water, food, etc.). Active testing of your water supply will enable your laboratory, working in conjunction with public health officials, to determine if there are any unique (and possibly lethal) disease organisms in your water supply. It is critical to be able to get the word out to your customers as soon as possible after discovering a health hazard in your water supply. In addition to your responsibility to protect public health, you must also comply with the requirements of the Public Notification Rule. Some simple methods include announcements via radio or television, door-to-door notification, a phone tree, and posting notices in public places. The announcement should include accepted uses for the water and advice on where to obtain safe drinking water. Call large facilities that have large populations of people who might be particularly	

		threatened by the outbreak, such as hospitals, nursing homes, the school district, jails, large public buildings, and large companies. Enlist the support of local emergency response personnel to assist in the effort.	
45. Do you have a procedure in place to respond immediately to a customer complaint about a new taste, odor, color, or other physical change (oily, filmy, burns on contact with skin)?	Yes ☐ No ☐	It is critical to be able to respond to and quickly identify potential water quality problems reported by customers. Procedures should be developed in advance to investigate and identify the cause of the problem, as well as to alert local health agencies, your state drinking water primacy agency, and your local emergency planning committee if you discover a problem.	

Now that you have completed the *Security Vulnerability Self-Assessment Guide for Small Water Systems Serving Populations Between 3,300 and 10,000*, review your needed actions and then prioritize them on the basis of the most likely threats.

Source: www.asdwa.org

APPENDIX 9.C WATER UTILITY RESPONSE, RECOVERY, AND REMEDIATION GUIDELINES

I. Contamination Event: (Articulated Threat with Unspecified Material)					
Event description: This event is based on the threat of intentional introduction of a contaminant into the water system (at any point within the system) without specification of the contaminant by the perpetrator.					
Initial notifications	Notify local law enforcement Notify local FBI field office Notify National Response Center	Notify local/state emergency management organization Notify ISAC	Notify other associated system authorities (wastewater, water) Notify local government official	Notify local/state health and/or environmental department Notify critical care facilities	Notify employees Consider when to notify customers and what notifications to issue Notify state governor
Response actions	**Source water**	**Drinking water treatment facility**	**Water distribution/storage**	**Wastewater collection system**	**Wastewater treatment facility**
	Increase sampling at or near system intakes Consider whether to isolate the water source if possible	Preserve latest full-battery background test as baseline Increase sampling efforts Consider whether to continue normal operations (if determination is made to reduce or stop water treatment, provide notification to customers and issue alerts) Coordinate alternative water supply	Consider whether to isolate the water in the affected area if possible	Assess what to do with potentially contaminated water within the system based on contaminant, contaminant concentration, potential for system contamination, and ability to bypass treatment plant If bypassed, notify local and appropriate state authorities, and downstream users; increase monitoring of receiving stream	Preserve latest full-battery background test as baseline Increase sampling efforts Consider whether to continue normal operations (if determination is made to reduce or stop water treatment, provide notification to customers and issue alerts)
Recovery actions	Recovery actions should begin once the contaminant is through the system.				
Recovery notifications	Notify customers Notify media Notify ISAC				

Appropriate utility elements	Sample appropriate system elements (storage tanks, filters, sediment basins, solids handling) to determine if residual contamination exists	Flush system on the basis of results of sampling Monitor health of employees	Plan for appropriate disposition of personal protection equipment (PPE) and other equipment
Remediation actions	On the basis of sampling results, assess need to remediate storage tanks, filters, sediment basins, solids handling	Plan for appropriate disposition of PPE and other equipment	If wastewaster treatment plant was bypassed, sample and establish monitoring regime for receiving stream and potential remediation based on sampling results

Note: Response, recovery and remediation actions may be tailored to a specified (identified) material if the physical properties for the material are known.

II. Contamination Threat at a Major Event

II. Contamination Threat at a Major Event					
Event description: This event is based on the threat of, or actual, intentional introduction of a contaminant into the water system at a sports arena, convention center, or similar facility.					
Initial notifications	Notify local law enforcement Notify local FBI field office Notify National Response Center Notify ISAC	Notify local/state emergency management organization Notify ISAC Notify wastewater facility Notify state governor	Notify other associated system authorities (wastewater, water) Notify local government official	Notify local/state health and/or environmental department Notify critical care facilities	Notify employees Consider when to notify customers and what notifications to issue
Response actions	**Source water**	**Drinking water treatment facility**	**Water distribution/storage**	**Wastewater collection system**	**Wastewater treatment facility**
	No recommended action to take	No recommended action to take	Coordinate isolation of water Assist in plan for draining the contained water Assist in developing a plan for sampling water for potential contamination based on threat notification Provide alternate water source	Coordinate acceptance of isolated water Monitor accepted water Assist in plan for draining the contained water Assist in developing a plan for sampling water for potential contamination based on threat notification	
Recovery actions	Recovery actions should begin once the contaminant is through the system.				
Recovery notifications	Notify customers in the area of the facility of actions to take Notify customers in affected area once contaminant-free clean water is reestablished Notify downstream users such as water suppliers, irrigators, and electric generating plants				
Water distribution/Storage	Consider flushing system via hydrants in distribution systems				
Remediation actions	**Water distribution/storage**	Assess need to decontaminate or replace distribution system components			
	Wastewater treatment plant	On the basis of sampling results, assess need to remediate storage tanks, filters, sediment basins, solids handling	Plan for appropriate disposition of PPE and other equipment	If wastewater treatment plant was bypassed, sample and establish monitoring regime for receiving stream and potential remediation based on sampling results	

III. Notification from Health Officials of Potential Water Contamination

Event description: This event is based on the water utility being notified by public health officials of potential contamination based on symptoms of patients.

	Source water	Drinking water treatment facility	Water distribution/storage	Wastewater collection system	Wastewater treatment facility
Initial notifications	Ask notifying official who else has been notified and request information on symptoms, potential contaminants, and potential area affected	Notify local law enforcement Notify local FBI field office Notify National Response Center Notify local/state emergency management organization	Notify other associated system authorities (wastewater, water) Notify local government official Notify state governor	Notify local/state health and/or environmental department Notify critical care facilities	Notify employees Consider when to notify customers and what notifications to issue Notify ISAC
Response actions	Increase sampling at or near system intakes Consider whether to isolate	Preserve latest full-battery background test result as baseline Increase sampling efforts Consider whether to continue normal operations (if determination is to reduce or stop water treatment, provide notification to customers and issue alerts) Coordinate alternative water supply (if needed)	Increase sampling in the area potentially affected and at locations where the contaminant could have migrated to; it is important to consider the time between exposure and onset of symptoms to select sampling sites Consider whether to isolate Consider whether to increase residual disinfectant levels	Increase sampling at pump stations, specifically in the area potentially affected Assess what to do with potentially contaminated water within the system according to contaminant, contaminant concentration, potential for system contamination, and ability to bypass treatment plant If bypassed, notify local and appropriate state authorities, downstream users (especially drinking water treatment facilities), and increase monitoring of receiving stream	
Recovery actions	Recovery actions should begin once the contaminant is through the system.				
Recovery notifications	Assist health department with notifications to customers, media, downstream users, and other organizations				
Appropriate utility elements	Sample appropriate system elements (storage tanks, filters, sediment basins, solids handling) to determine if residual contamination exists		Flush system according to results of sampling Monitor health of employees	Plan for appropriate disposition of personal protection equipment (PPE) and other equipment	
Remediation actions	On the basis of sampling results, assess need to remediate storage tanks, filters, sediment basins, solids handling, and drinking water distribution		Plan for appropriate disposition of PPE and other equipment	If wastewater treatment plant was bypassed, sample and establish monitoring regime for receiving stream and potential remediation based on sampling results	

Note: Patient symptoms should be used to narrow the list of potential contaminants.

IV. Intrusion through Supervisory Control and Data Acquisition (SCADA)					
Event description: This event is based on internal or external intrusion of the SCADA system to disrupt normal water system operations.					
Initial notifications	Notify local law enforcement Notify local FBI field office	Notify National Infrastructure Protection Center (NIPC) at 1-888-585-9078 (or 202-323-3204/5/6)	Notify other associated system authorities (wastewater, water) Notify employees	If the water is assessed to be unfit for consumption, consider when to notify customers and what notification to issue	
Response actions	**Source water**	**Drinking water treatment facility**	**Water distribution/storage**	**Wastewater collection system**	**Wastewater treatment facility**
	Increase sampling at or near system intakes Consider whether to isolate	Preserve latest full-battery background test result as baseline Increase sampling efforts Temporarily shut down SCADA system and go to manual operation using established protocol Consider whether to shut down system and provide alternate water	Monitor unmanned components (storage tanks and pumping stations) Consider whether to isolate	Temporarily shut down SCADA system and go to manual operation using established protocol Monitor unmanned components (pumping stations)—required only if wastewater SCADA system is compromised If SCADA intrusion caused release of improperly treated water, consider whether to continue normal operations (if determination is made to reduce or stop water treatment, provide notification to customers/issue alerts)	
Recovery actions	Recovery actions should begin once the intrusion has been eliminated and the contaminant/unsafe water (if this occurs) is through the system.				
Recovery notifications	Employees Local law enforcement Notify customers and media if the event resulted in contamination and the full range (see scenario I) of standard notifications were made				
Appropriate utility elements	With FBI assistance, make an image copy of all system logs to preserve evidence	With FBI assistance, check for implanted backdoors and other malicious code and eliminate them before restarting SCADA system		Install safeguards before restarting SCADA Bring SCADA system up and monitor system	
Remediation actions	Assess and/or implement additional protections for SCADA system			Check for an NIPC water sector warning, based on the intrusion that may contain additional protective actions to be considered; NIPC warnings can be found at www.NIPC.gov or https://www.infragard.org for secure access Infragard members	

V. Significant Structural Damage Resulting from an Intentional Act

Event description: This event is based on intentional structural damage to water system components to disrupt normal system operations.

	Source water	Drinking water treatment facility	Water distribution/storage	Wastewater collection system	Wastewater treatment facility
Initial notifications	Notify local law enforcement Notify local FBI field office Notify National Response Center	Notify local/state emergency management organization Notify state governor Notify ISAC	Notify other associated system authorities (wastewater, water) Notify local government officials	Notify local/state health and/or environmental department Notify critical care facilities	Notify employees Consider when to notify customers and what notification to issue
Response actions	Deploy damage assessment teams; if damage appears to be intentional, then treat as crime scene—consult local/state law enforcement and FBI on evidence preservation Inform law enforcement and FBI of potential hazardous materials Coordinate alternative water supply, as needed Consider increasing security measures Based on extent of damage, consider alternate (interim) treatment schemes to maintain at least some level of treatment				
Recovery actions	Recovery actions should begin as soon as practical after damaged facility is isolated from the rest of the utility facilites.				
Recovery notifications	Employees Local law enforcement			Notify local FBI office	
Appropriate utility elements	Dependent on the feedback from damage assessment teams			Implement damage recovery plan	
Remediation actions	Repair damage			Assess need for additional protection/security measures for damaged facility and other critical facilities within the utility	

INDEX

Note: **Boldface** numbers indicate illustrations and tables.

AACM Company, 3.2
Accuracy of sampling, 1.4
Adaptability, climate change and, 8.10
Africa, performance indicators for, 7.10–7.11
African Water Supply and Sanitation Utilities (WSSU), 7.11
Aggregate data, 1.7
Aggregate time series data, 1.9
Agricultural droughts, 8.5
  economic aspects of water shortage and, 2.18–2.20
Air temperature, 1.27
*Alegre, **Helena,*** 7.1
American Water Works Association Research Foundation (AWWARF), 3.2, 6.2, 7.13–7.14
Analysis of water savings, 1.30–1.36
AND mode, in fault tree analysis, 9.19, **9.20**
Apparent losses performance indicators, 7.29, **7.30, 7.31, 7.34**
AQUATOOL, 3.20–3.21, 3.22, 3.24
Arrangement of data, 1.8–1.9
Artificial intelligence, computer programs for integrated management, 3.1, 3.22
Asian Development Bank data book for performance indicators, 7.10, **7.11**
Atmosphere, climate change and, 8.1
Auditor use of performance indicators, 7.4
Austin, Texas, water system optimization and example of, 5.18–5.21, **5.18, 5.20**
Authorized consumption performance indicators, 7.28, **7.34**
Autocorrelation errors, 1.29–1.30
Automatic meter reading (AMR), 1.10
Availability performance indicators, 7.12
Availability defined, water distribution systems, 6.5, 6.9–6.11, **6.10**
Availability vs. demand, 2.7–2.8, **2.7, 2.8**
Average availability defined, 6.5
Average rates of water use in, 1.21–1.22, **1.21, 1.22, 1.23–1.24**

Benchmarking using performance indicators, 7.4, 7.7, 7.10
Benchmarking Water and Sanitation Utilities (BWSU), 7.10
Best linear unbiased estimates (BLUE), 1.25
Billing:
  bimonthly cycle of, 1.18–1.19
  data obtained from, 1.10
  monthly cycle of, 1.16–1.19, **1.16, 1.17**
Bimonthly billing cycle, 1.18–1.19
Biological threats to safety and security of water supply, 9.4, **9.5, 9.6**
Biospheres and climate change, marine and terrestrial, 8.3
Branched systems, design of water distribution systems and, 4.3–4.5, **4.4**
Breaks in water distribution system, 6.2

California Department of Water Resources, 3.18
California, seasonal water use, **1.14**
CALSIM, 3.18, **3.19**
Carbon dioxide and climate change and, 8.3, **8.6**
Categorical data, 1.7–1.8
Census data, **1.12**
Center for Advanced Decision Support for Water and Environmental Systems (CADSWES), 3.18, 3.24
Center for Low Cost Water and Sanitation (CREPA), 7.10
Center for Training, Research and Networking for Development (TREND), 7.10
Center for Water Policy Research, Australia, 3.2
Central limit theorem, 1.32
Chance constraint, in reliability and availability analysis, 6.28
Chemical threats to safety and security of water supply, 9.4, **9.5**
Chlorine content, 5.1
  booster station optimization, 5.22–5.25, **5.24, 5.25**
  optimization of, 6.27–6.28
Climate change effects on water management, 8.1–8.18
  atmosphere in, 8.1

Climate change effects on water management (*Cont.*):
biospheres in, marine and terrestrial, 8.3
carbon dioxide and, 8.3, **8.6**
climate system explained and, 8.1–8.3, **8.2**
cryosphere in, 8.3
defining change in, 8.3
droughts and, 8.5, **8.7**, 8.8
effects of change and, 8.5–8.13
energy balance of Earth and, 8.3, **8.4**
evaporation and transpiration in, 8.5
general circulation models (GCM) in, 8.3, 8.5
global climate graphs, **8.6**
greenhouse effect and, 8.3, 8.5
groundwater and, 8.9
hydrologic cycle in, 8.5
hydrosphere in, 8.3
literature related to, 8.1
numerical weather prediction models (NWPMs) in, 8.3
photosynthetic process in, 8.3
predicting change, 8.3–8.5, 8.3
regional effects of, 8.10–8.13
river basin/regional runoff effects of, 8.10–8.13, **8.12, 8.13, 8.14**
runoff, floods, droughts and, 8.8
sensitivity, adaptability, vulnerability assessment in, 8.10
snowfall and snowmelt levels and, 8.7–8.8
soil moisture and, 8.5–8.7
Special Report on Emission Scenarios (SRES) in, 8.5
stability of climate and, 8.3
storm frequency and, 8.8
streamflow variability and, 8.8–8.9
uncertainties in, 8.13
water management options and, 8.13, **8.14**
water resource systems effects and, 8.9–8.10
Climate system explained, 8.1–8.3, **8.2**
Cluster sampling, 1.4
Cohen condition, pump optimization, 5.15
Coliform bacteria in drinking water, 5.1
Commercial, institutional, industrial water use, 1.12–1.13
economic aspects of water shortage and, 2.18–2.20
Community water systems in U.S., **1.3**
Comparison of means method, 1.31–1.33
Component (mechanical) failure mode, in water distribution systems, 6.3, 6.4, 6.5–6.11
Components of water demand, 1.13–1.19
Computer programs for integrated management, 3.1–3.26
AQUATOOL in, 3.20–3.21, 3.22, 3.24
artificial intelligence in, 3.1, 3.22
CALSIM in, 3.18, **3.19**
conjunctive stream-aquifer management in, 3.18

Computer programs for integrated management (*Cont.*):
database and database management system (DBMS) in, 3.16
decision support systems (DSS) in, 3.1–3.2, 3.15–3.21, **3.15, 3.16**, 3.22–3.23, **3.23**
differential dynamic programming (DDP) in, 3.12, 3.13
discrete time optimal control in, 3.9
EPANET and, 3.6, 3.14
evolution of, 3.6, **3.7**
genetic algorithms in, 3.12–3.13, **3.12**
global minimum in, 3.13, **3.13**
GRG2 programming codes in, 3.11
groundwater management subsystems in, 3.10
heuristic search methods vs. optimization techniques, 3.13–3.14
hydroinformatics and, 3.6, **3.7**, 3.22
HYDRUS flow model in, 3.13–3.14
integrated hydrosystems management and, 3.2–3.6
interfacing optimization and simulation programs in, 3.10–3.14, **3.11**
linear programming (LP) in, 3.18
mathematical programming in, 3.11, 3.13–3.14
mixed integer linear programming (MILP) in, 3.18
modeling for, 3.6–3.7
nonlinear programming (NLP) in, 3.8, 3.11
optimization formulations in, 3.8–3.10, **3.8**
Power and Reservoir System Model (PRSYM) in, 3.17–3.18
PROSIM, 3.18
prospects for, 3.14
prospects of, 3.22–3.23
reservoir system operation for water supply and, 3.9–3.10
RiverWare in, 3.18–3.20, **3.20**, 3.22, 3.24
SANJASM, 3.18
simulated annealing (SA) in, 3.12–3.13
simulation in, 3.6–3.7, **3.7**, 3.10–3.14
state of practice in, 3.21–3.22
successive approximate linear quadratic regulator (SALQR) method in, 3.12, 3.13–3.14
supervisory control automated data acquisition (SCADA) in, 3.14, 3.22, 5.22
total water management (TWM) and, 3.3, **3.4**
Trinity River Basin, Texas, DSS, 3.16–3.17, **3.17**
TVA Environment and River Resource Aid (TERRA) DSS, 3.17, 3.22, 3.24
water distribution system operation and, 3.9
Water Resources Engineering Simulation Language (WRESL) in, 3.18
watershed approach to management in, 3.5
WaterWare in, 3.21, 3.22, 3.24
Conditional probability approach, in safety and security of water supply, 9.15, **9.17**

Confidence grading scheme, using performance indicators, 7.20, **7.24**
Conjunctive stream-aquifer management, 3.18
Conservation and saving water:
  analysis of, 1.30–1.36
  background of, 2.3–2.11
  end use accounting systems in, 1.34–1.36, **1.35, 1.36**
  importance of, 3.5–3.6
  metering and, 2.4
  outside effects or externalities in, 1.31
  price elasticity approach to, 2.3, 2.4–2.6, **2.6**
  risk-based approach to, 2.3, 2.8
  statistical estimation of, 1.31–1.33
  time series analysis of, 1.33–1.34
  total water management (TWM) and, 3.3, **3.4**
  use restrictions and, 2.4
Conservation equations for design of water distribution systems, 4.3
Constant failure rate, constant repair rate, water distribution systems, 6.10–6.11
Constant/common variance, 1.24
Constraints, on pump optimization, 5.4
Consumer price index (CPI), 1.27
Contamination threat at major events, **9.42**
Context information and performance indicators, 7.2–7.3, **7.3**, 7.41
Contingency water distribution systems, 6.5
Continuous data, sample size determination for, 1.5–1.7
Continuous variables, 1.7
Contractual agreements and use of performance indicators, 7.5
Cost of water, 1.27
  drought management and (*see* Drought management)
  price elasticity approach to, 2.3, 2.4–2.6, **2.6**
  risk-price relationship in, 2.11–2.17, **2.12, 2.13, 2.18, 2.19**
Costs performance indicators, 7.12
Critical paths, in fault tree analysis, 9.19
Cross sectional data, 1.8–1.9
Cryosphere and climate change, 8.3
Cumulative probability distribution, 6.28
Customer billing data, 1.10
Customer premise categories, 1.13
Customer satisfaction performance indicators, 7.11
Customer time series data, 1.9
Cut-set analysis, 9.13–9.15, **9.14**
  reliability and availability analysis, minimum for node/system, 6.14, **6.15**, 6.16
Cyber threats to safety and security of water supply, 9.2, 9.4

Damage assessment, drought management and, 2.20–2.22, **2.22**
Data arrangements, 1.8–1.9
Data reliability and accuracy performance indicators, 7.19–7.20
Data scales, 1.7–1.8
Data set development, 1.7–1.9
Database and database management system (DBMS), computer programs for integrated management, 3.16
Decision support systems (DSS), 3.22–3.23, **3.23**
  AQUATOOL in, 3.20–3.21, 3.22, 3.24
  computer programs for integrated management, 3.1–3.2, 3.15–3.21, **3.15, 3.16**
  conjunctive stream-aquifer management in, 3.18
  database and database management system (DBMS) in, 3.16
  linear programming (LP) in, 3.18
  mixed integer linear programming (MILP) in, 3.18
  Power and Reservoir System Model (PRSYM) in, 3.17–3.18
  RiverWare in, 3.18–3.20, **3.20**, 3.22, 3.24
  structure of, 3.15, **3.16**
  Trinity River Basin, Texas, 3.16–3.17, **3.17**
  TVA Environment and River Resource Aid (TERRA) in, 3.17, 3.22, 3.24
  Water Resources Engineering Simulation Language (WRESL) in, 3.18
  WaterWare in, 3.21, 3.22, 3.24
Defining climate change, 8.3
Demand (*see* Water demand)
Dependent variables, 1.8, **1.8**
Design of water distribution system, 4.1–4.13, 9.1–9.2, **9.2, 9.3**
  applications for, 4.8–4.10
  branched systems in, 4.3–4.5, **4.4**
  chance constraint in, 6.28
  conservation equations for, 4.3
  constraints in, 4.2–4.2
  demand patterns and, 4.2
  duality gap in, 4.6
  dynamic programming in, 4.3
  flow rate calculation in, 4.5
  general system design via NLP in, 4.6–4.7
  genetic algorithms in, 4.3, 4.4, 4.7–4.8
  geometric programming in, 4.3
  hierarchy of components in, **9.3**
  integer programming in, 4.3
  layout of system for, 4.2
  linear programming (LP) in, 4.3, 4.4, 4.5–4.6, 4.8
  looped pipe systems via linearization in, 4.5–4.6
  mathematical formulation of, 4.2–4.3
  New York City example of, 4.8–4.10, **4.9, 4.10**
  nonlinear programming (NLP) in, 4.3, 4.6–4.7, **4.7**, 4.8
  operational parameters for, 4.2–4.3
  operations defined for, 4.2
  optimization methods in, 4.3–4.8

Design of water distribution system (*Cont.*):
optimization-simulation model and NLP in, 4.6–4.7, **4.7**
pipe sizing calculations in, 4.5–4.6
problem definition in, 4.1–4.2
reliability-based, 6.26–6.27
simulated annealing in, 4.3, 4. 7–4.8
stochastic search techniques in, 4.3, 4.7–4.8
Determinants of urban water demand, 1.20, **1.21**
Differential dynamic programming (DDP), computer programs for integrated management, 3.12, 3.13
Digital Elevation Model (DEM), 3.22
Disaggregate data, 1.7
Disaggregate estimate of water use, 1.20
Discrete step function, 1.26
Discrete time optimal control, computer programs for integrated management, 3.9
Discrete variables, 1.7
Dispersion bands, performance indicators, 7.13
Displacement meters, 1.1–1.2
Distribution (*see* Design of water distribution systems)
Distribution input performance indicators, 7.28
Distribution repair definitions, 6.2–6.3
Double log models, 1.28
Drought function in calculations, 2.23
Drought management, 2.1–2.34
availability vs. demand, 2.7–2.8, **2.7, 2.8**
background of water conservation in, 2.3–2.11
chi square distributions in evaluating risk for, 2.26–2.31, **2.28, 2.29**
classification of droughts, 8.5
climate change and, 8.5, **8.7**, 8.8
damage assessment and, 2.20–2.22, **2.22**
demand adjustment flowchart, 2.26–2.27, **2.27**
demand models and, 2.6–2.8
drought function in calculations, 2.23
droughts defined and classified, 8.5
economic aspects of water shortage and, 2.18–2.20
emergency relief measures in, 2.19, **2.21**
evaluation of risk in, 2.13–2.17, **2.16**, 2.26–2.31, **2.28, 2.29**
forecasting drought in, 2.10–2.11
increased prices and, 2.22–2.23
lessons learned through, 2.1
metering and, 2.4
operation and management planning under, 2.17–2.31, **2.20**
options for, **2.2–2.3**, 2.3–2.4
Palmer Drought Severity Index (PDSI) and, 2.10
Phoenix Arizona, example of, 2.29–2.31, **2.30, 2.31**
price elasticity approach to, 2.3, 2.4–2.6, **2.6**
return periods and risk values, 2.17, **2.17, 2.20**
risk-based approach to, 2.3, 2.8
risk safety factor relationship in, 2.25–2.26, **2.25, 2.26**, 2.30, **2.31**, 2.32

Drought management (*Cont.*):
risk-price relationship in, 2.11–2.17, **2.12, 2.13, 2.18, 2.19**
severity of drought as risk index, 2.9–2.11
Sheer Steila Drought Index (DI) in, 2.10
Southern Oscillation Index (SOI) in, 2.10
Surface Water Supply Index (SWSI) in, 2.10–2.11, 2.23–2.26, **2.24**
uncertainty and risk in demand and, 2.23–2.26, **2.24**
use restrictions and, 2.4
willingness to pay levels in, 2.18
Duality gap, design of water distribution systems and, 4.6
Dutch contact club for water companies, use of performance indicators, 7.12
Dynamic programming, 5.2
design of water distribution systems and, 4.3
*Dziegielewski, **Benedykt***, 1.1

Early life period, and failure rate, 6.6
Economic aspects of water shortage, 2.18–2.20, 7.1
Economic indices in failure of water distribution systems, 6.5
Economy performance indicators, 7.12
Efficiency calculations, pumps, 6.27–6.28
*Ejeta, **Messele Z.***, 2.1, 3.1
Electronic remote meter reading (ERMR), 1.10
Emergency demand conditions, reliability and availability analysis, 6.2
Emergency relief measures, drought management and, 2.19, **2.21**
Emergency response planning, recovery, remediation, in safety and security of water supply, 9.11, **9.40–9.41**
End use accounting systems, 1.34–1.36, **1.35, 1.36**
End uses of water, 1.19
Energy balance of Earth and climate change, 8.3, **8.4**
Engineering approach to performance indicators, 7.12–7.13
Environment performance indicators, 7.12
Environmental Advisory Board (EAB), 3.3
EPANET, 5.16, **5.16, 5.17**, 5.21
chlorine booster station optimization in, 5.23
computer programs for integrated management, 3.6, 3.14
reliability and availability analysis, 6.28
water system optimization and, 5.2, 5.5, 5.12
Error components, in regression analysis, 1.30
Error terms, 1.27
Estimated generalized least squares (EGLS) regression, 1.30
Estimating water use, 1.1–1.2, 1.20–1.30, 1.31–1.33
Evaluation of risk, drought management and, 2.13–2.17, **2.16**, 2.26–2.31, **2.28, 2.29**
Evaporation, climate change and, 8.5
Event symbols, in fault tree analysis, 9.19, **9.21**

Every *k*th systematic samples, 1.4
Exponential models, 1.28
Externalities, 1.31

Failure density, 6.6
  water distribution systems, 6.6–6.9
Failure modes in water distribution systems, 6.1–6.5
Failure probability, in water distribution systems, 6.6
Failure rate, in water distribution systems, 6.6–6.9, **6.7**
Financial performance indicators, **7.23–7.24**, 7.51
Financial risk, 2.8–2.9
Financing bodies, use of performance indicators, 7.4
Floods, climate change and, 8.8, 155
Flow rate calculation:
  design of water distribution systems and, 4.5
  pump optimization, 5.3–5.5
  volume deficits in, 6.23–6.26
Force of mortality function (*see* Instantaneous failure rate)
Forecasting climate change, 8.3–8.5
Forecasting drought, 2.10–2.11
Fourier series, 1.26
Frequency and duration indices in failure of water distribution systems, 6.5

Gate symbols, in fault tree analysis, 9.19, **9.20**
General performance indicators, 7.12
General algebraic modeling system (GAMS), water system optimization and, 5.2
General circulation models (GCM), climate change and, 8.3, 8.5
Generalized least squares (GLS) regression, 1.30
Generalizing function performance indicators, 7.13
Genetic algorithms:
  computer programs for integrated management, 3.12–3.13, **3.12**
  design of water distribution systems and, 4.3, 4.4, 4.7–4.8
Geographic information systems (GIS), 1.11
Geometric programming, design of water distribution systems and, 4.3
Global climate change graphs, **8.6**
Global minimum, computer programs for integrated management, 3.13, **3.13**
*Goldman, **Fred E.***, 5.1
Great Man Made River Scheme, **3.23**
Greenhouse effect and climate change, 8.3
GRG2 programming codes, computer programs for integrated management, 3.11, 5.5, 5.7, 5.12
Groundwater and climate change, 8.9
Groundwater management, computer programs for integrated management, optimization, 3.10
Guaranteed standards schemes (GSS) and performance indicators, 7.5

Hazard function (*See* instantaneous failure rate)
Heavy metal content of drinking water, 5.1
Heteroskedasticity, 1.29–1.30
Heuristic search methods vs. optimization techniques, computer programs for integrated management, 3.13–3.14
Hierarchy of components, water distribution system, **9.3**
Homeland Security Act, safety and security of water supply, 9.7
Housing characteristics, 1.11–1.12
Hydrological droughts, 8.5
Hydrant leak repairs, water distribution systems, 6.3
Hydraulic availability analysis, water distribution systems, 6.12–6.13, **6.13, 6.14, 6.18**, 6.18, **6.19, 6.21**
Hydraulic performance indicators, 7.13
Hydroinformatics, 3.6, **3.7**, 3.22
Hydrologic cycle, 1.1–1.2
  climate change and, 8.5
Hydrologic effect of climate change, 8.5
Hydrosphere in climate change, 8.3
HYDRUS flow model, 3.13–3.14

Independence, in regression analysis, 1.25
Independent variables, 1.8, **1.8**
Index of reliability, water distribution systems, 6.4–6.5
Index, seasonal, 1.26
Industrial water use (*see* Commercial, institutional, industrial)
Infant mortality period, and failure rate, 6.6
Information sharing on safety and security of water supply, 9.11
Instantaneous failure rate (hazard function, force of mortality function) in, 6.6
Institutional water use (*see* Commercial, institutional, industrial)
Integer programming, design of water distribution systems and, 4.3
Integrated hydrosystems management, 3.2–3.6
  computer programs for, 3.1–3.26
  history of, 3.3–3.5
  importance of, 3.5–3.6
  total water management (TWM) and, 3.3, **3.4**
  watershed approach to management in, 3.5
International field test of IWA system of performance indicators, 7.31–7.36, **7.36**
International Water Association (IWA), 7.9–7.10, 7.15–7.35, **7.19–7.24**
International Water Supply Association (IWSA), 7.5
Interval data, 1.8
Intrusion through supervisory control and data acquisition (SCADA), **9.44**

Joint leak, water distribution systems, 6.3

KYPIPE simulation, 6.27

*Lansey, **Kevin E.***, 4.1, 6.1
Lead and Copper Rule, 5.1
Leak repair, water distribution systems, 6.2
Leaks in water distribution system, 6.2
Linear modeling, 1.28, 2.6–2.8
Linear programming (LP), 3.18
  design of water distribution systems and, 4.3, 4.4, 4.5–4.6, 4.8
Log linear models, 1.28
Logarithmic modeling, 2.6–2.8
Looped pipe systems via linearization, 4.5–4.6
Losses (*see* Performance indicators; water losses)

Macrodata, 1.7
Main breaks, water distribution systems, 6.3
Main leaks, water distribution systems, 6.2
Malaysian Water Association (MWA), use of performance indicators, 7.14
Maps, service area, 1.11–1.13
Mathematical in-core nonlinear optimization systems (MINOS), water system optimization and, 5.2
Mathematical programming:
  computer programs for integrated management, 3.11, 3.13–3.14
  water system optimization and, pump operation, 5.3–5.12
Maximum contaminant levels (MCLs), 5.1
*Mays, **Larry W.***, 2.1, 3.1, 5.1, 6.1, 8.1, 9.1
Mean time between failure (MTBF), water distribution systems, 6.9, 6.16
Mean time between repairs (MTBR), water distribution systems, 6.9
Mean time to failure (MTTF), water distribution systems, 6.6–6.9, **6.8**
Mean time to repair (MTTR), water distribution systems, 6.9, 6.16
Measurements of water, 1.1–1.2
Mechanical availability of piping components, water distribution systems, 6.16–6.17, **6.18**
Median household incomes, 1.27
Meteorological drought, 8.5
Meters and metering, 1.1–1.2, 1.10, 1.13
  drought management and, 2.4
Metropolis algorithm, pump optimization, 5.13
Microdata, 1.7
Minimum month method, 1.15–1.19, **1.16, 1.17**
Minimum operational pressure, 6.17–6.18
Mixed integer linear programming (MILP), 3.18
Modeling, 1.22–1.30
  chlorine booster station optimization in, 5.22–5.25, **5.24, 5.25**
  computer programs for integrated management, 3.6–3.7
Modeling (*Cont.*):
  drought management and, 2.6–2.8
  pump optimization operation, 5.3–5.5
  water system optimization and, 5.1–5.2
MODFLOW model, 3.18
Monthly billing cycles, 1.16–1.19, **1.16, 1.17**
Multiple regression techniques, 1.25
Multivariate regression models, 1.33

National Geographic Society, 3.3
Network deficit, in water distribution systems, 6.25
New York City, design of water distribution systems and, example of, 4.8–4.10, **4.9, 4.10**
Nodal and system availability, water distribution systems, 6.17
Nodal and system reliability, water distribution systems, 6.13–6.16
Nominal data, 1.7–1.8
Nonlinear programming (NLP), 3.11
  computer programs for integrated management, 3.8
  design of water distribution systems and, 4.3, 4.6–4.7, **4.7**, 4.8
  pump optimization, 5.21, **5.22**
  water system optimization and, 5.2
Nonordinal data, 1.7–1.8
Nonprobability sampling, 1.3
Nonrevenue water performance indicators, 7.27, 7.29, **7.30, 7.31, 7.34**
Normality, in regression analysis, 1.25
North American Classification system (NAICS) codes, 1.13
North Marin County pumping optimization, 5.19–5.21, **5.20**
North Marin Water District (NMWD), water system optimization and, 5.7–5.12, **5.9, 5.10, 5.11**
Notification of potential contamination, safety and security of water supply, **9.43**
Numerical weather prediction models (NWPMs) in climate change, 8.3

Objective oriented management and performance indicators, 7.7
Operating pressure performance indicators, **7.34**
Operation of water systems, 5.1–5.29
Operational performance indicators, **7.21**, 7.51
Operational parameters for design of water distribution systems and, 4.2–4.3
Operational pressure, minimum, 6.17–6.18
Optimal design (*see* Design of water distribution systems)
Optimal operation (*see* Operation of water systems)
Optimization formulations, computer programs for integrated management, 3.8–3.10, **3.8**
Optimization model, water distribution systems, 6.27–6.28

Optimization model flowchart, pump optimization, 5.7, **5.8**
Optimization of water system operations, 5.2
background and history of, 5.1–5.2
chlorine booster station optimization in, 5.22–5.25, **5.24, 5.25**
chlorine content and, 5.1
cost function in, 5.13–5.16
design challenges for future of, 5.25–5.26
dynamic programming in, 5.2
EPANET and, 5.2, 5.5, 5.12, 5.16, **5.16, 5.17**, 5.21
general algebraic modeling system (GAMS) in, 5.2, 84
GRG2 in, 5.5, 5.7, 5.12
Lead and Copper Rule and, 5.1
mathematical in-core nonlinear optimization systems (MINOS) in, 5.2
mathematical programming approach to, 5.3–5.12
maximum contaminant levels (MCLs) in, 5.1
Metropolis algorithm in, 5.13
models for, 5.1–5.2
nonlinear programming (NLP) in, 5.2, 5.21, **5.22**
North Marin Water District (NMWD) example of, 5.7–5.12, **5.9, 5.10, 5.11**
optimization and simulation model linkage in, 5.5, **5.6**
optimization of water distribution systems, 5.2
pumps, 5.1, 5.3–5.12
Austin, Texas example of, 5.18–5.21, **5.18, 5.19, 5.20**
Cohen condition in, 5.15
constraints in, 5.4
cost function in, 5.13–5.16
decision variables in, 5.13
EPANET in, 5.16, **5.16, 5.17**, 5.21
flowchart of optimization model in, 5.7, **5.8**
Metropolis algorithm in, 5.13
model formulation in, 5.3–5.5, 5.12–5.13
nonlinear programming (NLP) in, 5.21, **5.22**
North Marin County example of, 5.19–5.21, **5.20**
North Marin Water District (NMWD) example of, 5.7–5.12, **5.9, 5.10**, 5.11
optimization and simulation model linkage in, 5.5, **5.6**
penalty function method in, 5.5–5.7
periodic behavior in, 5.14–5.15, **5.14**
pressure head in, 5.6
quality of water and, 5.4
reduction technique as solution to, 5.5–5.7
simulated annealing in, 5.12–5.22, **5.17**
supervisory control automated data acquisition (SCADA), 5.22
tank capacity and, 5.14–5.15

Optimization of water system operations (*Cont.*):
Total Coliform Rule and, 5.1
U.S. Environmental Protection Agency (USEPA) and, 5.2
Optimization-simulation model and NLP, design of water distribution systems and, 4.6–4.7, **4.7**
OR mode, in fault tree analysis, 9.19, **9.20**
Ordinal data, 1.7–1.8
Ordinary least squares (OLS) regression, 1.24, 1.29–1.30,
Organization/personnel performance indicators, 7.12
Output indicators, 7.8–7.9
Outside effects, conservation and saving water, 1.31

Palmer Drought Severity Index (PDSI), 2.10
Panel data, 1.8–1.9
Penalty function:
performance indicators, 7.13
pump optimization, 5.5–5.7
Performance failure mode, water distribution systems, 6.3–6.4
Performance indicators as management support tool, 7.1–7.53
American Water Works Research Foundation (AWWARF) and, 7.13–7.14
apparent losses, 7.29, **7.30, 7.31, 7.34**
Asian Development Bank data book for, 7.10, **7.11**
assessment and global reporting in, 7.39, 7.40–7.42, **7.42**, 7.51–7.52
auditors' use of, 7.4
authorized consumption, 7.28, **7.34**
availability, 7.12
benchmarking using, 7.4, 7.7, 7.10
Benchmarking Water and Sanitation Utilities (BWSU) and, 7.10
classifications and definitions used in IWA manual, **7.25**, 7.26–7.27, **7.27**
complementary, 7.15
concept of, 7.2–7.3
confidence grading scheme in, 7.20, **7.24**
context information and, 7.2–7.3, **7.3**, 7.41
contractual agreements and use of, 7.5
costs, 7.12
customer satisfaction, 7.11
data availability, reliability, accuracy in, 7.38–7.39
data collection for, 7.39
data input example in, 7.40–7.42, **7.42–7.47, 7.49**
data reliability and accuracy, 7.19–7.20
definition of, 7.38, 7.40
dispersion bands and, 7.13
distribution input, 7.28
economy, 7.12
engineering approach to, 7.12–7.13
environment, 7.12

Performance indicators as management support tool (*Cont.*):
establishment of, 7.38–7.39
factors affecting, 7.2–7.3
financial, **7.23–7.24**, 7.51
financing bodies, use of, 7.4
forms for use of, 7.4–7.5, 7.4
general, 7.12
generalizing function and, 7.13
global system of, 7.2–7.3
guaranteed standards schemes (GSS) and, 7.5
hydraulic performance, 7.13
implementation of, phases for, 7.36–7.40, **7.37**
inputs and outputs in, 7.26–7.27, **7.27**
internal data flows for, 7.39
international field test of IWA system of, 7.31–7.36, **7.36**
International Water Association (IWA) and, 7.9–7.10, 7.15–7.35, **7.19–7.24**
International Water Supply Association (IWSA) and, 7.5
levels of, 7.15
mains length, **7.34**
Malaysian Water Association (MWA) use of, 7.14
matrix of confidence grades in, **7.24**
nonrevenue water, 7.27, 7.29, **7.30, 7.31, 7.34**
objective definition for, 7.43
objective oriented management and, 7.7
operating pressure, **7.34**
operational, **7.21**, 7.51
organization of IWA manual on, 7.20–7.26
organization/personnel, 7.12
output indicators in, 7.8–7.9
penalty function and, 7.13
personnel indicators, **7.19**
physical, **7.20**
pressurization time, **7.34**
private participation in management and, 7.7
production, 7.12
quality and quality certifying entities, use of, 7.4, 7.5, 7.12, 7.13, **7.22**
raw water, imported or exported, 7.27, 7.34
real losses, 7.29, **7.30, 7.31, 7.34**
reference PI system for, 7.43
relevancy of, selection of, 7.38, 7.41–7.42, 7.43
reliability, 7.13
results interpretation in, 7.39–7.40
running costs, **7.34**
Scandinavian cities group use of, 7.11–7.12
scope of application for, **7.6**, 7.37
service connections, **7.34**
SIGMA Lite software and, 7.10, 7.30–7.31, **7.35**, 7.40–7.42, **7.42–7.47**
software for, 7.39
state of the art in, 7.5–7.15
state variables in, 7.13

Performance indicators as management support tool (*Cont.*):
strategic performance assessment policy using, 7.37–7.38
supplied water, 7.28
supranational organizations using, 7.4
system input volume, 7.28
team profile for, 7.38, 7.43
transmission input, 7.28
treated water, imported or exported, 7.27, **7.34**
treatment input, 7.27
U.K. privatization and, lessons learned from, 7.7–7.9
unaccounted for water (UFW) in, 7.27
United States and, status of, 7.13–7.14
users, benefits, scope of application for, 7.3–7.5
utility information and, 7.2–7.3
variables and, 7.2–7.3, 7.13, **7.26, 7.34**
water abstracted, 7.27, **7.34**
water balance assessment using, 7.27, **7.28**, 7.29–7.30, 7.48–7.51, **7.50**
water charges, **7.34**
water losses, 7.28, **7.30, 7.31, 7.32–7.33**, 7.42–7.52, 7.47–7.48
water produced, 7.27, **7.34**
water resource, **7.19**, 7.51
Water Utility Partnership for Capacity Building in Africa (WUP) and, 7.10–7.11
World Bank benchmarking toolkit and, 7.10
worldwide use of, 7.14–7.15, **7.16–7.17, 7.18**, 7.31–7.36, **7.36**
Personnel performance indicators, **7.19**
Phoenix, Arizona, drought management and, example of, 2.29–2.31, **2.30, 2.31**
Photosynthetic process in climate change, 8.3
Physical performance indicators, **7.20**
Physical threats to safety and security of water supply, 9.4
Pipes:
mechanical availability of piping components in, 6.16–6.17, **6.18**
size calculation in design of water distribution systems, 4.5–4.6
Policymaking bodies, use of performance indicators, 7.4
Pooled time series cross section (TSCS) data, 1.9, 1.27–1.28
Pooled time series data, 1.8–1.9
Population and housing characteristics, 1.11–1.12
Population studies, water use, 1.3
Power and Reservoir System Model (PRSYM), 3.17–3.18
Precision of sampling, 1.4
Predicting climate change, 8.3–8.5
President's Commission on Critical Infrastructure Protection (PCCIP), 9.4

Pressure deficits, water distribution systems, 6.24–6.26
Pressure head, pump optimization, 5.6
Pressurization time, performance indicators, **7.34**
Price elasticity approach, 2.3, 2.4–2.6, **2.6**
Price of water, 1.27
Privatization and performance indicators, lessons learned from, 7.7–7.9
Probabilistic risk, 2.8–2.9
Probability sampling, 1.3
Production performance indicators, 7.12
Proportional data, sample size determination for, 1.5–1.7
Proportions, sampling for, 1.6
PROSIM, 3.18
Public Health, Security and Bioterrorism Preparedness and Response Act, 9.6–9.7, **9.7**
Public water supply use statistics, 1.2
Pumping system, 5.1, **9.23, 9.24**
  Austin, Texas example of, 5.18–5.21, **5.18, 5.20**
  Cohen condition in, 5.15
  constraints in, 5.4
  decision variables in, 5.13
  efficiency calculations for, 6.27–6.28
  fault tree for, **9.24**
  flow rate calculations in, 5.3–5.5
  flowchart of optimization model in, 5.7, **5.8**
  hydraulic availability analysis in, 6.12–6.13, **6.13, 6.14, 6.18**, 6.18, **6.19, 6.21**
  mathematical programming approach to operations of, 5.3–5.12
  Metropolis algorithm in, 5.13
  model formulation in, 5.3–5.5
  nonlinear programming (NLP) in, 5.21, **5.22**
  North Marin County example of, 5.19–5.21, **5.20**
  North Marin Water District (NMWD) example of, 5.7–5.12, **5.9, 5.10, 5.11**
  optimization and simulation model linkage in, 5.5, **5.6**
  penalty function method in, 5.5–5.7
  periodic behavior in, 5.14–5.15, **5.14**
  pressure head in, 5.6
  reduction technique as solution to, 5.5–5.7
  schematic diagram of, **9.23**
  simulated annealing in, 5.12–5.22, **5.17**
  supervisory control automated data acquisition (SCADA), 5.22
  tank capacity and, 5.14–5.15

Quality of water, 5.1
  performance indicators, 7.4, 7.5, 7.12, 7.13, **7.22**
  pump optimization, 5.4, 86

Rainfall, 1.27
RAM-W Risk Assessment Methodology, 9.8
Random effects models, 1.30
Random sampling, 1.13
Random variables, 1.7
Raw water, imported or exported, performance indicators, 7.27, **7.34**
Reading meters (*see* Meters and metering)
Real losses, performance indicators, 7.29, **7.30, 7.31, 7.34**
Recovery, remediation, in safety and security of water supply, 9.11, **9.40–9.41**
Reduction technique, pump optimization, 5.5–5.7
Regression analysis, 1.22–1.23
Regulatory agency use of performance indicators, 7.4, 7.5
Relationships in water use, 1.20–1.30
Reliability and availability analysis, 6.1–6.31
  American Water Works Association Research Foundation (AWWARF) in, 6.2
  availability defined for, 6.5, 6.9–6.11, **6.10**
  average availability defined in, 6.5
  breaks in, 6.2
  chance constraint in, 6.28
  chlorine application and, 6.27–6.28
  component (mechanical) failure mode in, 6.3, 6.4, 6.5–6.11
  conditional probability approach in, 9.15, **9.17**
  constant failure rate, constant repair rate in, 6.10–6.11
  constraints in, 6.27–6.28
  contingency in, 6.5
  cut-set and tie-set analysis in, 6.14, **6.15**, 6.16, 9.13–9.15, **9.14, 9.16**, 9.19
  demand variation failures in, 6.25
  economic indices in, 6.5
  emergency demand conditions and, 6.2
  EPANET and, 6.28
  failure density function in, 6.6
  failure density, failure rate in, 6.6–6.9, **6.7**
  failure modes in, 6.1–6.5
  failure probability in, 6.6
  fault tree analysis in, 9.15–9.23, **9.18, 9.22**
  framework for reliability-based design model and, 6.26–6.27
  frequency and duration indices in, 6.5
  hydrant leak repairs in, 6.3
  hydraulic availability analysis in, 6.12–6.13, **6.13, 6.14, 6.18**, 6.18, **6.19, 6.21**
  instantaneous failure rate (hazard function, force of mortality function) in, 6.6
  joint leak in, 6.3
  leak repair in, 6.2
  leaks in, 6.2
  main breaks in, 6.3
  main leaks in, 6.2
  mean time between failure (MTBF) and, 6.9, 6.16
  mean time between repairs (MTBR) and, 6.9
  mean time to failure (MTTF) in, 6.6–6.9, **6.8**, 6.16
  mean time to repair (MTTR) in, 6.9

Reliability and availability analysis (*Cont.*):
mechanical availability of piping components in, 6.16–6.17, **6.18**
minimum operational pressure in, 6.17–6.18
nodal and system reliability in, 6.13–6.16, 6.17
optimization model for, 6.27–6.28
path enumeration method in, 9.13–9.15, **9.14**
performance failure mode in, 6.3–6.4
performance indicators, 7.13
pressure deficits in, 6.24–6.26
pump efficiency and, 6.27–6.28
reliability analysis for, 6.11–6.17
reliability indices in, 6.4–6.5
repair and, distribution repair definitions in, 6.2–6.3
repairable systems and, 6.7, 6.9
safety and security of water supply, 9.11–9.23
service leak repairs in, 6.3
severity indices in, failures and, 6.5
simultaneous failures in, probability of
state enumeration method in, 9.12–9.13, **9.12, 9.13**
stationary availability, stationary unavailability in, 6.11
time to failure (TTF) analysis in, 6.5, 6.9
time to repair (TTR) analysis in, 6.5, 6.9
total network deficit in, 6.25
Tucson, Arizona example of, 6.17–6.23, **6.19, 6.20, 6.21**
unavailability calculation in, 6.10–6.11, **6.10**
valve leak repairs in, 6.3
volume deficits in, 6.23–6.26
water distribution systems, 6.11–6.17
Reliability-based design model, 6.26–6.27
Reliability-based optimization model, 6.27–6.28
Reliability indices, water distribution systems, 6.4–6.5
Remediation, in safety and security of water supply, 9.11, **9.40–9.41**
Repair, distribution repair definitions in, 6.2–6.3
Repairable systems, water distribution systems, 6.7, 6.9
Reservoir system operation for water supply, computer programs for integrated management, optimization, 3.9–310
Response planning, recovery, remediation, safety, and security of water supply, 9.11, **9.40–9.41**
Risk:
evaluation of, 2.13–2.17, **2.16**, 2.26–2.31, **2.28, 2.29**
risk–price relationship in, 2.11–2.17, **2.12, 2.13, 2.18, 2.19**
Risk-based approach to drought management and, 2.3, 2.8
Risk–price relationship, drought management and, 2.11–2.17, **2.12, 2.13, 2.18, 2.19**
River basin/regional runoff effects of climate change, 8.10–8.13, **8.12, 8.13, 8.14**
RiverWare, 3.18–3.20, **3.20**, 3.22, 3.24
RiverWare Rule Language (RWRL), 3.18
Running costs performance indicators, **7.34**
Runoff, climate change and, 8.8, 8.10–8.13

Safe Water Drinking Act of 1947 (SDWAA), 5.1, 5.2
Safety and security of water supply, 9.1–9.44
biological threats to, 9.4, **9.5, 9.6**
chemical threats to, 9.4, **9.5**
conditional probability approach in, 9.15, **9.17**
contamination threat at major events and, **9.42**
cut-set and tie-set analysis in, 9.13–9.15, **9.14, 9.16**, 9.19
cyber threats to, 9.2, 9.4
emergency response planning, recovery, remediation, 9.11, **9.40–9.41**
fault tree analysis in, 9.15–9.23, **9.18, 9.22**
Homeland Security Act and, 9.7
information sharing on, 9.11
intrusion through supervisory control and data acquisition (SCADA), **9.44**
notification of potential contamination, **9.43**
path enumeration method in, 9.13–9.15, **9.14**
physical threats to, 9.4, **9.4**
President's Commission on Critical Infrastructure Protection (PCCIP) and, 9.4
prior to September 11, 2001, 9.4–9.6
Public Health, Security and Bioterrorism Preparedness and Response Act and, 9.6–9.7, **9.7**
reliability assessment and, 9.11–9.23
response to September 11, 2001 and, 9.6–9.8
security tools for, 9.8–9.9
small drinking water systems and, vulnerability assessment, 9.9–9.11, **9.30–9.39**
state enumeration method in, 9.12–9.13, **9.12, 9.13**
structural damage and, **9.45**
terrorist threats to, **9.4**, 9.4–9.8
threats to, 9.2–9.4
typical water supply system and, 9.1–9.2, **9.2, 9.3**
U.S. Environmental Protection Agency (USEPA) and, 9.6, 9.8, **9.8, 9.9**
Visual Interactive Site Analysis Code (VISAC) and, 9.12
vulnerability assessment for, 9.8–9.11, **9.10**
vulnerability assessment in, 9.26–9.29
Vulnerability Self-Assessment Tool (VSAT) in, 9.9
Water Sector Critical Infrastructure Advisory Group and, 9.4–9.5
Safety factor and risk relationship, drought management and, 2.25–2.26, **2.25, 2.26**, 2.30, **2.31**, 2.32
Sakarya, ***A. Burcu Altan***, 5.1
Sampling for proportions, 1.6
Sampling of water users, 1.3–1.7

SANJASM, 3.18, 70
Saving water (*see* Conservation and saving water)
Scales, data, 1.7–1.8
Scandinavian cities group, use of performance indicators, 7.11–7.12
Seasonal (annual) cross sectional data, 1.9
Seasonal and nonseasonal use, **1.14**, 1.15–1.19, **1.16, 1.17**, 1.26
Seasonal indexes, 1.26, 26
Sectoral disaggregate estimate of water use, 1.20
Sectors of water users, 1.13–1.15
Security tools, for safety and security of water supply, 9.8–9.9
Semilogarithmic modeling, 2.6–2.8
Sensitivity to climate change, 8.10
Service Providers Performance Indicators and Benchmarking Network (SPBNET), 7.11
Service area data and maps, 1.11–1.13
Service connections performance indicators, **7.34**
Service leak repairs, water distribution systems, 6.3
Severity indices, water distribution systems, 6.5
Sheer Steila Drought Index (DI), 2.10
SIGMA Lite software, performance indicators, 7.10, 7.30–7.31, **7.35**, 7.40–7.42, **7.42–7.47**
Simple random sampling, 1.3
Simulated annealing:
- Austin, Texas example of pump operations, 5.18–5.21, **5.18, 5.20**
- cost function in, 5.13–5.16
- decision variables in, 5.13
- design of water distribution systems and, 4.3, 4.7–4.8
- EPANET and, 5.21
- Metropolis algorithm in, 5.13
- pump optimization, 5.12–5.22, **5.17**
- pumps, North Marin County example of pump operations, 5.19–5.21, **5.20**
- water system optimization and, 5.12–5.22, **5.17**

Simulation, computer programs for integrated management, 3.6–3.7, **3.7**, 3.10–3.14
Simultaneous failures in water distribution systems, probability of, 6.17
Size of samples, 1.4, 1.5–1.7
Small drinking water systems, safety and security of water supply, vulnerability assessment, 9.9–9.11, **9.30–9.39**
Snowfall and snowmelt levels, climate change and, 8.7–8.8
Soil moisture and climate change, 8.5–8.7
Southern Oscillation Index (SOI), 2.10
Special metering, 1.10
Stability of climate and climate change, 8.3
Standard error, sampling, 1.4–1.5
Standard error, 1.32
Standard Industrial Classification (SIC) codes, 1.13
State enumeration method, safety and security of water supply, 9.12–9.13, **9.12, 9.13**
State variables in performance indicators, 7.13
Stationary availability, stationary unavailability, water distribution systems, 6.11
Statistical comparison, 1.31–1.33
Stochastic search techniques, design of water distribution systems and, 4.3, 4.7–4.8
Storm frequency and interval, climate change and, 8.8
Strategic performance assessment policy using performance indicators, 7.37–7.38
Stratified random sampling, 1.4
Stratified sampling, 1.4
Streamflow variability and climate change, 8.8–8.9
Structural damage, safety and security of water supply, **9.45**
Successive approximate linear quadratic regulator (SALQR) method, computer programs for integrated management, 3.12, 3.13–3.14
Supervisory control automated data acquisition (SCADA):
- chlorine booster station optimization in, 5.23–5.25
- computer programs for integrated management, 3.14, 3.22, 5.22
- pump optimization, 5.22
- safety and security of water supply, **9.44**

Supplied water performance indicators, 7.28
Supranational organizations, use of performance indicators, 7.4
Surface Water Drinking Rule, 5.1
Surface Water Supply Index (SWSI), 2.10–2.11, 2.23–2.26, **2.24**
System availability, water distribution systems, 6.17
System input volume performance indicators, 7.28
System reliability, water distribution systems, 6.13–6.16
Systematic sampling, 1.4

Tank capacity, pump optimization, 5.14–5.15
TERRA, TVA Environment and River Resource Aid, 3.17, 3.22, 3.24
Terrorist threats to safety and security of water supply, **9.4**, 9.4–9.8.4
Threats to water supply safety, 9.2–9.4
Tie-set analysis, 9.13–9.15, **9.14, 9.16**
Time series analysis, 1.25–1.26
- conservation effects, 1.33–1.34

Time series data, 1.8–1.9
Time to failure (TTF) analysis, water distribution systems, 6.5, 6.9
Time to repair (TTR) analysis, water distribution systems, 6.5, 6.9
Total Coliform Rule, 5.1
Total network deficit, water distribution systems, 6.25
Total trihalomethane (TTHM) levels, 5.16

Total water management (TWM), 3.3, **3.4**
Transmission input performance indicators, 7.28
Transpiration, climate change and, 8.5
Treated water, imported or exported, performance indicators, 7.27, **7.34**
Treatment input performance indicators, 7.27
Trihalomethane (TTHM), 5.16
Trinity River Advanced Computing Environment (TRACE), 3.16
Trinity River Basin, Texas, DSS, 3.16–3.17, **3.17**
Tucson, Arizona, reliability and availability analysis example of, 6.17–6.23, **6.19, 6.20, 6.21**
Tung, Y.K., 6.1
TVA Environment and River Resource Aid (TERRA) DSS, 3.17, 3.22, 3.24
Two-tail test, 1.33, 33

U.K. privatization and performance indicators, lessons learned from, 7.7–7.9
U.S. Army Corps of Engineers (USACE), 3.3, 3.22
U.S. Census data, **1.12**
U.S. community water systems, **1.3**
U.S. Environmental Protection Agency (USEPA), 5.2, 9.6, 9.8, **9.8, 9.9**
U.S. Soil Conservation Service Monthly Report, 2.11
Unaccounted for water (UFW) performance indicators, 7.27
Unavailability calculation, water distribution systems, 6.10–6.11, **6.10**
Uncertainty and risk, drought management and, in demand and, 2.23–2.26, **2.24**
Union for African Water Suppliers, 7.10
United States use of performance indicators, status of, 7.13–7.14
Universidad Politécnica de Valencia (UPV), Spain, 3.20
Urban areas, determinants of demand in, 1.20, **1.21**
Use (*see* Water use)
Use restrictions, drought management and, 2.4
Useful life period, and failure rate, 6.6
Utility company, performance indicators, use of, 7.3–7.4
Utility information performance indicators, 7.2–7.3

Valve leak repairs, water distribution systems, 6.3, 101
Variables, 1.7, 1.24, **1.29**
performance indicators, 7.2–7.3, 7.13, **7.26, 7.34**
Variance, 1.24
Venturi meters, 1.1–1.2
Visual Interactive Site Analysis Code (VISAC), safety and security of water supply, 9.12
Volume deficits, water distribution systems, 6.23–6.26
Volumetric units of measure, 1.1–1.2
Vulnerability assessment, safety and security of water supply, 9.8–9.11, **9.10**, 9.26–9.29
Vulnerability Self-Assessment Tool (VSAT), safety and security of water supply, 9.9
Vulnerability to climate change, 8.10

Wastewater disposal, price of, 1.27
Water abstracted, performance indicators, 7.27, **7.34**
Water and Wastewater Utilities Indicators, 7.10
Water balance, 7.29–7.30
performance indicators, 7.48–7.51, **7.50**
Water charges performance indicators, **7.34**
Water demand analysis, 1.1–1.37
adjustment of demand, in drought management, 2.26–2.27, **2.27**
analysis of water savings in, 1.30–1.36
availability vs. demand, 2.7–2.8, **2.7, 2.8**
average rates of water use in, 1.21–1.22, **1.21, 1.22, 1.23–1.24**
components of water demand in, 1.13–1.19
customer premise categories in, 1.13–1.15
data arrangements in, 1.8–1.9
data set development in, 1.7–1.9
definition and measurement of water use, 1.1–1.2
determinants of, 1.20, **1.21**
modeling of water use in, 1.22–1.30
modeling of, 2.6–2.8
public supply water use, 1.2
sampling of water users, 1.3–1.7
seasonal and nonseasonal components in, 1.15–1.19, **1.16, 1.17**, 1.26
service area data and maps in, 1.11–1.13
variation failures in, 6.25
water use data in, 1.10–1.11
water use relationships in, 1.20–1.30
Water distribution system:
American Water Works Association Research Foundation (AWWARF) in, 6.2
availability defined for, 6.9–6.11, **6.10**
availability defined in, 6.5
average availability defined in, 6.5
breaks in, 6.2
chlorine application and, 6.27–6.28
component (mechanical) failure mode in, 6.3, 6.4, 6.5–6.11
computer programs for integrated management, optimization, 3.9
constant failure rate, constant repair rate in, 6.10–6.11
contingency in, 6.5
demand variation failures in, 6.25
design of, 4.1–4.13
economic indices in failure of, 6.5
emergency demand conditions and, 6.2
EPANET and, 6.28.28
failure density, failure rate in, 6.6–6.9, **6.7**
failure modes in, 6.1–6.5
failure probability in, 6.6

Water distribution system: (*Cont.*):
framework for reliability-based design model and, 6.26–6.27
frequency and duration indices in failure of, 6.5
hydrant leak repairs in, 6.3
hydraulic availability analysis in, 6.12–6.13, **6.13, 6.14, 6.18**, 6.18, **6.19, 6.21**
instantaneous failure rate (hazard function, force of mortality function) in, 6.6
joint leak in, 6.3
leak repair in, 6.2
leaks in, 6.2
main breaks in, 6.3
main leaks in, 6.2
mean time between failure (MTBF) and, 6.9, 6.16
mean time between repairs (MTBR) and, 6.9
mean time to failure (MTTF) in, 6.6–6.9, **6.8**
mean time to repair (MTTR) in, 6.9, 6.16
mechanical availability of piping components in, 6.16–6.17, **6.18**
minimum operational pressure in, 6.17–6.18
nodal and system reliability in, 6.13–6.16, 6.17
optimization model for, 5.2, 6.27–6.28
performance failure mode in, 6.3–6.4
pressure deficits in, 6.24–6.26
pump efficiency and, 6.27–6.28
reliability analysis for, 6.11–6.17
reliability and availability analysis of, 6.1–6.31
reliability indices for, 6.4–6.5
repair and, distribution repair definitions in, 6.2–6.3
repairable systems and, 6.7, 6.9
service leak repairs in, 6.3
severity indices in failure of, 6.5
stationary availability, stationary unavailability in, 6.11
time to failure (TTF) analysis in, 6.5, 6.9
time to repair (TTR) analysis in, 6.5, 6.9
total network deficit in, 6.25
unavailability calculation in, 6.10–6.11, **6.10**
valve leak repairs in, 6.3
volume deficits in, 6.23–6.26
Water losses performance indicators, 7.28, **7.30, 7.31, 7.32–7.33**, 7.42–7.52
Water management options and climate change, 8.13, **8.14**
Water produced performance indicators, 7.27, **7.34**
Water production records, 1.10
Water quality performance indicators, 7.4, 7.5, 7.13, **7.22**
Water resource performance indicators, **7.19**
Water resource systems:
climate change and, 8.9–8.10
performance indicators, 7.51
Water Resources Engineering Simulation Language (WRESL), 3.18
Water Sector Critical Infrastructure Advisory Group, 9.4–9.5
Water use, 1.1–1.2, 1.10–1.11
analysis of water savings in, 1.30–1.36
average rates of, 1.21–1.22, **1.21, 1.22, 1.23–1.24**
commercial, institutional, industrial, 1.12–1.13
components of water demand in, 1.13–1.19
data used in determining, 1.10–1.11
end use accounting systems in, 1.34–1.36, **1.35, 1.36**
estimating, 1.20–1.30, 1.31–1.33
modeling of, 1.22–1.30, 2.6–2.8
population studies in, 1.3
public and government facilities, 1.13
public supply, 1.2
relationships in, 1.20–1.30
seasonal and nonseasonal components in, 1.15–1.19, **1.16, 1.17**, 1.26
sectors of water users, 1.13–1.15
user sampling, 1.3–1.7
Water Utility Partnership for Capacity Building in Africa (WUP), **7.10–7.11**
Watermarque, 7.14
Watershed approach to management, 3.5
WaterWare, 3.21, 3.22, 3.24
Wear out life period, and failure rate, 6.6
Weighting, in regression analysis, 1.30
Willingness to pay, drought management and, 2.18
World Bank benchmarking toolkit, performance indicators, 7.10
Worldwide use of performance indicators, 7.14–7.15, **7.16–7.17, 7.18**, 7.31–7.36, **7.36**

Zero mean, 1.24